LIFE IN THE CHESAPEAKE BAY

THE JOHNS HOPKINS UNIVERSITY PRESS
Baltimore and London

LIFE IN THE

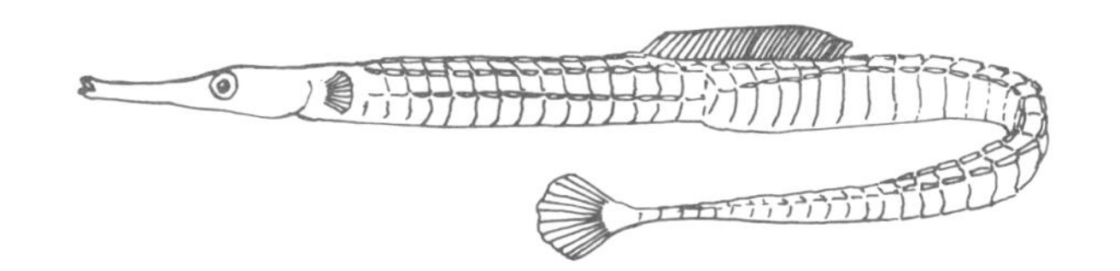

CHESAPEAKE

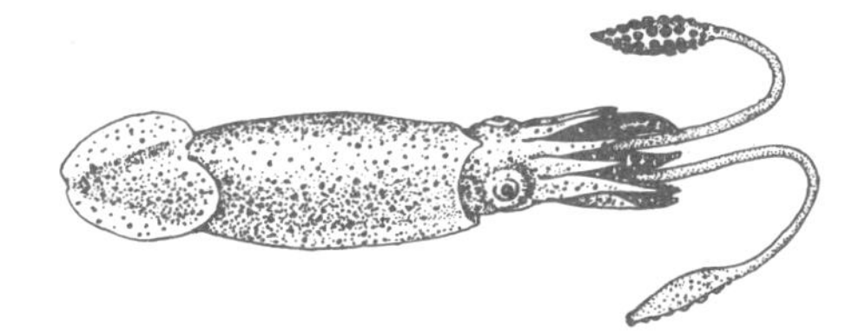

BAY

Second Edition

ALICE JANE LIPPSON
ROBERT L. LIPPSON

This book has been brought to publication with the generous assistance of the Baltimore Gas and Electric Company.

Printed in the United States of American on acid-free paper
06 05 04 03 02 01 00 99 98 97 5 4 3 2 1

Photo credits: Joyce R. Firstman: pp. 48–49, 216; Peter J. Gedeon: pp. 2–3, 214–215; Alice Jane Lippson: pp. 4, 22–23, 24, 50, 82–83, 112–13, 114, 158–59, 180–81, 228–29, 230; Robert L. Lippson, Ph.D.: pp. 84, 182, 258–59; Gordon Thayer, Ph.D., *National Marine Fisheries Service, Beaufort Laboratory, Beaufort, N.C.:* p. 160; Prints of photos by Alice Jane and Robert L. Lippson by Joyce R. Firstman.

The Johns Hopkins University Press
2715 North Charles Street
Baltimore, Maryland 21218-4319
The Johns Hopkins Press Ltd., London

Library of Congress Cataloging-in-Publication Data will be found at the end of this book.
A catalog record for this book is available from the British Library.

ISBN 0-8018-5476-8
ISBN 0-8018-5475-X (pbk.)

WE DEDICATE THIS BOOK
to our parents, who gave us the light,
and to our children and grandchildren, who are the light.

Contents

The Chesapeake Bay is the largest estuary in the United States and one of the largest in the world, with more than 10 million people living along or near its shores. The Bay is heavily utilized for sport and recreation. Many visitors from other areas, as well as people who live nearby, come to its tidewaters for fishing, hunting, swimming, sailing, waterskiing, crabbing, or beachcombing. Whoever comes to the Bay, for whatever activity, encounters its life in some form or other, and most cannot be without curiosity. This book is for the curious—from the casual beachcomber intrigued by the shape of a shell to the more serious student of estuarine ecology. It will also be helpful in the identification of plants and animals from the coastal lagoons and estuaries of the mid-Atlantic coast. Many of the same organisms found in the Chesapeake Bay and discussed in this book occur from Cape Hatteras to Cape Cod.

When we wrote the first edition of this guidebook, we concentrated on the marine animal and plant life likely to be encountered in the Chesapeake Bay. This new edition of *Life in the Chesapeake Bay* covers 116 additional species. It now includes a broad coverage of birds that are a part of the estuarine system. We have also significantly expanded the chapter on marshes, newly retitled "Wetlands." In addition, we have included several species of reptiles and mammals closely associated with Chesapeake Bay habitats. Over the past 10 years we have continued our studies of the plant and animal communities of the Chesapeake Bay. We have ranged the Chesapeake and its tributaries on numerous cruises aboard our boat, *Odyssey,* and have gained a greater understanding of the role birds play in an estuary. There are many species of waterfowl that feed on submerged aquatic vegetation, the produce of wetlands, and invertebrates and fish. There are other species of birds, such as herons and egrets, pelicans and cormorants, and ospreys and eagles, which prey mostly on fish and invertebrates. Many of these fish and invertebrate feeders have increased in abundance and have extended their range significantly, particularly in the case of the brown pelican. Then, too, bottlenosed dolphins have increased in numbers and occurrence in the Bay. Several reasons for the growing numbers of some species come to mind: reduced hunting pressure, habitat improvements, better water quality, enlightened resource management, and environmental education. And there are no doubt other reasons, including favorable weather conditions and little-understood cycles within a species or group of species. Regardless of the causes, many of the top predators, that is, the fish eaters, have prospered. By inference, the Bay's fish population has also increased, which augurs well for the Chesapeake if we maintain our vigilance.

More than 2,000 plants and animals have been identified from the Chesapeake Bay region, but only a fraction of these will commonly be seen by the layman. Many are rare, some so rare that they have been observed only once; many are minute, too small to be identified without the aid of a microscope; and many are not, strictly speaking, marine organisms but are, rather, animals such as waterfowl, mammals, or amphibians closely associated with Chesapeake Bay habitats but not actually living in the water.

This newly revised guidebook covers marine animals, along with selected birds, seaweeds, and wetland plants of the Chesapeake Bay and the mid-Atlantic area. It discusses and illustrates 70 species of plants; more than 200 species of invertebrates, such as insects, crabs, clams, oysters, worms, and jellyfishes; more than 100 species of fishes; several reptiles; 65 species of birds; and freshwater and marine mammals. Most of the animals and plants included in this book are likely to be seen by an observant person walking along the water's edge, treading a marsh, peering down from a jetty or pier, searching through a

Preface

bushel of oysters, or bringing up a fishing line or crab trap. Some deeper-water species, such as the chartreuse-eyed mantis shrimp or the spotted brief squid, are less likely to be encountered under ordinary circumstances but are included because of their uniqueness, abundance, or interesting characteristics.

This guidebook is for anyone who comes in contact with the tidewaters of the Chesapeake: the year-round Bay resident—adult or child—the summer visitor or camper, the hunter and fisherman, the conservationist, the student with a project to complete. The science teacher will find this book valuable in the classroom and for field trips. The contents, sufficiently comprehensive to be of special aid to the professional scientist, provide a single practical identification guide covering invertebrates, fishes, and other species, with a focus on estuarine organisms of the Bay.

Field guidebooks are usually organized on a phylogenetic and taxonomic level, that is, progressing from plants to animals and from the simplest forms to the most advanced organisms. This is a logical method, easily understood by scientists, but the organizational subtleties are not obvious to many nonscientists. This book is organized according to broad, easily recognized habitats and emphasizes ecological relationships rather than classical phylogenetic hierarchies. We feel that this approach is more meaningful and useful to the greater part of the intended audience.

Ecology is a dynamic science. Owing to the research efforts of numerous scientists, our understanding of the biology of individual species, their taxonomic status, and their relationship with other organisms and their environment is subject to frequent revision and expansion. This book should be regarded as a snapshot—a moment in time—of accrued information on estuarine and estuarine-associated organisms. The reader should be aware that new information, constantly coming to the fore and thereby increasing our knowledge, is the essence of science.

We take pleasure in acknowledging all those who have helped us with this second edition. Dr. Edward Christoffers ferreted out difficult-to-find and necessary information for us. Richard Kleen, a "big man" in the birding world, made sure that the sections on birds would fly. Our longtime colleague, Dr. Victor Kennedy, kept us current on new work in crustacea and mollusks. Dr. Anson "Tuck" Hines did the same from his lab on the western side of the Bay. We thank Dr. Dale Calder, Dr. Fred Holland, and Dr. Austin Williams for updating us on the taxonomy of hydroids, polychaetes, and crustaceans, respectively. Dr. Max Hensley, our mentor in matters herpetological, was always available and helpful when we called on him for help. We thank Dr. Frank Schwartz for allowing us to use some of his work on the turtles of Maryland. Tim Goodger waded through the wetlands chapter and kept us from going over our heads. Dr. John McDermott helped us by providing information on nemertean and polychaete worms. Dr. Richard Snider was a fount of information on insects. Dr. T. Wayne Porter, our old friend and professor, gave us his always helpful and valued guidance. John Page Williams shared his keen observations on the Bay's natural history with us. We deeply appreciate all the kindnesses and help that our good friend Sara V. "Sally" Otto has given to us over the years. Capt. Ed "Snoozer" Watson, another wonderful friend, has helped us navigate around many a rocky shoal, and we are grateful to him. We acknowledge, as well, all those who contributed to the success of the first edition: the staff at the National Marine Fisheries Laboratory at Oxford, Maryland; the photographic assistance given us by Mrs. Joyce Firstman and Ted Stadel, which continues as a major contribution in this edition; and the scientific knowledge imparted to us by our colleagues Nancy Kirk Mountford, Dr. Edythe Humphries, Dr. Gordon Thayer, Dr. Robert J. Diaz, and Dr. William Hargis.

Robert Harington, Barbara Lamb, and Carol Ehrlich of the Johns Hopkins University Press have been a delight to work with during the production of this book. We thank them for their patience, good nature, and consummate professionalism.

Finally, we wish sincerely to acknowledge all our scientific colleagues whose years of research on the Chesapeake Bay and other coastal waters have provided the great wealth of information on the plants and animals we discuss in this book.

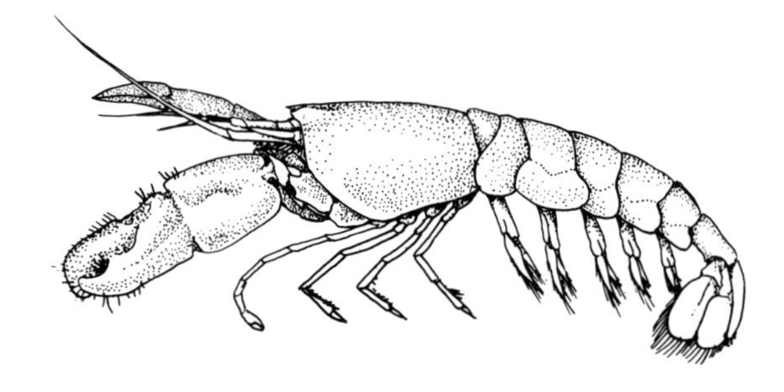

LIFE IN THE CHESAPEAKE BAY

ECOLOGY OF THE CHESAPEAKE BAY

The Chesapeake Bay was formed some 12,000 years ago when the last great ice sheet melted, raising the level of the sea and flooding the valley of the ancient Susquehanna River. The path of this old riverbed formed the deep channels of the present Chesapeake Bay, which is 180 miles long and runs in a north-south direction roughly parallel to the Atlantic seacoast, passing through two states, Maryland and Virginia.

This vast body of water is highly branched and its shoreline is deeply and irregularly incised by many embayments and tributary streams of various sizes. Nineteen principal rivers and 400 lesser creeks and streams are tributaries of the Bay, creating more than 4,600 miles of tidal shoreline. The main stem of the Bay and all its tributaries upstream to the limits of their tidewaters make up the Chesapeake Bay estuarine system (p. 6). The western shore rivers are generally larger than those on the Eastern Shore and are the principal source of freshwater gathered from broad watersheds extending up into the Appalachian mountain ranges to the west and north. Three of these rivers—the Susquehanna, the Potomac, and the James—contribute 80 percent of the total freshwater entering the Bay. The Eastern Shore rivers cut across the low, flat countryside of the Delmarva Peninsula and are characterized by large expanses of marshlands, which support great numbers of migratory waterfowl during autumn and winter.

Compared to the oceans, the Chesapeake Bay is very shallow, the average depth of the main stem being less than 30 feet and the average depth of the entire system, including all tidewater tributaries, only 20 feet. Generally, the bottom of the main stem and tributaries slopes gradually from both shores to deeper channel waters. Broad shoals and extensive flats occur in the Bay. Deep holes also occur in various spots; the deepest, 174 feet, is located just southeast of Annapolis, Maryland. The vast expanses of relatively shallow water in the Bay support a wide vari-

1. ECOLOGY OF THE CHESAPEAKE BAY

ety of bottom life that thrives at depths of less than 20 feet. The Chesapeake's world-famous oyster and soft-shelled clam harvests are attributable to the amount of suitable shallow-water habitat present in the Bay. The deeper areas, too, are important to the ecology of the Bay; many animals seek the warmer, more stable environment that prevails in these areas during the winter months, when the shallower waters become colder in response to abrupt changes in winter air temperatures.

The tides rise and fall twice daily in the Chesapeake. The magnitude of this rise and fall varies along the length of the Bay, the tidal range being about three feet at the mouth, gradually decreasing to a foot in the vicinity of Annapolis, and from there to the head of the Bay increasing again to two feet. At the head of the larger tributaries, such as the Potomac River, the tidal range may increase to as much as three feet. The height of the tide varies with the time of month and the season. The greatest tidal amplitudes are usually associated with high winds and maximum monthly (spring) tides. The regular rise and fall of waters in the Chesapeake create many areas of intertidal habitat which are occupied by communities of organisms especially adapted to living under these conditions.

The Chesapeake Bay lies within a temperate geographic zone with seasonal changes in water temperature more marked than those found in estuaries to the south but not as extreme as those found toward the north. Most biota of the Bay respond to seasonal temperature cycles and to day length. In spring, there is a general resurgence of activity: migratory species such as fishes and crabs move toward warming shallows from deeper channels or from the ocean, where they have overwintered; anadromous herrings and shads move in from the sea and upstream to freshwater streams to spawn; and bottom-living invertebrate communities begin to grow and reproduce. Throughout spring and summer many fish enter the Bay to feed on the abundant schools of smaller prey fishes, such as anchovies and menhaden. In fall and winter, there is a general movement of many migratory species out of the Bay and a decline in activity and growth of bottom fauna. However, certain species, such as waterfowl, have a reversed cycle, with greatest growth and abundance occurring during the colder months. Thus, an awareness of the seasonal changes in community life in each habitat is imperative for anyone studying Bay life.

HOW SALTY IS THE WATER?

One of the most important factors in understanding Bay ecology is the comprehension of salinity (salt content) distribution throughout the Bay system. Unlike the ocean, where salt content varies little over a relatively large area, the Chesapeake Bay, like all estuaries, contains waters that range from fresh to nearly as salty as ocean waters. An estuary is defined as a somewhat restricted embayment in which the flow of freshwater mixes with high-salinity ocean water. At the head of the Bay and at the head of each Bay tributary stream (the geographical fall line), tidal influence is apparent, but little or no ocean-derived salt is present. The salinity increases gradually downstream; midway down the Bay salinity concentration averages about 15 parts of salt to 1,000 parts of water (15 ppt), a concentration approximately half that of ocean salinity, which is generally about 30–35 ppt. Salinity also increases from the surface to the bottom, and deeper waters may be 2–3 ppt saltier than surface waters. A number of physical and geographic factors contribute to this phenomenon, the most significant being that saltier ocean water is heavier and intrudes into the Bay along the bottom, whereas lighter freshwater flows downstream in the surface layer. Salinity distribution is also affected by two other factors: the Coriolis effect,

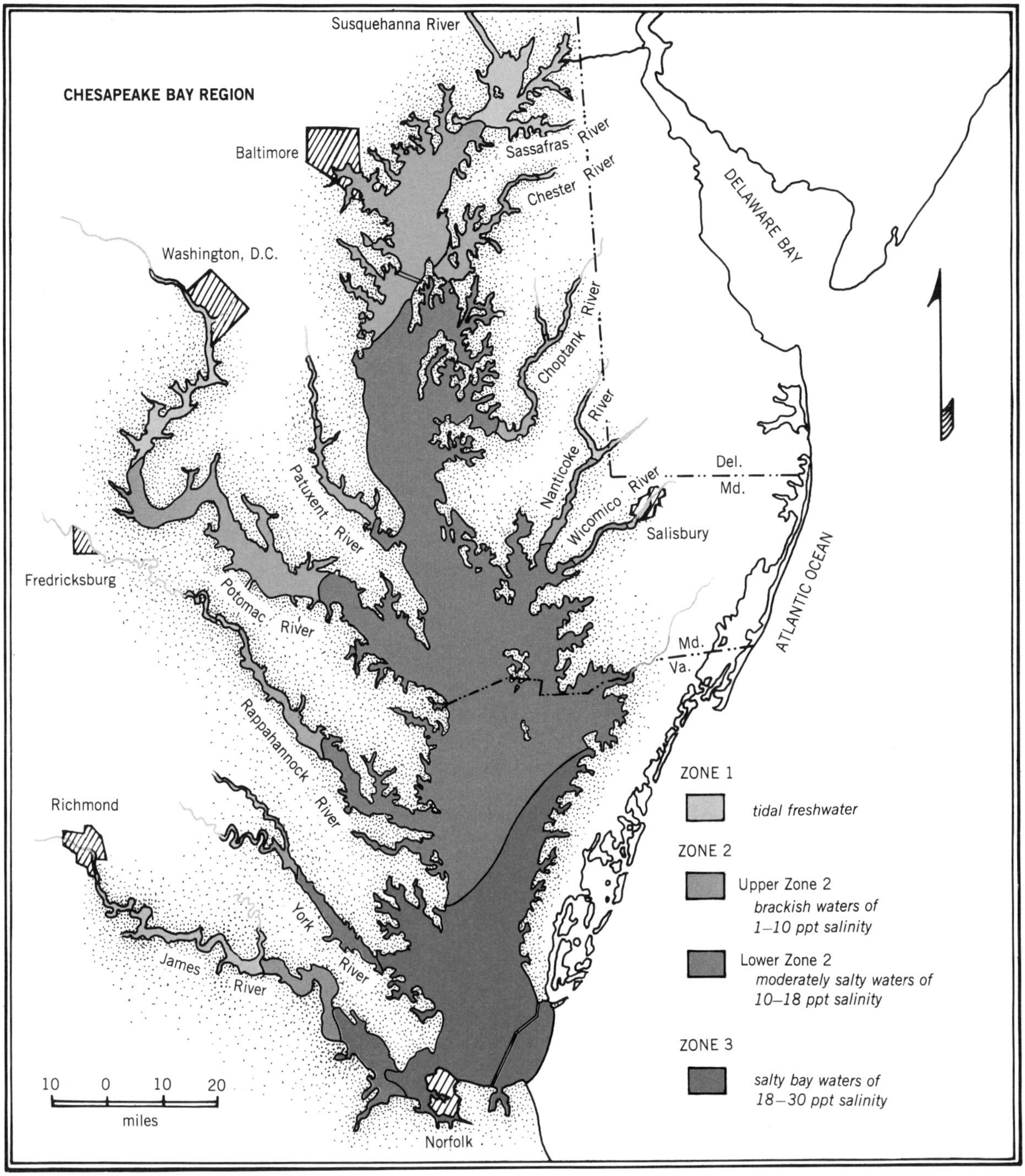

CHESAPEAKE BAY REGION

Susquehanna River
Baltimore
Sassafras River
Chester River
DELAWARE BAY
Washington, D.C.
Choptank River
Patuxent River
Nanticoke River
Wicomico River
Del.
Md.
Salisbury
Fredricksburg
Potomac River
ATLANTIC OCEAN
Md.
Va.
Rappahannock River
Richmond
York River
James River
Norfolk

10 0 10 20
miles

ZONE 1
tidal freshwater

ZONE 2
Upper Zone 2
brackish waters of 1–10 ppt salinity
Lower Zone 2
moderately salty waters of 10–18 ppt salinity

ZONE 3
salty bay waters of 18–30 ppt salinity

caused by the earth's rotation on its axis; and the higher discharge of freshwater from western shore tributaries, which results in saltier waters along the eastern shore of the Bay. In spring, when freshwater flows are highest, owing to spring rains and melting snow, salinity in a given area may run about 2 ppt less than average, and in the autumn, when freshwater flows are normally lowest, salinities may run 2–6 ppt higher.

Because of the wide range of salinities, Chesapeake Bay waters provide perfectly suitable habitats both for freshwater species in the upper reaches of the main stem of the Bay and its tributaries and for typical ocean species toward the mouth. Those species with the greatest tolerance for salinity changes (called euryhaline species) penetrate far into the estuary. Thus, certain euryhaline freshwater species, such as yellow perch or the brightly colored pumpkinseed sunfish, may be found well downstream into salinities as great as 10 ppt, and euryhaline marine species, such as spot or Atlantic croaker, may be found upstream in water that is nearly fresh. Certain species found in the Chesapeake are true estuarine species. They are distributed throughout the Bay itself but do not extend into freshwater or into the sea.

No matter what the habitat—a beach, a marsh, an intertidal mud flat—salinity determines, to a large extent, the kinds of species that live there. It is, therefore, important to know the general salinity of any area you observe. To help the reader, we have divided the Chesapeake Bay region into three zones according to salinity. Zone 1 covers tidal freshwaters at the head of the Bay and its tributaries. Many short creeks and inlets off the Bay proper or off larger rivers do not have freshwater input and are, therefore, without a Zone 1 region. Zone 2 covers waters ranging from slightly brackish to moderately salty, with average maximum salinities of about 18 ppt. This zone encompasses the largest area and is inhabited by the typical estuarine biota of the Chesapeake. It has been subdivided into upper Zone 2, where average salinity ranges from 1–10 ppt, and lower Zone 2, from 11–18 ppt. Zone 3 covers the lower Bay regions, with salinities from 18 ppt to ocean salinities;. Many ocean animals typical of the Atlantic seacoast inhabit Zone 3 along with many estuarine species.

Most of the organisms found in the Chesapeake Bay occur in other bays, inlets, and tidal tributaries of the mid-Atlantic coast, and salinity zones similar to those of the Chesapeake Bay occur in these bodies of water. A species list of all the plants and animals we discuss, along with their distribution by salinity zones, as well as a bird species list and their seasonal occurrence, is provided at the end of this book. In addition, there is a selected reading list for those who desire further references to seacoast biota and their habitats.

TYPICAL HABITATS OF THE BAY AND ITS TRIBUTARIES

Animals that live in the Chesapeake system have evolved and adapted over the ages to live in a variety of habitats. Some species are restricted to a specific habitat; others are ubiquitous. The little oyster crab lives only within a living oyster, whereas the indomitable blue crab is as often seen swimming in the deep waters of the open bay as skulking among the waving strands of a grass bed.

Organisms are often categorized by the broad general habitat they occupy. Pelagic plants and animals live in the open water and benthic plants and animals live at, or in the bottom. The pelagic group is further divided into nekton, free swimmers (primarily fishes) capable of strong self-directional movement, and plankton, composed of feeble or nonswimming organisms that are easily conveyed by the tides and currents. Plankton contains the most abundant and diverse life forms in the Chesapeake Bay. Most are microscopic plants and animals that are unseen, or, if seen, rarely noticed by the ordinary person. They are small in size but not in importance, constituting, as they do, the primary food source for most higher life in the Bay. The minute plant forms called phytoplankton include single-celled algae, such as diatoms, and dinoflagellates. Some species are so minute that they can be observed only under the highest-powered microscopes; others aggregate and can readily be seen as floating masses of green scum or as reddish brown patches on the water's surface. Phytoplankton is consumed in vast quantities

The waters of the Chesapeake Bay and its tributaries. The region is zoned according to the saltiness of the water. Because biotic communities change from zone to zone, the reader should use this map to identify the salinity zone of any locale before attempting to identify organisms found there.

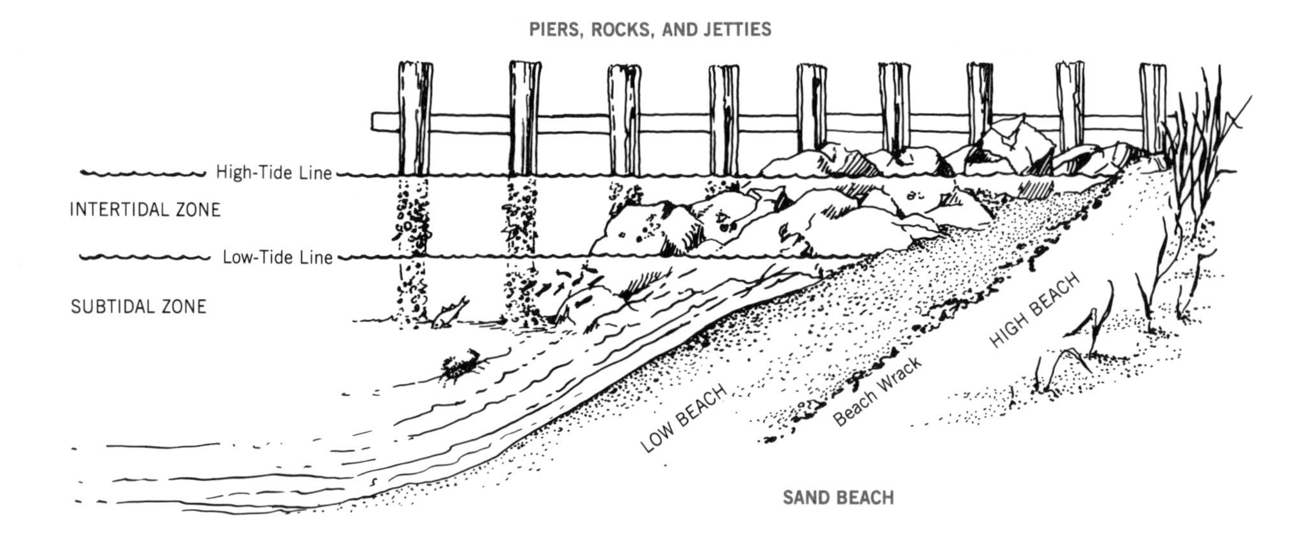

by herbivorous microscopic planktonic animals called zooplankton. Most zooplankton is microscopic, but larger free-floating animals, such as jellyfishes, amphipods, shrimps, and certain worms, are also considered part of the zooplankton. The most important group of zooplankton in the Bay are copepods, tiny crustaceans related to shrimp and crabs. Copepod populations are prolific; hundreds can occupy a quart of water dipped from the Bay. Some copepods are large enough (1 mm or more) to be seen by the naked eye. Raise a clear container of Bay water to the light and you will oftentimes be able to see the small creatures jerkily propelling themselves through the water. There are thousands of species of microscopic plants and animals in the Chesapeake, but only brief mention is made of them in this book as most readers will be unable to see them clearly enough to identify them without a microscope.

The benthos is made up of the plants and animals that live in or on the bottom. Benthic plants include many microscopic algae—the green, red, and brown seaweeds found along shoreline waters—and many species of rooted aquatic plants that grow in shallow waters. Benthic animals creep or crawl along the bottom, burrow into the mud or sand, or attach themselves to any hard surface, be it rock, rope, piling, or shell. The attached forms are referred to as sessile animals.

Animals that burrow in the bottom are called infauna, whereas sessile animals and animals that move over the bottom are referred to as epifauna. Designations are not rigid, since some infauna emerge from the bottom sediments to crawl along the bottom or even become part of the plankton at certain times during their life cycles. The common clamworm, for example, may burrow into bottom sediments, crawl among the seaweeds and barnacles of a pier piling, or swarm to the surface in late spring or summer to spawn. Many benthic invertebrates produce free-swimming or -floating pelagic larvae. Larvae of worms, barnacles, and mollusks may compose the greatest portion of the zooplankton at certain times of the year. The biological associations that exist in the Chesapeake Bay, whether pelagic or benthic, are largely dependent on the habitat type, which is delineated by type of bottom sediment, salinity, water depth, or other physical and chemical features.

We have distinguished eight easily recognizable habitat types in the Chesapeake Bay. There is a chapter for each of the eight habitats with descriptions of the biological community and illustrations of the animals found there. Many species, such as waterfowl, shorebirds, birds of prey, blue crabs, and fish that roam the Bay, are common to more than one type of habitat; others are restricted to a single habitat.

Sand Beaches

Sand beaches in the Chesapeake Bay are not buffeted by waves and winds as are the turbulent ocean beaches. They are not as broad and do not form the high, protective dune lines of the seacoast. However, like the ocean beaches, they present different beach zones according to the tides. The intertidal zone existing along the lower reach of the beach is intermittently submerged as the tide ebbs and flows, whereas the upper beach zone is wet only during the very highest tides.

The fauna is not generally as diverse on sand beaches as it is in areas with muddier bottoms. Oftentimes the most abundant indications of life are heron tracks or empty sea shells and other dead remains blown up onto the beach, frequently concentrated in the beach wrack, a windrow of debris along the beach.

Intertidal Flats

Intertidal flats occur along the shore where the bottom is alternately exposed and covered by the tides. The bottom may be very soft and oozy, composed of fine silts and muds, or it may be relatively firm, with much sand intermixed with the muds. The landward boundary may be a mud bank or may gradually rise to a marsh habitat. Sandy-mud flats often merge with a sand beach. A flat with a deep slope toward the channel will be only a narrow band at low tide, whereas one with a gentle slope will extend for hundreds of feet toward the channel. In some smaller creeks or streams, the bottom may be totally exposed during low tide. Intertidal flats harbor a wide diversity of plants and animals buried in the muds or crawling over the surface, including bacteria, algae, worms, snails, and little buglike crustaceans called amphipods. At low tide, intertidal flats offer up their riches to a variety of shorebirds, herons, gulls, and terns. At high tide, the intertidal flat becomes a shallow inshore habitat for pelagic marine life moving shoreward with the rising waters.

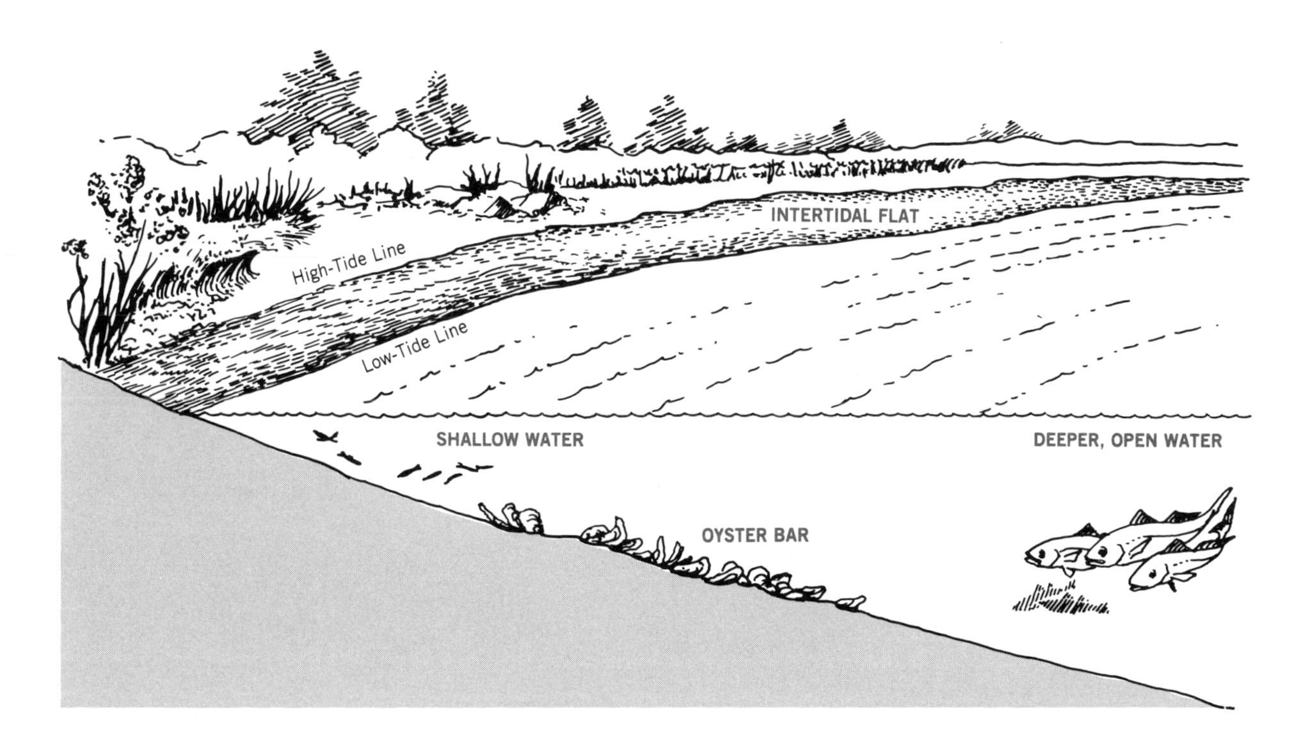

Piers, Rocks, and Jetties

The hard surfaces of pier pilings, rocks, jetties, and other similar structures provide a suitable habitat for many attached plants and sessile animals, which in turn provide food and haven for other animals. The composition of the pier or jetty community varies according to the tidal range. Animals at the highest levels, moistened only by spray, differ from those found within the intertidal and subtidal zones. Biota typical of the pier or rock habitat are also common on any submerged firm substrate, such as crab traps, sunken logs, ropes, oyster shell beds, and larger rocks or pebbles. Small fish and invertebrates of the shallow waters become closely associated with this community as they feed on the rich variety of food attached to the hard substrate.

Shallow Waters

The shallow subtidal zone, only a few feet deep, is a habitat often encountered by the swimmer, crabber, or boater. Schools of small fishes may often be seen darting away from a leg movement, a little nip may be felt on the toe, or a stream of bubbles may be seen rising from a tiny hole in the bottom. Ducks and geese dive and "tip up" to feed on aquatic vegetation and invertebrates. Most of the benthic life here is similar to that found in the exposed intertidal zones, although some shallow-water animals are found only subtidally. The pelagic life is varied and abundant, but because many of the creatures are too small to be noticed or too rapid in their movements to be readily identified, they are unknown or unrecognized even to those who often wade or swim in this habitat.

Seagrass Meadows and Weed Beds

In many places shallow-water areas contain broad expanses of aquatic plant beds. These plants in themselves create a special and separately considered habitat for a number of diverse organisms.

Aquatic plants take hold in softer mud and silt bottoms and blanket the bottom with their waving leaves, affording protection and food to a variety of waterfowl, invertebrates, and small fishes. Many invertebrates dwell

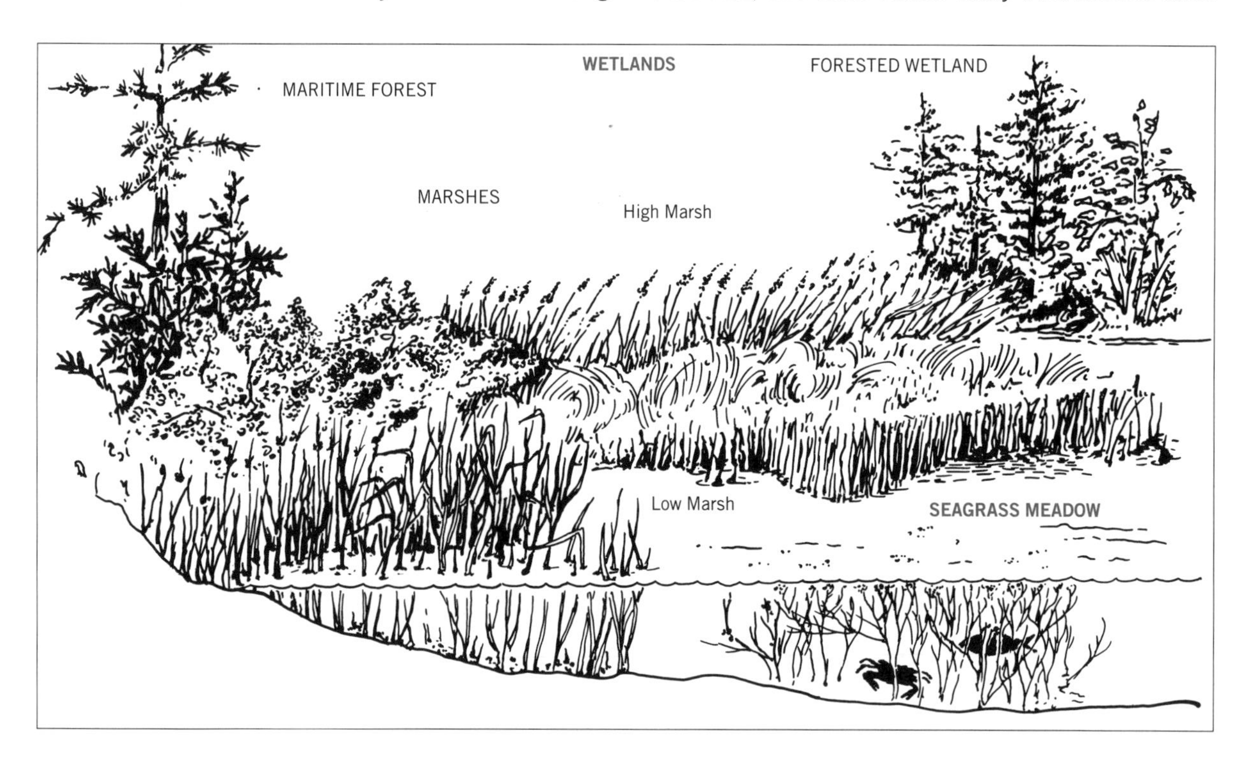

among the holdfast roots, attach to the leaves, or bury themselves in the substrate. The most common submerged aquatic plants are described in this book principally because of their importance as habitats for animals. In the fresher waters of Zone 1 and Zone 2, in the upper Bay and tributaries, stands of sago pondweed, redhead grass, horned pondweed, and widgeon grass form weed beds, whereas in the mid and lower Bay, meadows of widgeon grass and eelgrass predominate. Submerged plant beds often signal their presence by tips of leaves or seed heads breaking the surface. The waving plants can also be seen when passed over in a boat, and the beds are often exposed during the very lowest tides.

Wetlands

Wetlands include a number of different habitats discussed in this book. They range from the low fresh- and saltwater marshes at the edge of the water to high marshes, maritime forests, and forested wetlands. Close to shore, emergent species of aquatic plants take hold in the mud bottoms to form low-marsh areas. Low-marsh plants give way at higher elevations to the more diverse marsh plant life that forms the high marsh. The high marsh is inundated only intermittently, during highest tides. The high-tide marsh merges with upland terrestrial plants where the narrow bands of salt-tolerant and moisture-tolerant plants growing just beyond the high marsh are often referred to as maritime forests, or bay edge forests. Forested wetlands grow along the flood plains of upper tributaries; here the trees and the lush understory of shrubs and vines are adapted to growing in waterlogged and flooded bottomlands. Each type of wetland has its own community life—some creatures attach themselves or cling to grasses, while others, such as beavers, build dams and impoundments in forested wetlands. Mussels burrow in the marsh soil, and birds feed and nest in wetland habitats. When wetlands are inundated by seasonal flooding or daily tidal fluctuations, mobile animals such as crabs and fishes move in to feed.

Oyster Bars

Oyster bars cover extensive bottom areas throughout the mid and lower Chesapeake Bay. The oysters provide a substrate for numerous sessile invertebrates, many of which are the same species found on the piers or rocks.

Mobile epifaunal invertebrates, such as worms, snails, and small crustaceans, creep in and out of the nooks and crannies of the oyster shells, grazing on the attached animals and sometimes preying on the oysters themselves by boring into their shells. Many small fishes are part of the oyster community, preying on the easily obtainable food and using empty oyster shells to attach and shelter their eggs and larvae. The major oyster bars of the Chesapeake are subtidal and therefore are not as easily observed by the Bay visitor. The oyster bar habitat provides a rich and distinct community important to Bay ecology; although not as accessible as the shoreline habitats, it should not be ignored by students of the Bay.

Deeper, Open Waters

As the water deepens, the benthic flora and fauna change. In these deeper reaches of the Chesapeake Bay and its tributaries light penetration is diminished and the seaweeds and rooted aquatic plants disappear. Bottom sediments gradually become finer, and the soft, silty ooze supports fewer benthic invertebrates. The bottom water and the soft sediments are often depleted of oxygen in the deepest channels, particularly during the summer. However, in the flowing waters above the bottom there is a thriving community of pelagic fishes and invertebrates. Schools of small forage fishes, such as menhaden and anchovies, and large predatory fishes, such as bluefish, striped bass, and seatrout, roam this habitat. Loons and sea ducks dive for fish and small clams in the deeper waters. Jellyfishes, blue crabs, squid, and arrow worms are also found in the open waters of the Bay. Here, as in the oyster bar habitat, evidence of life is often gleaned not by direct contact but through secondary evidence: an unusual fish is caught by an angler; the water surface is broken by innumerable ripples, evidence of a school of menhaden; or the winglike "fins" of a cownose ray create a momentary swirl on the surface.

CLASSIFICATION OF ANIMALS AND PLANTS

To some, the biologist's favorite occupation seems to be one of categorizing organisms and labeling them with obscure scientific names, much to the consternation of the layman. This process—called taxonomy, the science of plant and animal classification—is important, however, because it provides some order and logic for understanding the relationships among the various groups and species of organisms. Furthermore, it means that every described species has a name that is understood by scientists throughout the world, no matter what language they speak. Many of the species in the Chesapeake Bay have a colloquial or English name, referred to as a common name. Some less common species have not yet acquired colloquial names, so we have given them what we consider to be appropriate common names. The first time a species is mentioned in this book, both common and scientific names are given; thereafter, only the common name is used.

Because this guidebook is organized according to habitat and because many of the same organisms may be found in more than one habitat, it will be helpful to discuss briefly the scientific classification of animals in the Chesapeake Bay. Later, readers can refer back to this section to help determine what general kind of plant or animal they are dealing with.

A hierarchy of categories, the Linnean System, is the most widely accepted system of classification, and it proceeds from kingdom, through phylum, class, order, family, and genus, to species. The characteristics of the major phyla occurring in the Chesapeake Bay are briefly described below, but be advised that it is often difficult for a novice to determine the phylum of a particular animal, especially some of the many wormlike creatures that are found in abundance in the Bay. However, in many instances only a single species of the phylum is found in the Bay, making identification somewhat more certain. The descriptions of some classes, the major subgroups within a phylum, are also discussed under some of the larger phyla. The first scientific name of a specific organism, the genus, is always capitalized; the second scientific name, the species designation, is not capitalized. Most scientific names are of Latin origin. Thus, *Callinectes sapidus,* the blue crab, when translated from the Latin becomes "beautiful savory swimmer." Some few animals are referred to in this book by generic names only.

Phylum Porifera—Sponges

Sponges are the simplest form of multicelled animals, with no organs or differentiated tissues. A simple skeletal structure of tiny glassy or calcareous spicules and spongin fibers provides a framework for the living cells. Water, bringing food and oxygen to the living cells, enters through tiny pores on the surface of the sponge, passes through a network of canals and chambers, then exits through other larger pores. Sponges in the Chesapeake Bay are not like large, soft, commercial bath sponges but are, rather, low encrusting forms that grow over rocks, stones, or pilings, or are irregularly shaped masses with knobbed, fingerlike projections. Their surface appears "suedelike" owing to the tips of the tiny skeletal spicules projecting through the skin. Often, broken pieces are detached and wash up along the shore. Sponge identification is generally based on microscopic examination of spicule shapes. Consequently, the collector will be able to identify most Chesapeake Bay sponges only to a general type, although a few species, such as the redbeard sponge, with its bright red to orange color and intertwining fingers, are easily recognized. Sponges may be confused with other encrusting growths, such as bryozoans or hydroids.

SPONGES

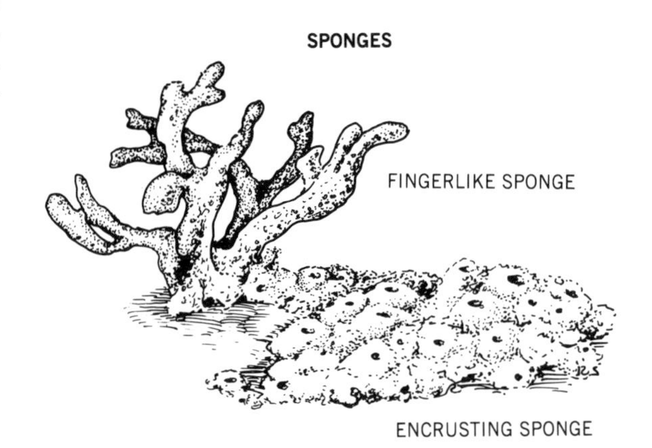

Phylum Cnidaria—Hydroids, Jellyfishes, Sea Anemones, and Corals

Cnidaria includes a number of varied classes of marine invertebrates all sharing certain characteristics. Cnidarians are more advanced than sponges. They are composed of two specialized layers of cells separated by a gelatinous noncellular layer, the mesoglea. They have only

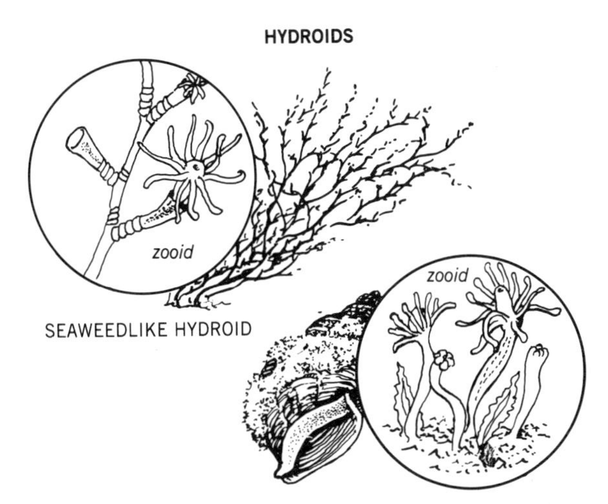

one opening—a mouth, which is connected to the gastrointestinal cavity. Radial symmetry, in which the body parts are arranged around a central axis, is characteristic of this phylum. Cnidarians are unique in the animal world in possessing stinging cells, or nematocysts. Another interesting characteristic of the phylum is the development of two different life forms—a sessile polyp stage and a free-floating medusa stage. Some species alternate from one stage to the other, whereas others have only a polyp or a medusa stage. Three classes of cnidarians occur in the Chesapeake Bay.

Class Hydrozoa—Hydroids. Hydroids are abundant in the Chesapeake Bay but often go unnoticed because they appear simply as thin, fuzzy coatings on piers, rocks, and shells or as waving fronds mistaken for seaweeds. Of the many types of hydroids in the Bay, some species form both polyps and hydromedusae (tiny floating jellyfishes) and others occur only as polyps. Most hydroids are colonial, that is, they are a collection of individual animals, called zooids, each with a mouth and tentacles, but with interconnected digestive cavities. Hydroid colonies attach to the substrate by means of a horizontal root system. In some hydroids the zooids arise directly from this base; in others, they emerge from branching stems, stems that are supported by a chitinous envelope around the soft tissues.

Class Scyphozoa—True Jellyfishes. Among Scyphozoa the medusa stage dominates, and the medusa bell is far larger and more noticeable than are hydromedusae. The mesoglea is thick and gelatinous. Tentacles containing many stinging cells extend from the rim of the medusa bell, and the mouth is edged with oral folds. The polyp stage is a tiny, inconspicuous form, whose primary function is to produce new medusae. There are only four species of jellyfishes in the Chesapeake. The most abundant and well known is the infamous sea nettle, which no one ever forgets after having encountered its stinging tentacles.

Class Anthozoa—Sea Anemones and Corals. Anthozoans have no medusa, only a polyp, stage. Anthozoan polyps differ from hydroid polyps in that the gastrointestinal cavity is divided by longitudinal septa into radiating compartments. Nematocysts are found on the feeding tentacles surrounding the mouth and the internal septa as well.

Sea anemones are solitary, soft-bodied polyps, topped with a ring of tentacles surrounding a slitlike mouth. Some anemones are attached to firm substrates by an adhering flat basal disk; others bury themselves in bottom substrates. There are only a few species of sea anemones in the Chesapeake Bay, and although none achieve the size or brilliant colorations of the tropical forms, they are nevertheless quite beautiful and, when observed underwater, look like graceful, swaying flowers. Unfortunately, most of

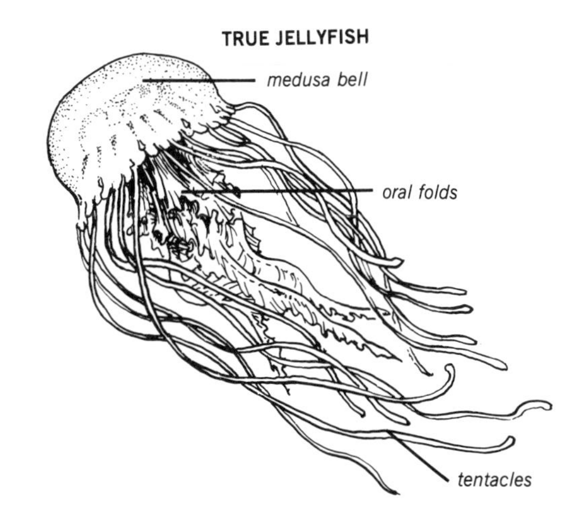

us see them out of the water, when they are withdrawn into small, inconspicuous blobs of jelly.

The coral rocks sold in shell and aquarium stores are the skeletal remains of living stony corals. Corals are closely related to sea anemones, but, rather than being solitary polyps, they are a colony of polyps each connected to the other over the surface of their skeletal base. The distinctive indentations of a coral skeleton mark the location of individual polyps. A single species of stony coral, the star coral, is found in the Chesapeake Bay.

Soft corals, similar to sea anemones and hard corals, are distinguished by having eight pinnate (featherlike) tentacles. A soft coral is a colony of small polyps supported by a skeletal mass composed of calcareous spicules or horny material. Each polyp is connected to others by a network of gastrodermal extensions that perforate the skeleton. Unlike the living tissue of hard corals, which is completely above the skeletal base, the living tissue of soft corals lies within the skeleton. A single species of soft coral, the whip coral, occurs in subtidal waters of the lower Chesapeake.

SEA ANEMONES AND CORALS

SOFT CORAL

SEA ANEMONE

HARD CORAL

Phylum Ctenophora—Comb Jellies

Ctenophores are similar to jellyfishes, with radial symmetry and a jellylike body tissue similar to mesoglea. However, they have no nematocysts, and their bodies are divided internally by eight ciliated bands of tissue edged with fused cilia, or combs. The combs beat rhythmically to propel the animal through the water, a swimming method different from that of jellyfishes, which propel themselves by pulsating the medusa bell. Only two species of ctenophores occur in the pelagic habitats of the Chesapeake Bay: the nut-shaped sea walnut, and an ovoid pink comb jelly.

COMB JELLY

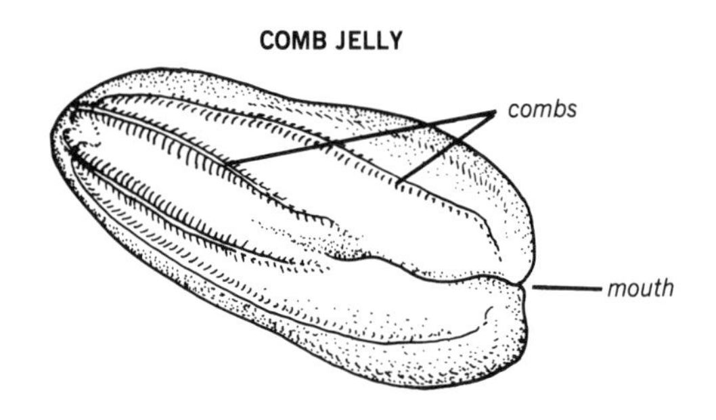

Phylum Platyhelminthes—Flatworms

There are many different types of "worms" or wormlike creatures in the Chesapeake Bay, and it is important to distinguish which general group or phylum a specimen belongs to if one has any hopes of being able to identify the species. Each phylum has distinctive characteristics that allow most laymen to identify the type of "worm" they are dealing with. Platyhelminthes is a very primitive group and includes parasitic tape worms and flukes as well as the free-living turbellaria, the only group that concerns us here. Turbellarian flatworms are abundant in the Bay. Most are very tiny and are hidden in bottom sediments, but a number are large enough to be easily visible. Almost any-

FLATWORM

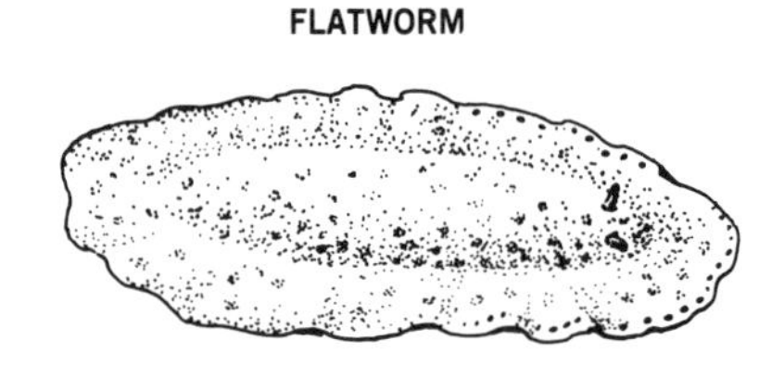

one who has picked up a shell or a stone from the water has probably unknowingly seen a flatworm, which appeared only as a flat gelatinous form. Flatworms are very thin and leaf-shaped, soft and unsegmented. They are one of the earliest group of animals to develop bilateral symmetry, a mouth (but no anus) and a digestive cavity, which in the larger free-living forms is highly diverticulated. They are often distinctly marked with tiny eyespots or other color-patterns.

RIBBON WORM

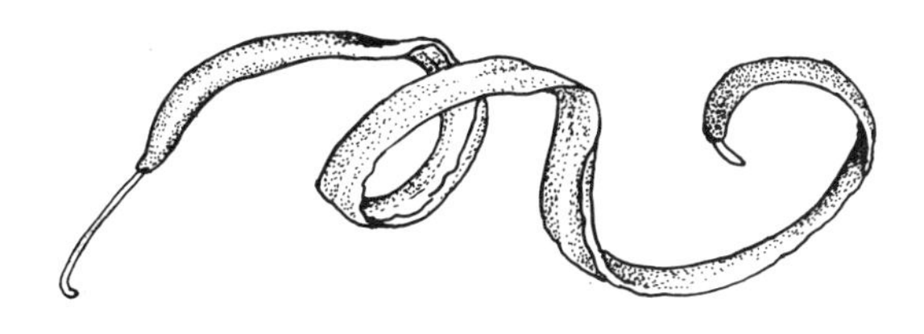

Phylum Rhynchocoela—Ribbon (Nemertean) Worms

As their name implies, ribbon worms are elongate, unsegmented, and often flattened worms without appendages. They are soft-bodied and break into fragments when handled. Ribbon worms are contractile, lengthening and shortening as they squirm and sometimes twisting themselves into a tangle. Ribbon worms are also known as proboscis worms because of a unique whiplike projection from their head. The proboscis is used for defense and capture of prey and may actually be longer than the worm itself. The proboscis retracts through a pore into a fluid-filled cavity surrounded by muscle. When the muscle contracts, the proboscis is quickly ejected. There are many ribbon worms throughout the Chesapeake Bay, buried in the mud, hiding under stones, among seaweeds or fouling growths on piers and jetties. Some are symbiotic and live in close association in hard clams and other mollusks and on blue crabs.

Phylum Annelida—Segmented Worms

Annelid worms are one of the most diverse and abundant groups of invertebrates in the Chesapeake Bay. The well-known terrestrial earthworm belongs to this phylum. Annelids are far beyond flatworms or ribbon worms in evolutionary development. They have a mouth and anus, a specialized head region with a brain, specialized organs for grabbing and tearing prey, and body cavities with organs analogous to our own. Some are extremely resistant to pollution and occur in contaminated sediments where little else lives. Most of the marine annelids in the Bay belong to the class Polychaeta, meaning "many hairs" and referring to the many bristles arising from special appendages called parapodia. The parapodia extend from each side of the body, one pair to each body segment. They are of diverse form and shape and may be lobed, jointed, feathery, stalked, or paddlelike. The head regions of polychaetes are also quite varied and are equipped with structures such as specialized tentacles, intricately feathered extensions, and beaklike jaws. Species identifications are often based on minute differences in parapodia or head structures, and determination of species generally requires the use of a microscope. Polychaetes of many types are found in almost every habitat of the Bay. There are burrowers, tube-builders, crawlers, and swimmers. Over 110 species of polycheate worms have been identified in the Chesapeake Bay, but we have included only those that have characteristic shapes or structures, build specialized tubes, or occupy unique habitats.

Other classes of annelids found in the Chesapeake include Oligochaeta (earthwormlike annelids) and Hirudinea (leeches). Little is known about leeches in the Bay, and this group is not covered here. Oligochaetes differ from polychaetes in that they have no parapodia or head appendages. They may have a few setae (or bristles) arising directly from each body segment in a bundle of two or more, but most oligochaetes are relatively smooth. Oligochaetes are primarily terrestrial or freshwater animals, but a number of

BRISTLE WORM (POLYCHAETE)

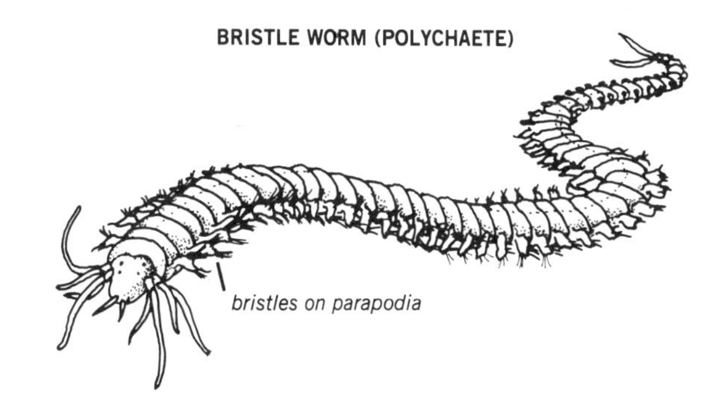

typically brackish-water species will be found in tidal fresh- and low-salinity waters of the upper Bay and its tributaries. Marine oligochaetes are generally small, threadlike worms and are difficult for nonspecialists to identify.

PHORONID WORMS

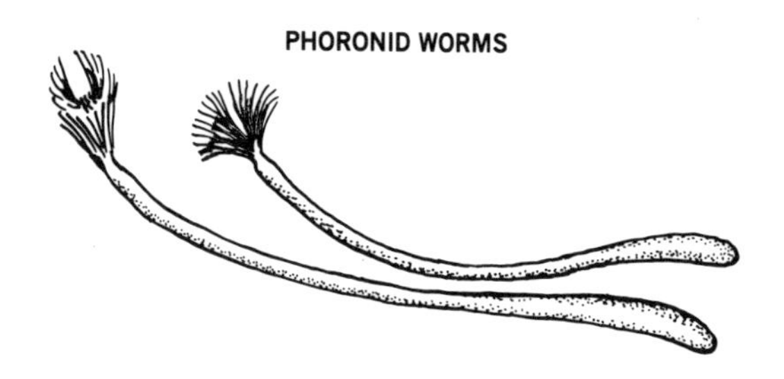

Phylum Phoronida—Phoronid Worms

Phoronida is a small phylum containing only 15 known species of wormlike animals, all marine and all living in parchmentlike tubes. Phoronids are more closely related to bryozoans (phylum Ectoprocta) than they are to annelids, although they look like some polychaete worms because of the tentacles that extend from their heads. The arrangement of two rows of tentacles in the shape of a horseshoe distinguishes them from all other wormlike creatures. Unlike annelids, they are unsegmented and have a rather smooth, soft body with a somewhat bulging end.

Phylum Mollusca—Mollusks

Clams, oysters, and snails are easily recognizable representatives of this diverse and important phylum. Of the six classes of mollusks, four occur in the Chesapeake Bay: gastropods (snails), with a single shell; bivalves (clams and oysters), with two valves or shells; chitons, with eight shell plates; and cephalopods, with an internal vestigial shell and here represented by the squid.

Class Gastropoda—Snails. The gastropoda contain the largest number of mollusk species. Representatives of this large and diverse group include periwinkles, whelks, slipper shells, marsh snails, and nudibranchs, as well as land snails and land slugs. The distinctive shapes and colors of the shells make identification relatively easy and enjoyable for the amateur. Gastropods have well-developed heads with tentacles and eyes and a large muscular foot adapted for gliding over almost any surface.

Class Pelecypoda—Clams, Oysters, and Mussels. The bivalves are also well represented in the Bay and include not only oysters and several species of clams but also mussels and arks. They differ from gastropods in that they have two valves or shells and a wedge-shaped foot adapted for burrowing into soft bottom sediments rather than for gliding over surfaces. They have two basic modes of feeding: suspension feeding, in which food particles are filtered from the water by means of gills; and deposit feeding, in which the bivalve sucks up organic material from the soft muds by means of long, flexible inhalant siphons.

MOLLUSKS

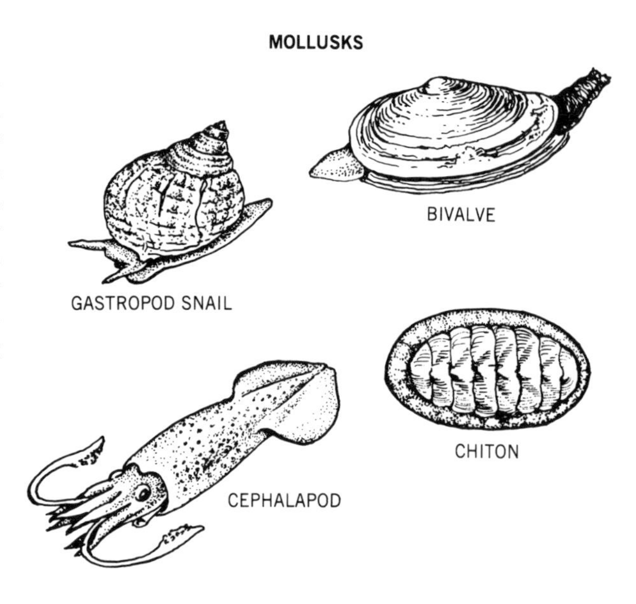

Class Polyplacophora—Chitons. This group of mollusks, the chitons, are elliptical-shaped animals with eight connected shell plates on top of their body and a flat, suckerlike foot on their lower, or ventral, surface. This foot enables the chiton to cling tightly to rocks, shells, and other hard surfaces, from which they are very difficult to dislodge. Their distinctive body structure makes it impossible to mistake chitons for any other animal. They graze like gastropods, moving slowly over the rocks and scraping off the algal film with small, rasping teeth. A single chiton, the common eastern chiton, is found in the Chesapeake Bay, in high-salinity areas only.

Class Cephalapoda—Squids. The squids, octopuses, and other members of the class Cephalapoda bear little resemblance to snails and clams, yet they are indeed close relatives. Cephalopods are the most highly developed of all mollusks, with large heads and eyes and a crown of tentacles homologous to the foot of other mollusks. The chambered nautilus is a cephalopod with an external shell; squid and cuttlefish have only a remnant of an internal shell; the octopus has no shell. Schools of brief squid are occasionally abundant in higher-salinity Chesapeake waters. This species is the only cephalopod occurring in the Bay.

Phylum Arthropoda—Jointed-legged Animals

Phylum Arthropoda contains more species and more individuals than any other animal group in the world and accounts for approximately 80 percent of the known animal species, largely because of the thousands of insect species that belong to this phylum. Arthropods have an external skeletal framework rather than an internal one, as do human beings and many other animals. Their bodies are covered with a hard, chitinous shell, or exoskeleton, to which their musculature is attached. Ligamentous joints allow movement of each body segment and appendage. In order to grow, arthropods must molt, leaving the old shell behind and forming a new, larger shell. There are two major types of arthropods, characterized by type of anterior appendages and also by certain other features.

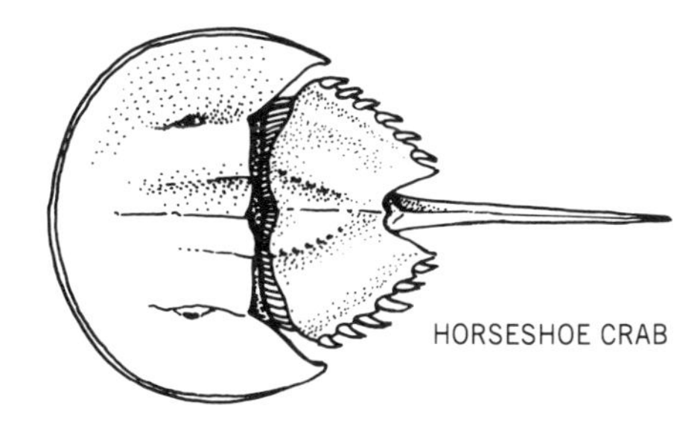
HORSESHOE CRAB

Both types are represented in the Chesapeake Bay. Horseshoe crabs, sea mites, and sea spiders belong to the first

INSECT
DRAGONFLY

major group, which also includes terrestrial spiders and scorpions. In other words, the horseshoe crab is more closely related to the common garden spider than to a blue crab, although it certainly resembles neither. The second major group includes class Crustacea, class Insecta, and a number of lesser classes. Insects are primarily terrestrial, although several species are found in tidal fresh- and low-salinity Chesapeake waters. Crustaceans are primarily aquatic and comprise crabs, lobsters, and shrimps in addition to innumerable microscopic animals of the zooplankton community, such as copepods. Crustaceans occupy every habitat of the Chesapeake Bay.

Barnacles are a sessile group of crustaceans. They were once thought to be mollusks because of the calcareous shells they construct and occupy. The animal within the shell, however, is jointed-legged, and its early stages of development are typical of other crustaceans. Two groups of small crustaceans, isopods and amphipods, are not familiar to most casual observers, but they are abundant and ubiquitous inhabitants of the Bay and, once noticed, will become familiar to any collector. Many common species are large enough to be identified easily. Isopods are buglike creatures with bodies flattened from top to bottom; amphipods are more shrimplike in appearance, with bodies flattened from side to side. Mantis shrimp are curious creatures living in the muddy bottoms. One species is found in the Chesapeake Bay. A number of distinctly different kinds of shrimps inhabit the Bay, including mysids, penaeids, snapping shrimps, and mud shrimps. Blue crabs, mud crabs, and oyster crabs are found throughout the Bay; lobsters do not occur in the estuary, but they are found offshore in the cooler, deeper, and saltier waters of the Atlantic Ocean.

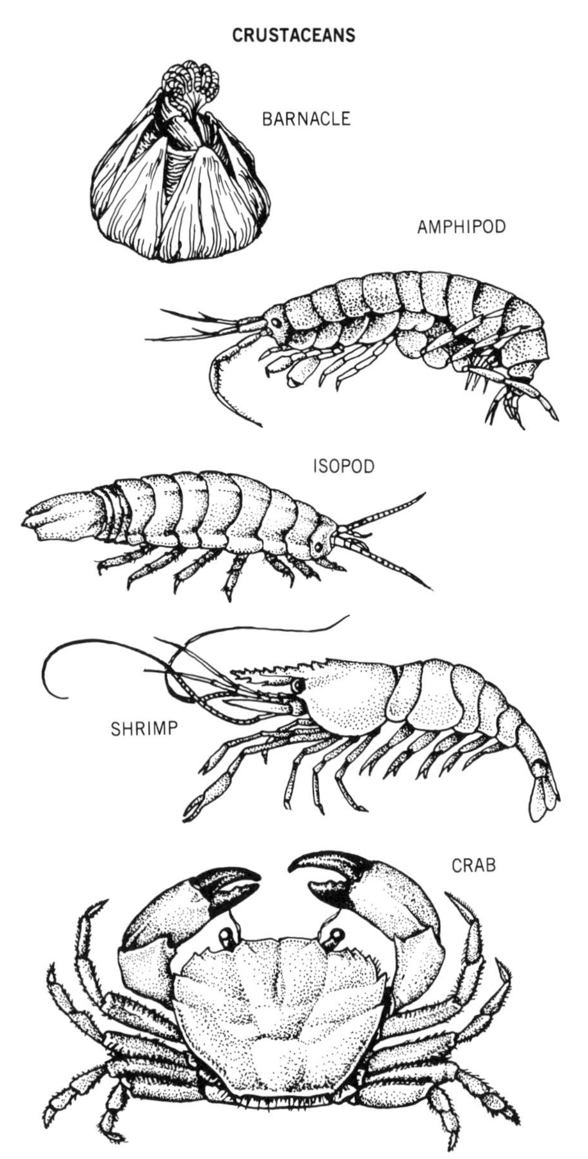

Phylum Bryozoa—Bryozoans

Bryozoans, or moss animals, as they are also called, are colonial, sessile animals, similar to hydroids but with individual zooids that are not connected to each other as they are in hydroids, and with lophophore (horseshoe-shaped) tentacles. Bryozoans are more evolutionarily advanced than hydroids. Because details of their body shapes and structures are not easily determined without the use of a microscope, most bryozoans are difficult, if not impossible, for the nonspecialist to identify to species. Bryozoans are present in many shapes, structures, and sizes: some are microscopic and grow along a creeping stolon; some are flat, hard, lacy-looking crusts growing on stones, shells, and vegetation; some are erect and appear like branched seaweeds or hydroids; and some are rubbery, lumpy masses very similar to certain sponges. All types are represented in the Chesapeake Bay.

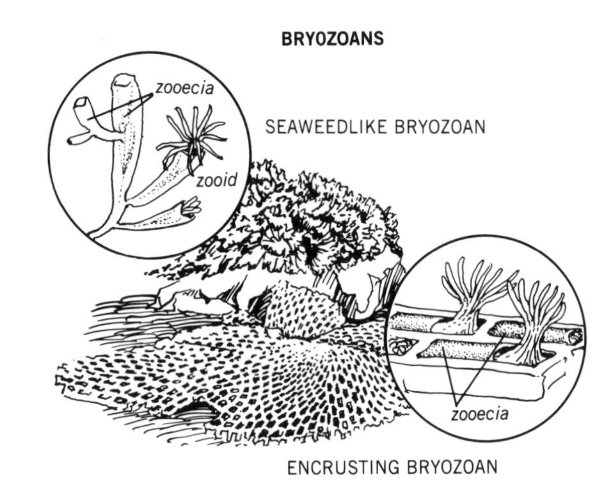

Phylum Echinodermata—Echinoderms

This group of exclusively marine animals includes the familiar sea stars, sea urchins, and sand dollars. Echinoderm means "spiny skinned" and refers to the prickly internal calcareous skeleton, which projects through the skin. The echinoderm body is divided into five parts around a central axis, a structure that is termed pentamerous radial symmetry. This five-sided symmetry is obvious if one looks at a sea star or the surface of a sand dollar. Echinoderms are considerably more advanced evolutionarily than radially symmetrical animals such as jellyfishes or sea anemones. Their method of locomotion—by means of hundreds of tiny tube feet—is unique and is accomplished by the progressive suctional grasping and releasing by the tube feet. Sea stars, or starfish, are particularly adept at opening clams and oysters by means of their tube feet, which can exert a powerful adhering force. Sea stars, brittle stars, and sea cucumbers are found only in the higher-salinity waters of the lower Chesapeake Bay.

ECHINODERMS

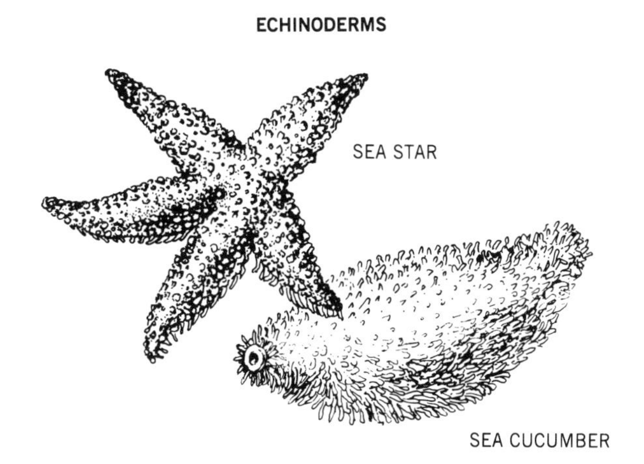

Phylum Chaetognatha—Arrow Worms

Arrow worms are small planktonic marine animals resembling fish rather than worms. They are often abundant in open waters of the mid and lower Bay but are inconspicuous because of their small size (an inch or less) and transparent bodies. They generally school offshore, out of view of the shoreline observer. At first glance, arrow worms look like fish with fins along their bodies and at the tip of their tail. A closer look, however, discloses a head with hooked claws, used for capturing small zooplankton and other prey, and tiny eyespots rather than the large, distinct eyes of fish.

CHAETOGNATH

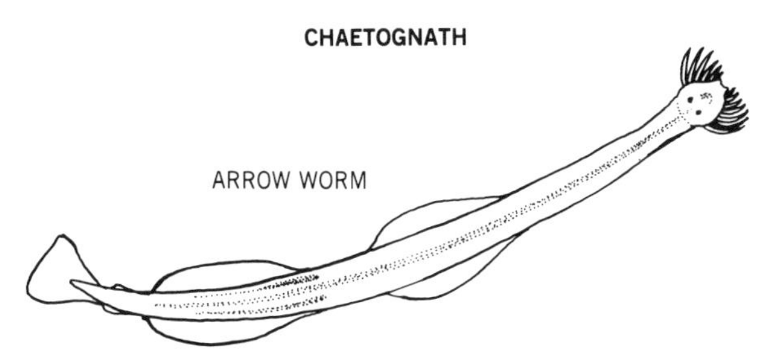

HEMICHORDATE

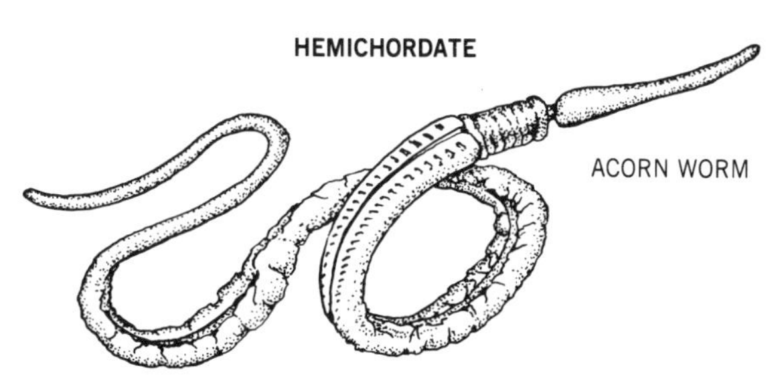

Phylum Hemichordata—Acorn Worms

Hemichordata is a small phylum of wormlike marine creatures that are interesting to scientists because their larvae have characteristics of both invertebrate larvae and fish larvae. Some zoologists believe this to be a transitional phylum between invertebrates and vertebrates. However, this advanced status is not apparent as one studies an adult worm. The only acorn worm in the Chesapeake Bay is easily recognizable by its three distinct and different-colored body segments: a white proboscis, an orange collar, and a brownish body. It lives in U-shaped burrows in the sandy-mud shallows of the mid and lower Bay.

Phylum Chordata—Nerve Cord Animals

Chordata is a very large phylum comprising two subphyla of invertebrates which have a notochord, a flexible, rodlike structure considered to be the primitive backbone, and the subphylum Vertebrata which comprises the vertebrates of the world: fish, birds, amphibians, reptiles, and mammals—and thus man. The invertebrate subphylum Urochordata has common representatives in the Chesapeake Bay.

Subphylum Urochordata—Tunicates. Tunicates show chordate characteristics—for example, the primitive notochord—only in their larval stages. Tunicates of the Chesapeake Bay, as adults, are small sessile animals with a tough outer skin, or tunic. The ubiquitous sea squirt of the Chesapeake is a tunicate. It is hard to believe that an undistinguished gelatinous mass like the sea squirt may actually be a precursor of the whole spectrum of higher animals with backbones.

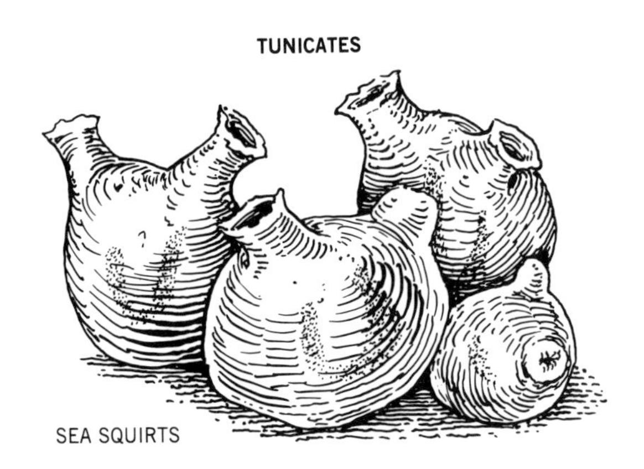

Subphylum Vertebrata—Vertebrates. Vertebrates are animals with a backbone and a brain case, or cranium. The backbone is made up of a series of separate segments called vertebrae. The central cord of the nervous system runs through the backbone, which supports and protects it. Vertebrates have two pairs of limbs, their skeletons articulated to the backbone. Five classes of vertebrates are included in this book, cartilaginous and bony fishes, reptiles, birds, and mammals. Fishes are classified into those with cartilaginous skeletons, such as sharks, rays, and skates, and those with calcified skeletons, the true bony fishes. The limbs of fishes, analogous to our arms and legs, are the paired pectoral fins, extending from each side, and the paired pelvic fins, extending from the belly. Most fishes also have several simple fins, including one or more dorsal or backfins; a caudal or tail fin; and an anal fin. Reptiles, including turtles and snakes, have a dry skin covered with scales or scutes. Most reptiles, such as turtles, alligators, and most lizards, have two pairs of limbs; however, snakes as well as a few species of lizards are legless. Reptiles are fertilized internally, and most lay large, well-protected eggs that have an abundant yolk supply. They are air breathers and are equipped with lungs rather than gills.

In many ways, birds are similar to reptiles, from which they are believed to have evolved. They breathe air with the use of lungs; their feathers, unique in the animal world, were derived from reptilian scales; and they are egg layers, as are the reptiles. Birds have modified their four limbs into a pair of wings and a pair of legs; their bones are hollow or very light; and they have exchanged the heavy-toothed jaws of reptiles for the ability to fly. Mammals are warm-blooded vertebrates with several more highly evolved skeletal structures than are found in other vertebrates. Their bodies are covered with hair in the adult stage or at some time during their embryonic development, and they generally have two pairs of limbs—although the hind limbs have been lost in some species such as dolphins. Mammals give birth to live young and nurse them. Mammals are quite diversified in their appearance, feeding, and habitat requirements. They range from the birdlike bats and armor-plated armadillos to plankton-feeding great whales.

● With some understanding of the ecological relationships among species, habitats, and salinity and with some knowledge of the types of animals that inhabit the Chesapeake Bay, you, reader, should be equipped with the necessary tools to identify most of the creatures you encounter. Always try to be aware of the general location within the Bay system where an organism was found and under what circumstances. Was it pulled out of sandy bottom muds? Was it attached to some other object? Was it floating free or swimming? Such awareness will reward the curious observer, who, with each visit to the shore, will gain a deeper knowledge and appreciation of the marvelous and myriad life that exists in the Chesapeake Bay.

VERTEBRATES

SHARK

CARTILAGINOUS FISH

BONY FISH

first dorsal fin
second dorsal fin
lateral line
gill flap (operculum)
caudal fin
anus (vent)
anal fin
pectoral fin
pelvic fin

BIRD

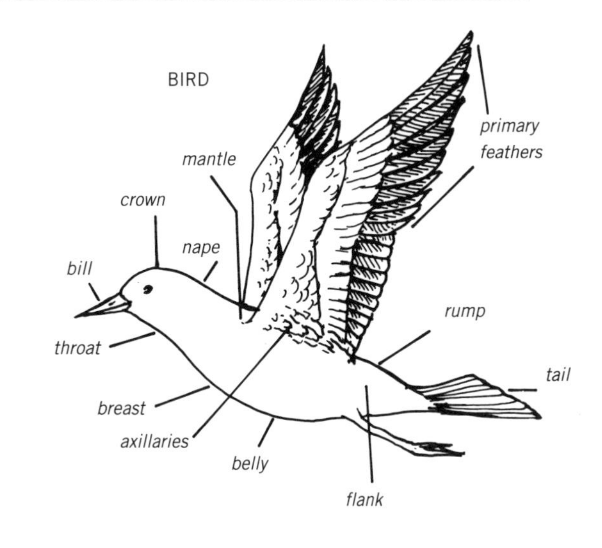

REPTILE

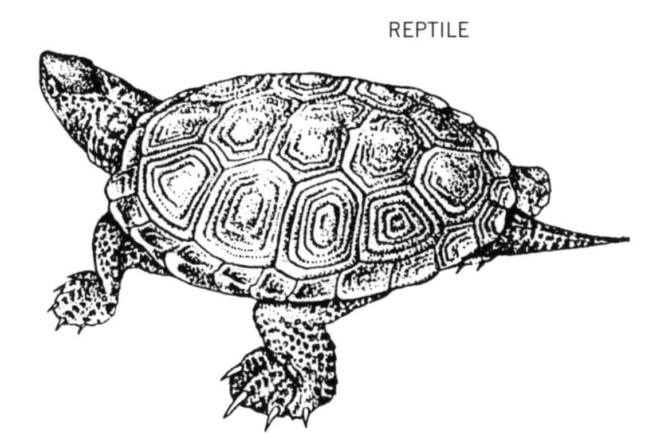

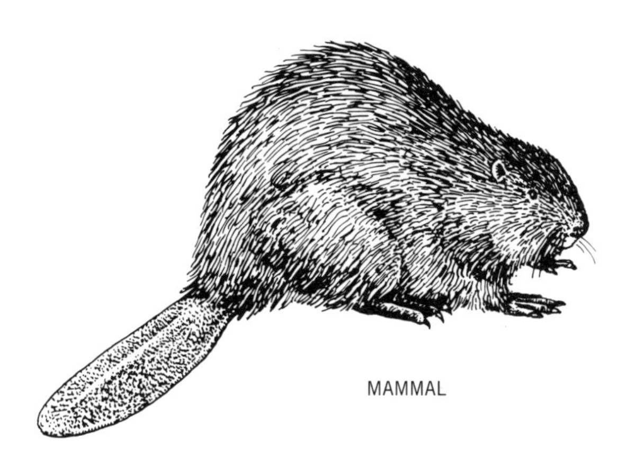

MAMMAL

SAND BEACHES

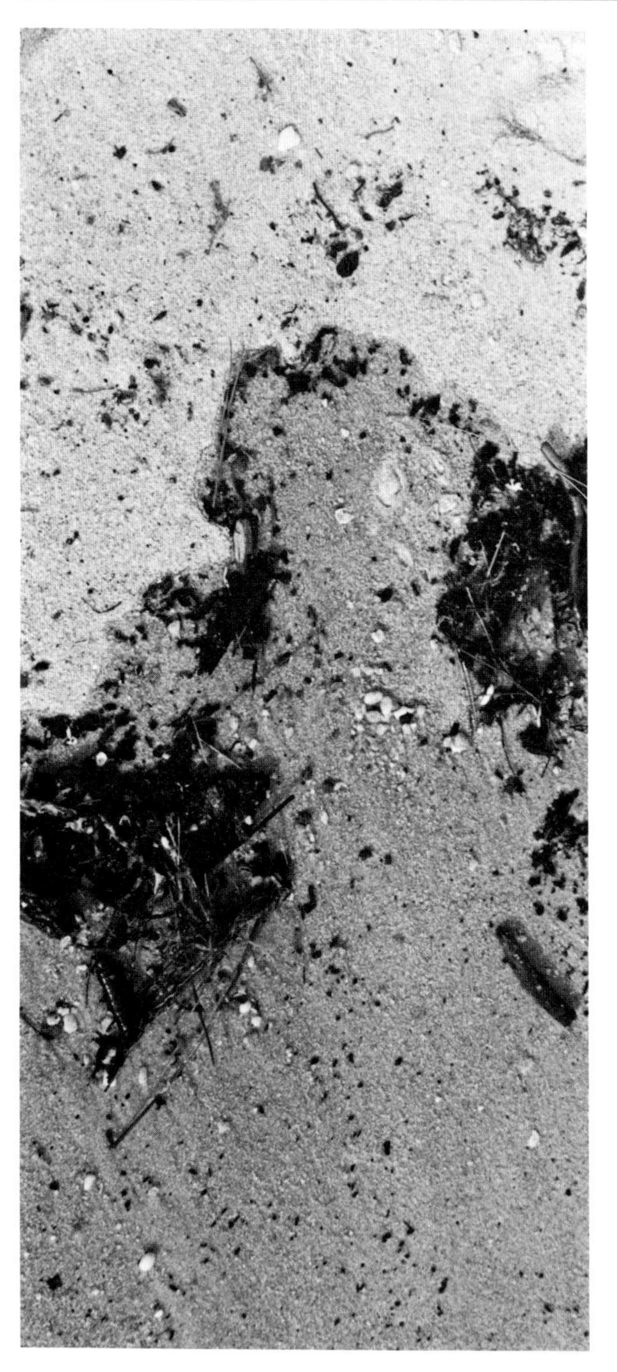

Sand beaches form the most extensive habitat along the Atlantic seacoast, but in the Chesapeake Bay they are found only where conditions are suitable for building and maintaining this type of shoreline. Waves, currents, sediments, and the slope of the shore all influence the make-up of a shoreline. A true sand beach slopes gently toward the water and is composed of sand grains or, sometimes, pebbles. The beach is usually located in a relatively exposed area vulnerable to high winds and waves that clean and scour the sand. Offshore, the prevailing currents must be strong enough to move and deposit sand particles toward the area to maintain the beach.

In the Chesapeake, the typical broad ocean-type beach, so well known to the seaside visitor, is found mostly in the lower Bay, where much of the shoreline is exposed to wide-open waters, which are strongly affected by ocean currents and waves. The exposed sand beaches of the middle and upper Bay are generally narrow and with less slope. Sandy-mud shorelines are far more characteristic of the head of the Bay and the upper tributaries. Here, wave action is generally less severe, and the tons of fine sediments continually emptying into the headwaters from the rivers and streams tend to be deposited along the shores.

The sand beach is continuous with and adjacent to many other habitats. Often, as the beach slopes toward the water, sediments become finer and combine with accumulated organic material, gradually changing the beach into a sand-mud flat. The beach is often bordered by a marsh; rocks or jetties jut out from the sand; or a pier crosses over the beach and extends to the water—the various habitats with their individual plant and animal communities are often close together.

Frequently, terrestrial animals wander onto the beach in search of food. Raccoons, opossums, foxes, and other small mammals prowl the beach at night in search of prey. Shorebirds skirt the edges of the breaking waves hunting

2. SAND BEACHES

for bits of food, pecking at live animals burrowed in the sand or stranded on the beach by the receding tides. Often, the only signs of Bay life are the empty carcasses of crabs, dried egg cases of skates, fish skeletons, or fragments of sponges and mollusk shells. These remains of marine life have often been moved many miles from their origin by tidal currents and storms.

DWELLERS ON THE BEACH

Marine forms that have emerged from the water to live on a sandy beach have had to adapt to a harsh environment. They are exposed to drying, or desiccation, by the heat of

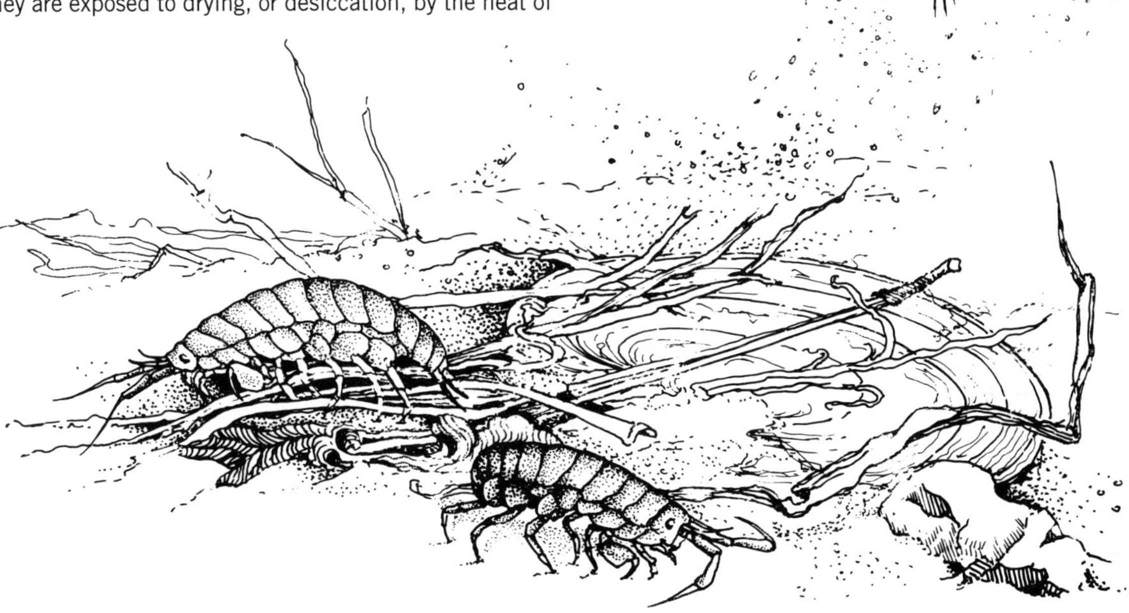

A beach hopper, *Talorchestia longicornis* (to 1 inch), jumps high over two scavenging beach fleas, *Orchestia platensis* (to ½ inch).

the sun and the winds; to a substrate that constantly shifts; and to only intermittent exposure to their once natural habitat—the water. Only a few species can live along the highest region of the beach, beyond the range of the tide, where conditions are harshest for marine animals, but these species are well adapted to their environment and their populations are often impressively large. Most high-beach animals are active at night, when temperatures have moderated. On the lower beach, in the intertidal zone, conditions are less severe, enabling many more marine species to exist. Incoming tides keep the sand moist and cool and transport food to the animals dwelling there. Beach animals are generally fast-moving and respond quickly to inundation and recession of the waves.

Much of the life of the intertidal beach is hidden below the surface of the sand and goes unnoticed unless purposely sought. Even then, there is a whole world of life that will never be seen by the searcher, although he may hold a teeming mass of it in a handful of damp sand. In the seemingly infinitesimally small spaces between the sand grains there is often a thriving community of microscopic animals and plants called the interstitial community. Certain tiny algal plants and bacteria may be so numerous that they color the sand a yellowish brown. A variety of minuscule animals, such as protozoans, gastrotrichs, roundworms, and copepods, crawl or swim between the sand grains, moving upward as the tide comes in and retreating deeper as it ebbs.

The large marine animals of the beach that can readily be seen and identified include beach fleas, sand diggers, horseshoe crabs, ghost crabs, and mole crabs. All are crustaceans with hard outer shells, which protect them from drying out, and all have appendages especially adapted for digging rapidly into the sand. Beach fleas and sand diggers are beach-dwelling amphipods belonging to certain amphipod groups that are distributed on beaches throughout the world.

The sand digger, *Neohaustorius schmitzi* (to ⅕ inch), moves water currents (blue arrows) in one direction to bury itself and in the opposite direction to feed.

Beach Fleas

Beach fleas, or beach hoppers, as they are also called, are the marine animals most likely to be seen on the drier sandy beaches of the Chesapeake Bay above the high-tide line, where the accumulated debris deposited by past tides reposes. Their vision is acute, and the slightest disturbance causes them to hop about the beach, much like a flea. Two species are common in the Bay. One, *Talorchestia longicornis,* we will call the beach hopper, to distinguish it from *Orchestia platensis,* the beach flea. Beach hoppers are often found above the tide line of less exposed beaches, sometimes hundreds of feet back from the water. They are nocturnal, tunneling into the sand by day and emerging at night to move down to the water's edge in search of food. Juvenile beach hoppers live closer to the water than the adults, in damp intertidal sands, where, during the day, they may often be seen by the thousand springing about at the water's edge.

Beach fleas generally live under beach wracks of dead and dying seaweeds. Like adult beach hoppers, they are nocturnal, but they will jump about during the day if the wrack is disturbed.

The paler, sand-colored beach hopper is a relatively

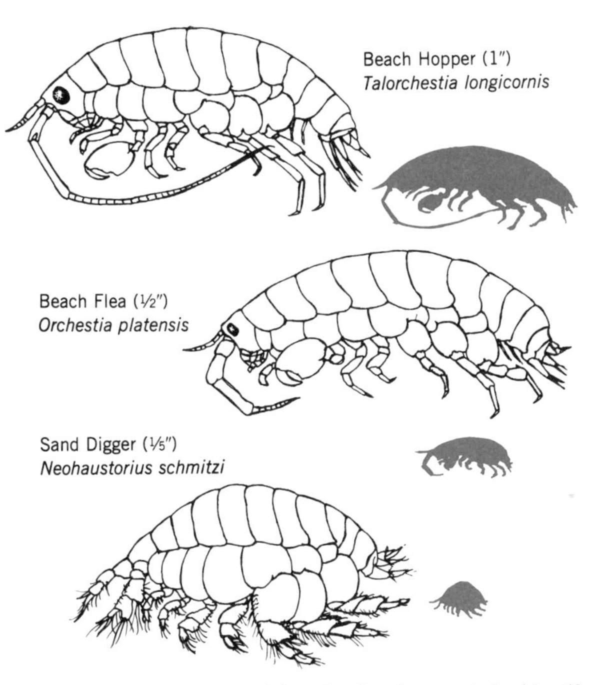

The size, body shape, and length of antennae help identify common beach amphipods. Silhouettes show actual sizes.

large amphipod, about an inch long, whereas the darker beach flea is half an inch long, or less. The very long antennae of beach hoppers, upon close inspection, help to distinguish them from beach fleas.

Sand Digger Amphipods

Sand digger amphipods are found low on the beach—near the water, where the sand is continuously wet—and beyond, in shallow subtidal waters. *Neohaustorius schmitzi* is one of the most widespread species of sand digger amphipods and is found throughout the Bay.

Sand diggers have bodies especially adapted to burrowing in loose sand. Their burrowing method depends on the sand being waterlogged, thereby allowing efficient expulsion of the sand grains. As sand diggers excavate one or two inches into the sand, their underbody forms a funnel, which moves water and food particles to their mouth parts. Their mouth parts are equipped with innumerable fine setae, which filter fine organic food particles out of the water. At times, they emerge from the bottom sands and swim to the surface, but then return almost immediately and rebury. The digger amphipods have enlarged plates along the sides of their bodies and are much smaller than the beach amphipods (less than one-fourth inch long). These plates form the sides of the water funnel. Sand diggers have large digging appendages and pale eyes so tiny as not to be visible.

Beach Tiger Beetles

Beach tiger beetles, as a group, are active scurriers and strong fliers. When approached, they wait, then fly off in a flash, alighting some distance away, generally facing the interloper. They usually are colored metallic green, bronze, or blue, and often have white or yellow spots.

The northeastern beach tiger beetle, *Cicindela dorsalis,* inhabits open, sandy beaches in several areas of the Chesapeake Bay from near the mouth of the Patuxent River to the mouth of the Bay. The adult beetle, about one-half to three-fifths inch long, has white to light tan wing covers, called elytra, sometimes embellished with fine, dark lines, and a shiny, bronze-green head and thorax. The adults are most active on bright, sunny days along the water's edge, where they feed, mate, and bask. They prey primarily on amphipods and flies; they also feed on dead crabs such as mole crabs and blue crabs, and on the occasional fish washed up on the beach.

Female beach tiger beetles typically deposit eggs just below the surface of the sand, usually above the high-tide line. The highly predaceous larvae hatch in late July and have been called "sit and wait predators." The larvae live in cylindrical, vertical burrows with their sand-colored heads bent at right angles to their bodies, poised to seize any unwary insect or amphipod that comes their way. They are equipped with two strongly curved hooks directed forward on the fifth abdominal segment. These hooks, when firmly pressed into the sides of the burrow, prevent the larva from being yanked out of its underground retreat by a strong, agile amphipod or insect. Depending on when the larvae hatch, they may develop into adults the following year or they may persist in the larval form for two years. As the larva molts and increases in size, the burrow is deepened from an average 4 inches to as much as 14 inches.

Adult tiger beetles have few natural enemies, although occasionally common grackles and beach wolf spiders will feed on them. The larvae, however, are often attacked by an antlike parasitic wasp, which stings and

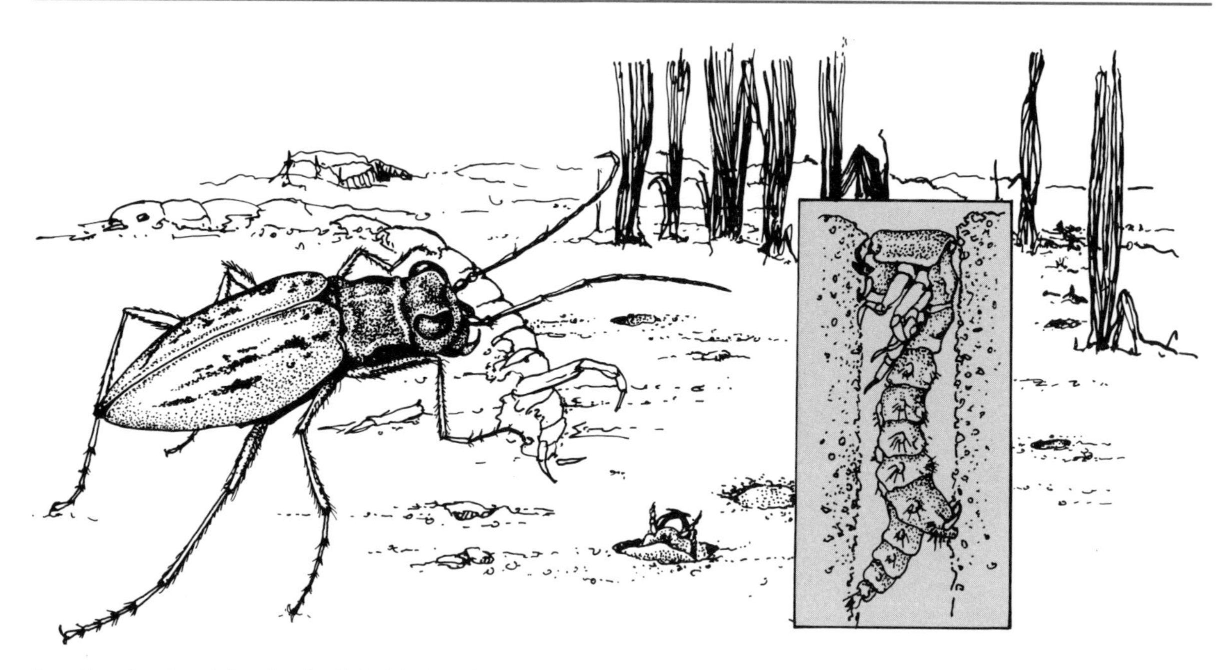

A northeastern beach tiger beetle, *Cicindela dorsalis* (to 3/5 inch), feeds on a beach amphipod. Insert shows a tiger beetle larva fastened within a burrow with its head plate closing the entrance.

paralyzes the larva and deposits an egg on it. When the egg hatches, the larval wasp is supplied with a ready source of food.

The northeastern beach tiger beetle formerly occurred in great swarms on sandy beaches from New England to Virginia. The large populations to the north of the Chesapeake Bay have all but disappeared. Both the alteration of beaches and heavy use by expanding human populations have probably contributed to the decrease in tiger beetle populations. The northeastern beach tiger beetle is listed as a threatened species under the Federal Endangered Species Act throughout the Chesapeake Bay and should not be disturbed.

Horseshoe Crabs

Atlantic horseshoe crabs, *Limulus polyphemus,* are misnamed, for they are not closely related to crabs. One of the most primitive crustaceans, they are literally unchanged since Devonian times, some 360 million years ago. Their lineage leads to terrestrial spiders and scorpions rather than to present-day crabs, making them cousins to the black widow spider rather than the blue crab.

Horseshoe crabs are unmistakable. They look like no other creature, with their brownish green, high-domed body, spiked tail, and widely spaced primitive eyes. Turn a horseshoe crab over and its five pairs of dark brown legs and leaflike gills will wave furiously as it sticks its tail into the sand and uses it as a lever to right itself. Horseshoe crabs are ubiquitous in the middle and lower Bay, in both shallow and deep waters and over all types of bottoms. Horseshoe crabs molt, as do all crustaceans. Their shell splits along a predetermined suture line on the front rim and the crab emerges and crawls away, now larger by a fourth than the discarded molt. The shell it leaves behind is often the only sign we see of this crab, as the larger horseshoe crabs usually congregate in deeper Bay waters during most of the year. Juvenile horseshoe crabs, however, remain in shallow waters, where trails etched on the bottom can often be seen in quiet coves. Follow one of these trails and you may discover a young sand-colored horseshoe crab well camouflaged on the bottom.

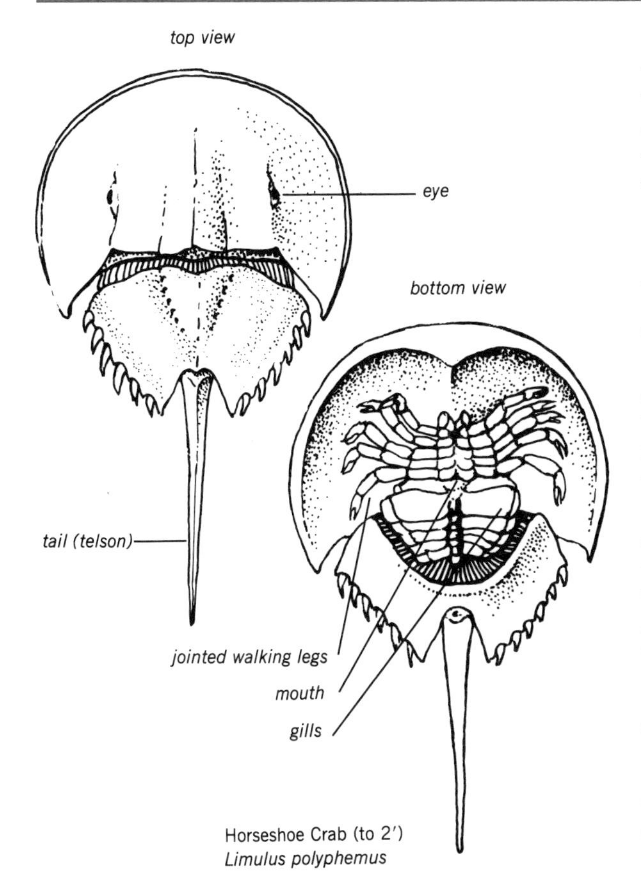

Horseshoe Crab (to 2′)
Limulus polyphemus

In May or June, during high spring tides, which are influenced by the full moon, a visitor to the Bay may witness the phenomenal annual spawning ritual of the horseshoe crab. During the spawning ritual, the smaller male clings to the back of the larger female as she crawls from the water to lay her strands of dark greenish eggs in shallow nests dug in the sand. Screaming gulls, terns, and shorebirds often join the melee, gorging on the freshly laid eggs. As the waves wash in they bathe the eggs with sperm released by the males and simultaneously cover the nests with a protective layer of sand. After spawning, most of the horseshoe crabs retreat with the ebbing waters, but some are stranded and remain behind, dug into the sand until the next tide. The buried eggs deposited in the high-tide zone require approximately two weeks to develop, and when the next spring tides return the young are ready to hatch. The turbulent waves and grinding sands break the eggs open, releasing hundreds of baby horseshoe crabs. As the tide recedes the young crabs are swept back to the water, where they will remain for nine or ten years, molting and growing until they are mature enough to return to the beach for spawning.

Ghost Crabs

On lower Bay beaches, particularly those closest to the ocean, the ghost crab, or sand crab, *Ocypode quadrata,* may be seen scurrying across the sand on the tips of its pointed legs, dodging and darting, then suddenly disappearing into a hole in the sand. The hole is the entrance to

a long burrow dug to a depth of three or four feet. The ghost crab is aptly named in Latin, as *Ocypoda* means "swift-footed." It is commonly called the ghost crab because it disappears so quickly from sight—there one moment, but gone in another, as it darts into a burrow or lies flattened just under the sand, its pale color camouflaging it perfectly. The two dark spots of its long stalked eyes provide the only clue to its position.

The marks of ghost crabs on a seemingly empty beach are easy to recognize, as the crabs leave distinctive criss-cross tracks. Oftentimes, a burrow entrance is edged with a fan-shaped scattering of sand clumps brought up and discarded there by the burrowing occupant.

Ghost crabs are more terrestrial than any other species of crab in the Chesapeake Bay, and their burrows can be found hundreds of feet from the water's edge. The older ghost crabs generally burrow highest up on the beach; young ghost crabs, which are a mottled gray and brown and much darker than adults, usually burrow closer to the water's edge, sometimes intertidally. Ghost crabs may be encountered on the beach at any time, but they are generally in or near their burrows during the day. As darkness falls, large numbers of ghost crabs scurry back and forth along the beach searching for food. Ghost crabs periodically moisten their gills and developing eggs, which are carried under their bodies, but rarely enter the water, preferring to brace themselves sideways into the wet sand and allow the waves to wash over their backs. Toward dawn, ghost crabs begin to move up on the beach to dig new burrows or return to old ones. During early morning, they can often be seen scooting in and out of their burrows, repairing and modifying the long and sometimes extensive tunnel. By afternoon they often plug the burrow entrance with sand and retreat into the cooler depths to wait nightfall again. Ghost crabs are only active on Chesapeake region beaches from spring to fall; in winter they remain deep in their burrows. Ghost crabs are not large—their distinctive square-shaped carapace is generally not over three inches wide—but they hold themselves high on their legs and look a great deal more imposing than their size warrants.

Ghost crabs are closely related and similar in shape to the fiddler crabs found along muddy banks and marshes, but fiddlers are much darker and smaller and occupy a completely different habitat.

Mole Crabs

The busy and delightful little mole crab, *Emerita talpoida* (*talpoida* means "molelike"), lives close to the breaking waves on many of the same beaches as the ghost crab. Mole crabs are numerous on ocean-side beaches, but within the Chesapeake Bay they are generally restricted to highly exposed beaches near the mouth. It is fascinating to watch the pattern of activity of a community of mole crabs on a beach. As a wave retreats, innumerable mole crabs will suddenly emerge from the soaked sand, race toward the water, and then instantaneously dig into the sand again as another wave rolls in. Mole crabs follow the tides, moving en masse up and down the slope of the beach and always settling within the zone of breaking waves.

Mole crabs are beautifully adapted to these somewhat perilous conditions. Their egg-shaped, compactly rounded bodies enable them to disappear quickly into the sand.

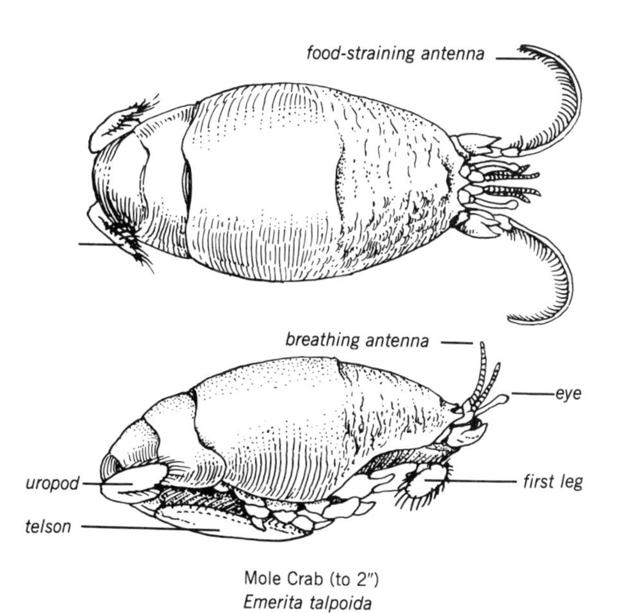

Mole Crab (to 2")
Emerita talpoida

Their shape bears little resemblance to that of a "typical" crab; their legs are small and are either folded close to their body or are digging furiously to bury before the next wave arrives. They have no claws with which to bite, nip, or threaten. Mole crabs are pale gray to light sand-colored, and they have two small dark eyes set at the end of slender pedestals and two types of antennae. The first set of antennae extend out between the eyes and are used for breathing. When a mole crab is buried and anchored firmly by its tailplate (or telson), it closes the breathing antennae together to form a water funnel, which projects above the sand. Water currents moving through this funnel keep the gills free from sand grains. As the waves recede over the buried mole crab, the second set of antennae—long, feeding antennae—unfold into the water and sift food particles through hundreds of fine antenna hairs. When the antennae net is filled with bits and pieces of food, the antennae are curled back to the mouth and the food is ingested.

A mole crab is easy to capture as it pops out of the sand. Watch the point where it reburies itself, then quickly scoop the crab out with your hand. If you search for mole crabs in spring or early summer, you may find three or four smaller crabs clinging tenaciously to the carapace of a larger one, an inch or so long. The smaller ones are the almost parasitic males, which mature at a size of less than one-eighth inch and rarely grow much larger than one-third inch.

In winter, mole crabs, unlike ghost crabs, leave the beach and retreat to deeper offshore waters.

SIGNS OF LIFE

Shells

Shells of mollusks are the most common and recognizable signs of marine life on the shores of the Chesapeake Bay. Shell collecting is popular with everyone. At home the shells can be examined at leisure and, with a little guidance, can easily be identified. Shell identification is fun, and a deeper appreciation of Bay life comes from knowing what kinds of animals the shells represent. The shells of the common mollusks of the Chesapeake Bay are included here with a guide to their identification and habitat.

Shells from burrowing, mud-dwelling clams, from marshland snails, or from deeper-water shell beds are often transported by prevailing currents to the beach. A shell from any location or habitat in the Bay may ultimately be cast ashore or moved to a completely different type of locale.

Shells can be separated into two major types: bivalves, of the class Pelecypoda, in which two separate shells are hinged together protecting the soft-tissued animal within; and univalves, of the class Gastropoda, in which most often a single-spiral shell surrounds the snail's body.

Bivalve shells may remain hinged together or may separate from their mate. In most instances, both shells (valves) of a species are essentially mirror images of the other, although some few species, such as jingle shells and oysters, have valves of slightly different shape and structure. Oftentimes, characteristics on both the inside and the outside of a shell must be studied in order to identify the species. The valves are usually hooked together by hinges bound together by a dark brown ligament. The hinges often have various types of interlocking ridges, or hinge teeth. The presence or absence or the number, shape, and size of these teeth are frequently characteristic of the species. Cardinal teeth are larger and central; lateral teeth are longer and narrower, extending either posteriorly or anteriorly. A rounded, cup-shaped chondrophore may also extend from the hinge.

Live bivalves open or clamp their shells shut by means of strong muscles attached to the interior concave surface of the valves. Distinctive muscle scars mark the sites of these muscle attachments. In some species, a pallial line between the muscle scars is formed where the muscular edge of the mantle, the soft outside body covering of the internal animal, is attached to the valve. Where a siphon has passed through the mantle, the pallial line curves in a U-shaped marking called the pallial sinus. On the outside of the valve, just above the hinge, the beak (or umbo) projects inwardly and usually anteriorly. Most bivalves have concentric growth lines, which originate at the beak and spread over the shell. In some species these lines are well sculptured and distinct, while in others they are barely visible. In certain species, such as scallops, sculpturing is radial and lines or ribs radiate from the beak. The outside surface of the shell is often covered with a sheath, the periostracum, which on some shells is thin and transparent, giving a gloss to the shell, and in others is dark and thick, presenting a coarse or rough surface. Weathered shells may have lost all signs of the periostracum and the sculptured surfaces may be smoothed, pitted, or eroded.

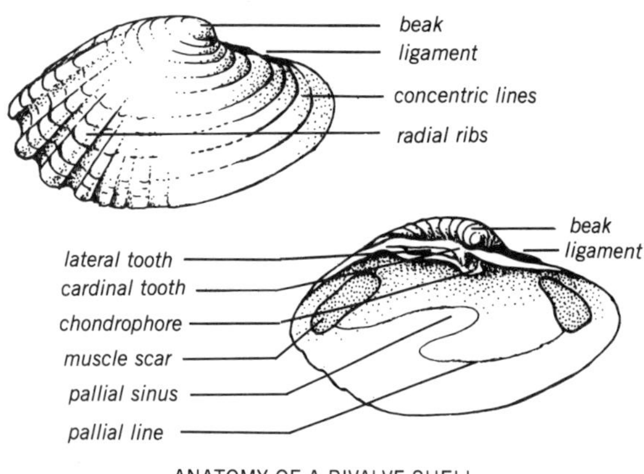

ANATOMY OF A BIVALVE SHELL

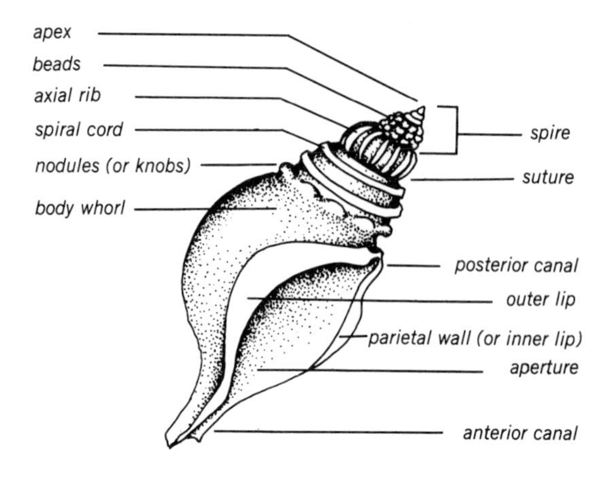

ANATOMY OF A GASTROPOD SHELL

Typical univalve or gastropod shells are single-spiral shells composed of a number of whorls coiled about an internal axis. The largest whorl at the base of the shell, the body whorl, contains the live animal and has an opening, the aperture, through which the anterior parts of the snail, such as the head, foot and mantle, may emerge. The aperture is bordered by a parietal wall, or inner lip, and an outer lip, and is often extended basally into an anterior canal, which encases the extended siphon in the living snail. In some snail species a similar posterior, or anal, canal is located at the top of the aperture. Most marine gastropods are right-handed (or dextral), with the outer lip of the aperture on the right side; this can easily be determined when

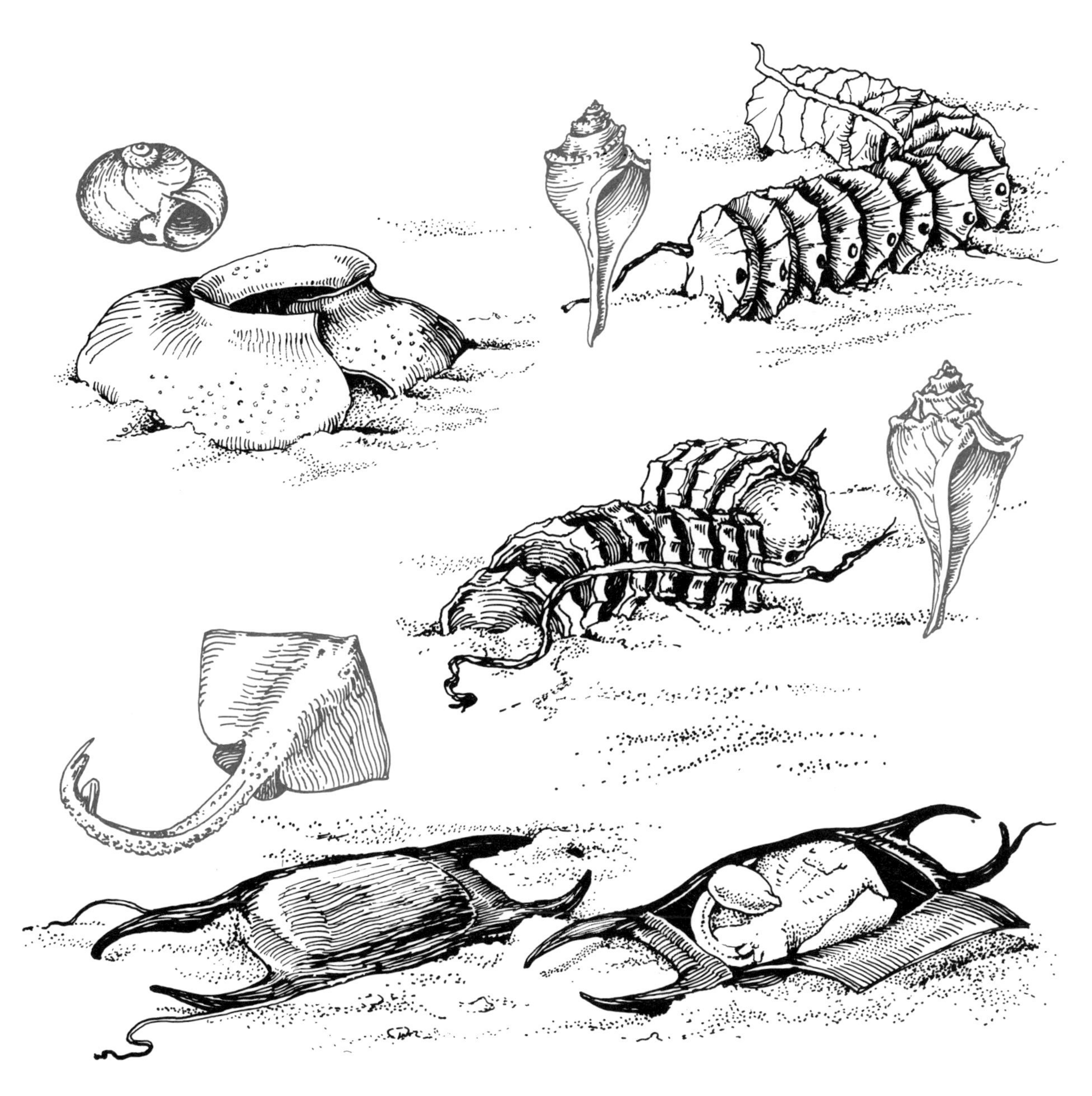

The egg cases of marine snails and skates commonly found on lower Chesapeake Bay beaches. The sand collar produced by the shark eye, *Neverita duplicata,* may be 5 to 6 inches in diameter. The flat-edged egg cases of the knobbed whelk, *Busycon carica,* and the knife-edged egg cases of the channeled whelk, *Busycotypus canaliculatus,* are 1 inch in diameter. The cutaway drawing shows how a baby clearnose skate, *Raja eglanteria,* lies within its protective coat, a mermaid purse 2 inches long; it is, however, rare to find an occupied skate egg case on the beach.

the shell is held with the tip of the spiral at the top and the aperture facing the viewer. Some few species of gastropods are left-handed (or sinistral), with the outer lip of the aperture to the left when held in the same position as above. The smaller whorls above the body whorl form the spire, and the tiny last few at the narrow tip form the apex. The continuous juncture between the whorls is the suture, which may be a thin line or a deeply grooved channel, irregular, wavy, or just slightly impressed. The surface of the shell may be variously sculptured, with axial ribs parallel to the axis of the shell or with spiral cords following the circular path of the whorls. Beads, nodules, and other markings may also be present on the whorls.

Egg Cases

The curiously shaped egg cases of shark eye snails, whelks, and skates are often cast up on beaches close to the mouth of the Chesapeake Bay. Shark eye snails produce strangely shaped sand collars. Few observers would guess that the inverted, cup-shaped collar or ring made of sand is of animal origin. Sand collars are usually dry and fragile and are easily broken after exposure to the sun and wind. The shark eye snail constructs these collars of sand and mucus and deposits its hundreds of tiny eggs in the semielastic, transparent ring.

Whelks construct long strands of easily recognizable coin-shaped, rubbery capsules. Each string of capsules was originally attached to a stone or shell and contained 20 to 100 minute eggs. These strings are often dislodged and carried ashore, where they are found empty and brittle on the beach, the baby whelks having long since emerged from a small hole at the edge of the capsule. Sometimes, however, some of the minuscule snails are still within the egg cases and can be shaken out of the open holes. Each species of whelk forms a capsule of a distinctive shape; those of the knobbed whelk are flat, like a coin, whereas the channeled whelk's egg capsule is knife-edged.

Mermaid purses are the dark brown leathery egg cases of skates. They are often entangled by their hooks in beach debris. The baby skates develop within their podlike containers, well protected by the tough casing. The hooks and tentacles on mermaid purses are of different sizes and shapes according to the species of skate. By the time a case is thrust upon the beach it is flat, split at one end and devoid of its occupant, a miniature skate, relative of rays and skarks.

BIRDS OF THE BEACH

Beach fleas, sand digger amphipods, mole crabs, and beach wrack cast upon the shore provide an array of food that attracts insects, nocturnal foraging raccoons and ghost crabs, and, above all, many birds. Gulls, terns, and several types of sandpipers fly over the beach, alight, and begin the unrelenting search for an unseen sand dweller, horseshoe crab eggs, or a scrap tossed up by the surf.

Shorebirds, such as the ruddy turnstone, sanderling, willet, and black-bellied plover, are wading birds. They are small to medium-sized and are usually compactly built. Some species are distinctly marked and easy to identify; others are quite difficult to determine because of seasonal changes in plumage and the often subtle differences in the length of bills and legs.

Gulls and terns, ever-present gleaners along the waterfront, are vocal, active birds frequently seen picking through stranded seaweed at the high-tide line, soaring overhead searching for small fish, or bobbing on the surface of the water just beyond the surf line. Terns and gulls appear to be similar at first glance; however, terns are generally more graceful and streamlined than gulls and fly with deeper wingbeats. Terns often fly with their bills pointed downward, while gulls tend to fly with their bills pointed horizontally. Their wings and bills, compared to those of gulls, are usually more pointed, and most species of terns have a forked tail, unlike gulls, which have squared or rounded tails. Terns plunge headlong into the water after a fish; gulls wheel and skim the water's surface when they see a meal—however, they occasionally plunge dive for a fish. Gulls and terns do share a definite similarity, however, in that varying plumages of the juveniles as well as seasonal changes in plumage make identification extremely difficult, sometimes even for expert birders.

Shorebirds

The ruddy turnstone, *Arenaria interpres,* is a medium-sized, stocky wader with a short, slightly uptilted, wedge-shaped bill. Turnstones have brown backs and red-orange legs with a bold breast band that looks like a vest, sometimes referred to as a calico pattern. In flight, they are well marked with white and dark accents on the wings, back, rump, and tail. Ruddy turnstones nest in the Arctic tundra during the summer and are transients in the mid-Atlantic and the Chesapeake Bay in spring, fall, and winter. Turn-

Shorebirds along a beach in the fall. Ruddy turnstones, *Arenaria interpres* (9½ inches), poke among pebbles; a willet, *Catoptrophorus semipalmatus* (15 inches), stands alert while another takes flight, flashing its black and white wings. Sanderlings, *Calidris alba* (8 inches), probe at the water's edge; and a black-bellied plover, *Pluvialis squatarola* (11½ inches), in its winter plumage walks behind.

stones regularly winter along the mid-Atlantic coast south to South America. They are very abundant birds, but are usually seen only in small flocks on beaches and mud flats and around rock jetties, flipping over stones and shells seeking small crustaceans, snails, and worms. They also eat the eggs of other birds, especially terns. Ruddy turnstones frequent the lower Chesapeake, particularly near Cape Charles and along the Chesapeake Bay Bridge Tunnel islands.

The sanderling, *Calidris alba,* is slightly smaller than the ruddy turnstone and is the palest of the sandpipers, a large and varied family of wading birds. The plumage is brown in the spring and light gray with white underparts in the winter. They have a prominent white stripe on the upper part of the wings, which is displayed during flight, and a black bill and legs. Sanderlings nest in the high Arctic during the summer months but are present in the Bay area in spring, fall, and winter. As fall commences, sanderlings move south as far as the West Indies and along the coast of South America. Sanderlings are the small, pale shorebirds that veer in quickly to the beach in small, tight flocks; dash to the surf just in front of a receding wave; probe the wet sand for worms, amphipods, and small mollusks; and scurry back up the beach, like mechanical toys, in front of an advancing wave. Sanderlings, like ruddy turnstones, are most often seen on lower Bay beaches.

The willet, *Catoptrophorus semipalmatus,* is a large sandpiper with long blue-gray legs and a straight bill. The plumage is a mottled gray-brown with a brown belly. It is a rather nondescript bird until it takes flight and shows its prominent white and black wing pattern and voices its noisy "pil-will-willet." Willets nest along the Eastern Shore from the Dorchester marshes south during mid-May to late July. They breed in tidal salt marshes, in saltmarsh cordgrass and saltmeadow hay. The nest is a grass-lined hollow on the floor of the marsh in which the female lays an average of four olive-colored eggs, marked with brown. The male will incubate the eggs at night and occasionally at midday. The female willet abandons the nest two or three weeks after the chicks hatch, while the male cares for the young for two more weeks. Willets feed on a varied diet which includes insects, worms, small mollusks, fiddler crabs, and fish. They often wade into the water and bob on the surface close to shore. The willet is generally a nervous species and is often the first bird to sound an alarm call.

The black-bellied plover, *Pluvialis squatarola,* is the

largest plover in the Bay and is distributed nearly worldwide. Plovers are wading birds, as are sandpipers; however, plovers are more compactly built and have thicker necks and shorter bills. The black-bellied plover has a characteristic hunched stance and lethargic behavior, which makes it appear dejected. This species has a black belly and sides; its back is speckled only when in breeding plumage, a phase seen in the spring and occasionally in the fall in the Chesapeake Bay area. Typically, along this part of the Atlantic coast, the black-bellied plover is in its winter plumage, which is grayish with a lighter belly. The black wing pits, or axillaries, contrasted against the white underwings, are a key characteristic of this bird when in flight. This large plover, known as the grey plover in Europe, nests in the dry tundra from May through August. Plovers run in short bursts, probe for a sandworm or small crab, and then resume their stop-and-start patrol of the beach.

Gulls and Terns

The herring gull, *Larus argentatus,* is one of the largest of the gulls. As is true of most gulls, its plumage gradually changes from an overall speckled brown in the juvenile to the distinct plumage of the breeding adult. The herring gull is a four-year gull, which means that it requires four seasons to attain breeding plumage, while most other gulls reach reproductive maturity in two or three years. The breeding adult is pale gray with a white head. It has a yellow bill with a red spot near the tip of the lower bill, and pink legs and feet. Herring gull chicks are often seen pecking at the red spot on the bill, which may be an innate begging behavior to stimulate feeding from the adult. This gull is very common along the Atlantic coast and throughout the Chesapeake Bay. It is often seen in great numbers on newly rising mountains of garbage and trash in landfills, as well as along the waterfront almost anywhere. The herring gull is a colonial nester and may build its nest in a variety of habitats. It lays an average of three olive to light blue eggs, which hatch in 24 to 28 days. Herring gulls are opportunistic feeders, feeding on insects, the eggs and young of other species, or small fish at the surface. They will often drop clams and oysters from the air onto a large rock, a pier, or a road in order to break the shell and feed on the meat inside.

The Caspian tern, *Sterna caspia,* is a massive bird and the largest tern in North America. It measures a maximum of 23 inches from the tip of the bill to the tip of the tail, which makes it just slightly smaller than the familiar herring gull. Readily identifiable by its red-orange bill marked

A first-year juvenile herring gull, *Larus argentatus* (25 inches), calls raucously behind a mature adult in full breeding plumage; a third-year herring gull is at the water's edge. A Caspian tern, *Sterna caspia* (21 inches), stands alert in the foreground.

The breeding plumage (*top row*) and winter plumage (*middle row*) of the royal tern, *Sterna maxima* (20 inches), the Caspian tern, and the least tern, *Sterna antillarum* (9 inches). In flight, the tips of the underwings of the royal tern are mostly pale, those of the Caspian tern dark; the tips of the upperwing of the least tern show a dark wedge.

with black near the tip, a black cap during breeding season, and black legs, it can be confused only with the royal tern. The Caspian tern, distributed throughout most of the world, breeds in small colonies along the Atlantic coast on flat sand, gravel beaches, shell banks, and occasionally in marshes. Often the nest is a simple scrape or depression lined with moss, grass, or seaweed. The two to three pinkish buff eggs may sometimes be placed in the crevice of rocks or concealed in beach rubble. This fiercest, largest, and least gregarious of the terns takes care of its young longer than any other tern species, feeding them five to seven months after they are able to fly. The Caspian tern is not abundant, but it is distributed throughout the Bay. This large "sea swallow," like most other terns, points its bill downward when searching for fish; it hovers before diving and then plunges into the water to capture its prey.

The royal tern, *Sterna maxima,* so similar in appearance to the Caspian tern, can be distinguished by a shaggy black cap during breeding season, a more deeply forked tail, and lighter underwings. It is slightly smaller and slimmer than the Caspian. The royal tern breeds along the mid-Atlantic coast, and perhaps in the lower Bay on Fishermans Island, in large, crowded colonies on open sand beaches and sand bars, and in isolated, sparsely vegetated areas. The female lays an average of one buff-white egg in a simple scrape. The fledglings separate from the parents and form a group known as a creche. Individual parents are able to recognize their own young, however, and continue to feed them, although the young are also able to feed themselves. Birds that live in large, loose colonies, such as eiders, flamingos, and penguins, often form creches. This form of grouping apparently reduces the risk of any one single chick's being preyed upon. The parents are also set free to spend more time foraging. Royal terns frequent the Bay from Tilghman Island to the south. They feed on crabs, squid, shrimp, and fish. The royal tern generally does not leave the coastal edges; the Caspian tern, on the other hand, can often be found inland.

The least tern, *Sterna antillarum,* is robin-sized and is the smallest tern in North America. The breeding adult, like the other terns, has a gray back and wings, a black cap and nape, a white forehead, and a forked tail. However, the bill is pale yellow to light orange with a black tip, and the legs are yellow rather than black. The immature bird is mottled, with a black eye stripe and black legs. Mature birds, in flight, have a distinctive black edge on their wings.

Least terns breed in their second year all along the Atlantic coast, including the Chesapeake Bay. Typically, they construct a shallow unlined nest on a sand bar or beach in which they lay one to three olive-buff eggs marked with brown. They occasionally build their nests on flat rooftops. The eggs hatch in about three weeks, and less than a month later the hatchlings are ready to begin flying. Least terns sometimes regulate the temperature of the developing eggs by wetting their plumage and shaking their wings to disperse droplets of water on them. Their peak breeding season in the Chesapeake Bay area is June to mid-July. Least terns will often be seen resting on pilings and buoys throughout the Bay area. They have very swift wingbeats, and when searching for food they tend to hover longer than other terns before plunging into the water for a meal of fish, small crustaceans, or insects. They also feed by skimming the surface of the water. They are common throughout the Chesapeake Bay during the spring and summer and migrate in the fall as far south as Brazil.

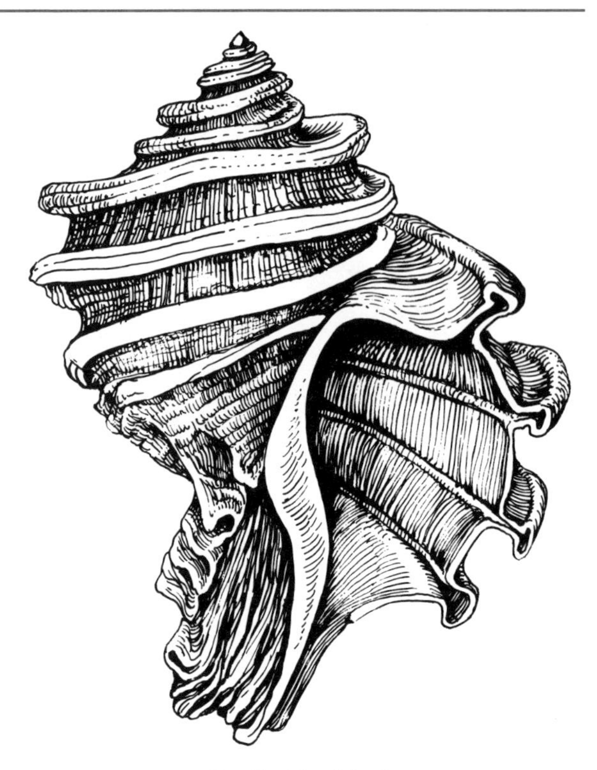

Four-lined Fossil Snail
Ecphora quadricostata

The first fossil described from North America, and Maryland's state fossil. Fragments of this shell are often found on certain Chesapeake Bay beaches.

ANCIENT RELICS OF THE SEA

Some 25 million years ago the coastal area of the mid-Atlantic, westward to Washington, D.C., and beyond, was submerged. During this time in geological history, known as the Miocene Epoch, a great variety of marine animals swam, crept, and burrowed over what is now farmland, woods, and cities in Maryland and Virginia. As the seas retreated and the coastal plain emerged, rich fossil-bearing formations containing historic evidence of ancient plants and animals were revealed.

We have included a brief discussion of fossils in this book because they are abundant, easily observed, and collectible in certain areas of the Bay and because they provide us with interesting examples of what marine life looked like long ago.

The first fossil described from North America came from the Miocene deposits of Maryland. The ancient four-lined fossil snail, *Ecphora quadracostata,* was described and illustrated in a publication in 1658. Walking along the beach in areas where fossil deposits are constantly being eroded, it is not unusual to find a scallop shell millions of years old mingled with the shells of living species such as the oyster and soft-shelled clam.

How, then, can a fossil be distinguished from a living species? In the case of the ancient scallop shell, which looks very much like scallops found in the sea at present, it is a relatively simple matter, for there are no large, living sea scallops in the Bay. However, fossil snails such as the fossil turret snail, *Turritella plebia,* and the fossil ark, *Anadara staminea,* do resemble living species found in the Bay. Generally, fossil shells are brittle and will break or crumble readily; the shells of recently dead specimens are stronger. It should be pointed out that this test fails miserably if you consider the fragile shells of such contemporary species as the soft-shelled clam, the razor clam, or the

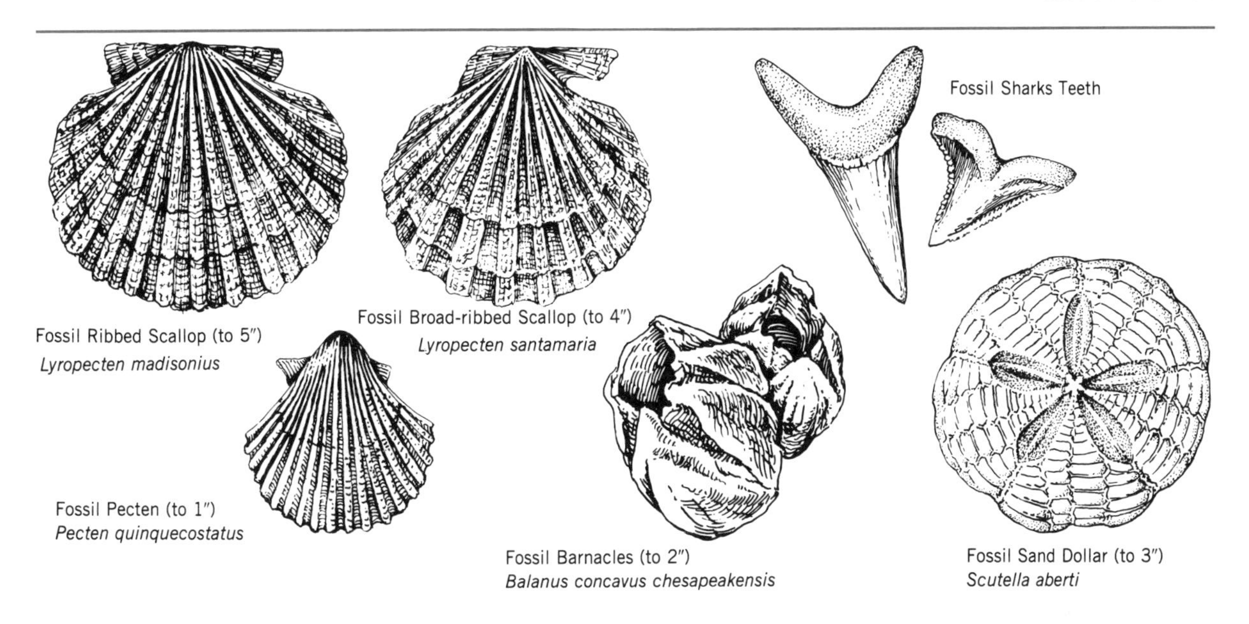

Some common fossils of the Chesapeake Bay. These species are easily identifiable as fossils. Scallops, pectens, giant barnacles and sand dollars do not now inhabit the Bay. Fossil shark's teeth are characteristically dark colored; teeth of dead contemporary sharks are ivory colored.

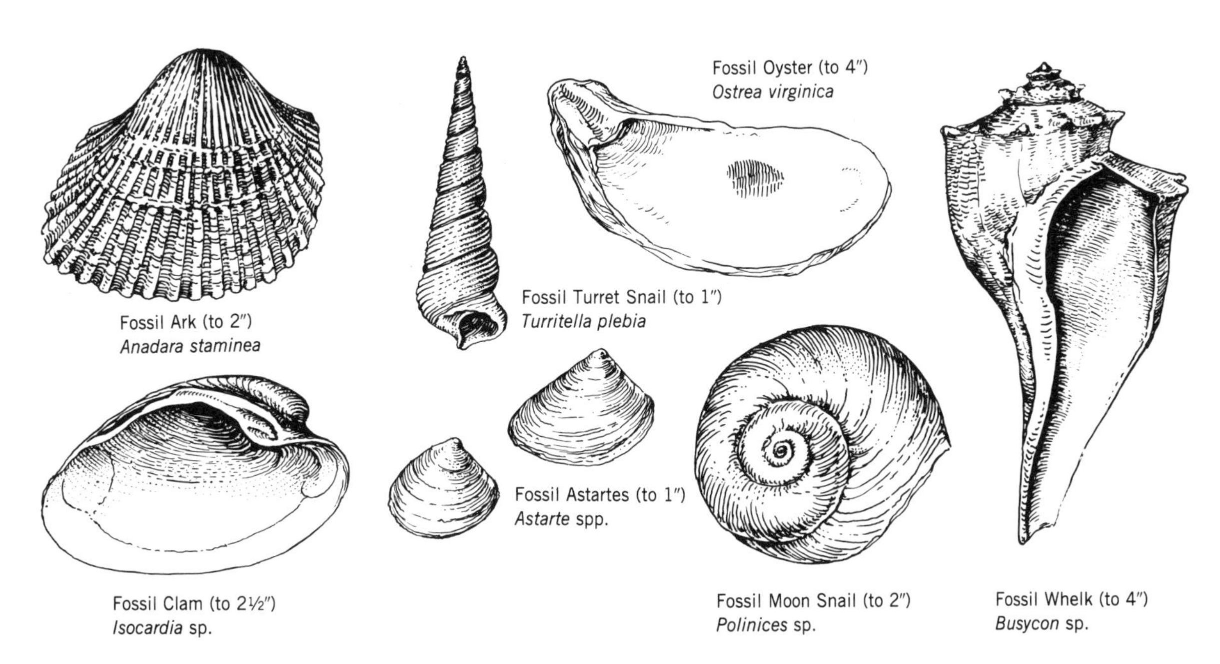

Common fossils that are difficult to identify because similar species currently inhabit the Chesapeake Bay. Of this group, the fossil turret snail can often be identified by its large size; similar living species are minuscule (see Shell Guide).

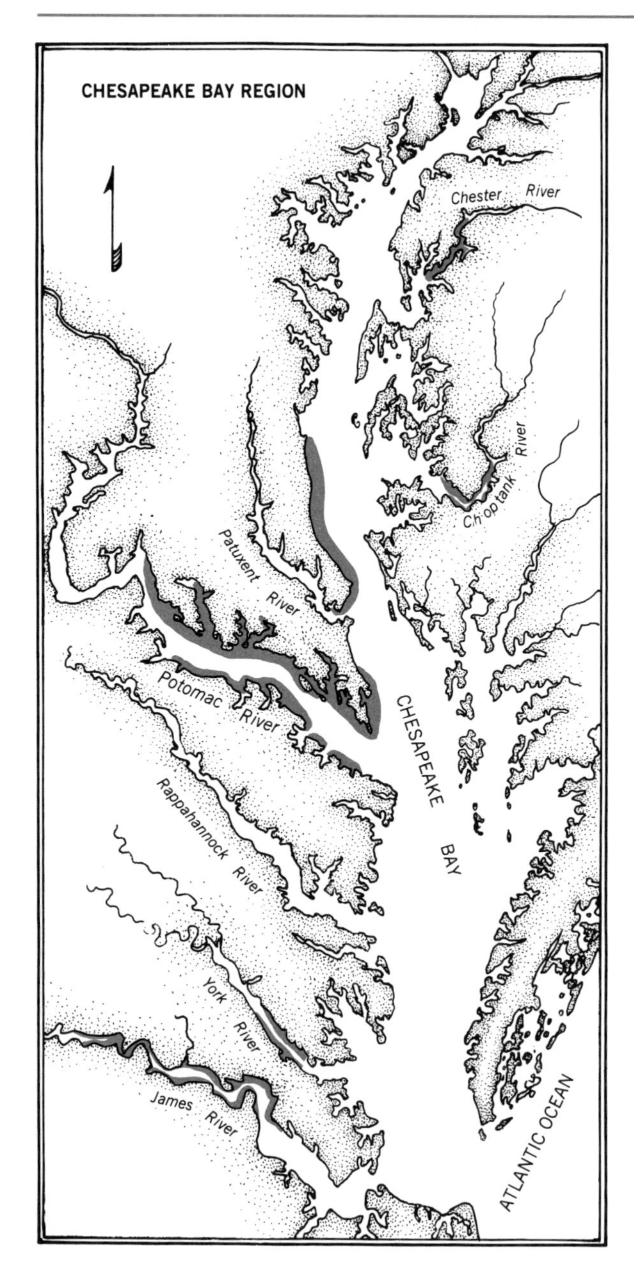

Shoreline areas of the Chesapeake Bay where fossil-bearing formations are located.

macoma clam. Examine your specimen carefully, looking particularly for evidence of muscle tissue adhering to the inner part of the shell, for remnants of the leatherlike hinge, and for any indication of an outer covering, the periostracum, which may be peeling in places. You can be sure that you have a living species if any of these conditions exists.

There are other fossils along the shores of the Chesapeake which are not mollusks, such as the fossil Chesapeake barnacle, *Balanus concavus chesapeakensis,* and the fossil sand dollar, *Scutella aberti,* and also sharks' teeth and vertebrate remains. Molluscan species are by far the most abundant, but fossils of other groups are not rare.

Fossil outcrops occur in several areas along the shores of the Bay and its tributaries. One of the best-known areas for collecting Miocene fossils, Calvert Cliffs, lies along the western shore of Maryland's Chesapeake Bay, from Chesapeake Beach southward to Solomons Island, at the mouth of the Patuxent River. The most abundant fossil in this formation is the large, fossil ribbed scallop shell, *Lyropecten madisonius.* Another scallop, the fossil broad ribbed, *Lyropecten santamaria,* is also quite common, as are a number of pelycypod species—*Anadara, Astarte, Isocardia,* and the fossil scallop, *Pecten quinquecostatus.* This is also a good area in which to find sharks' teeth.

The Maryland shore of the Potomac River from Point Lookout up to the Port Tobacco River is an excellent collecting locality for *Turritella, Ecphora,* the fossil oyster, *Ostrea virginica;* the fossil barnacle, *Balanus;* the fossil moon snail, *Polinices;* the fossil whelk, *Busycon;* and many other mollusks.

In the lower Chesapeake Bay of Virginia, a large area on the western shore, bounded by the James and York rivers and known as the Yorktown deposit, contains abundant fossils, including scallop species, teeth and vertebrae of sharks, whale bones, and many molluscan species.

● It is ironic that the most accessible and familiar Chesapeake Bay habitat, the place people most often visit to picnic, swim, or just walk along the shoreline, is usually the least rewarding from the viewpoint of encountering marine life. But for those who want to explore other habitats sand beaches allow easy passage, to the plants and muds of a fringe marsh bordering the beach, or to the water's edge and the easily plucked-out riches of the shallow waters.

Shell Guide

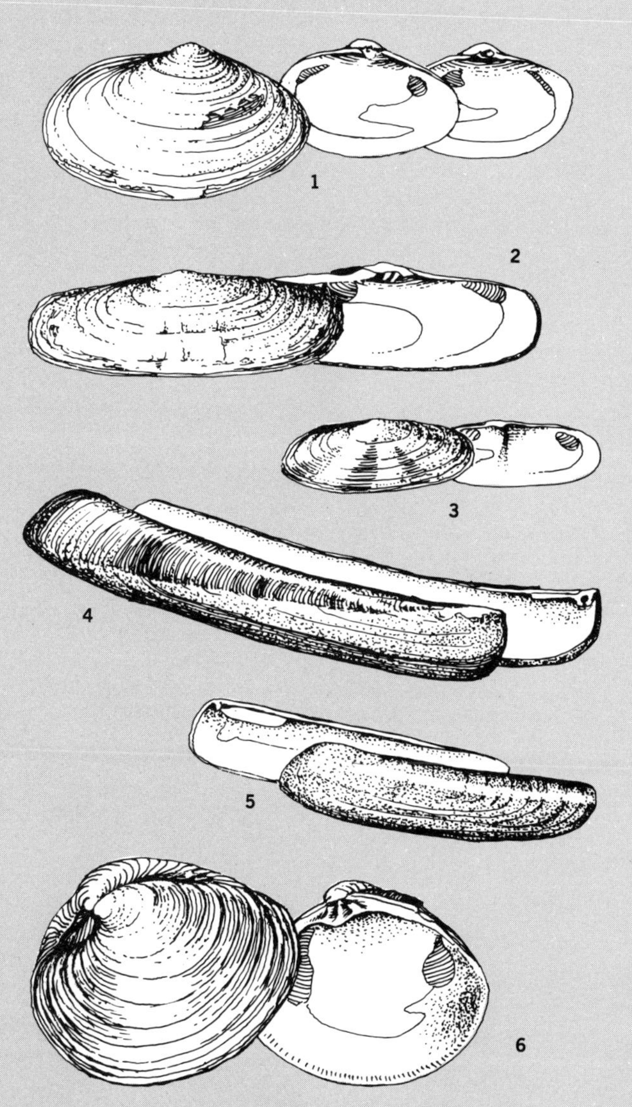

CLAMS OF THE BAY

1. **Soft-shelled Clam** (Manninose, longneck clam) *Mya arenaria.* Buried in mud, sand, intertidal to subtidal, Zones 2 and 3. Shell thin; hinge of one valve with projecting tooth, other with socket. Chalky white with gray to yellowish periostracum. To 4 inches.

2. **Stout Razor Clam** *Tagelus plebius.* Buried in sandy-mud, sand, mud, intertidal to subtidal, Zones 2 and 3. Shell strong, rectangular, length about three times width. White with thick olive-green to brown periostracum. Hinge with two small projecting teeth. To 3½ inches.

3. **Purplish Razor Clam** *Tagelus divisus.* Buried in sandy-mud, sand, mud, intertidal to subtidal, Zone 3. Smaller and less rectangular than stout razor clam. Whitish with purplish rays; thin, glossy brown periostracum. Hinge with two small teeth; obscure rib crossing inside of shell. To 1½ inches.

4. **Common Jackknife Clam** (Straight razor clam) *Ensis directus.* Buried in sandy-mud, intertidal to subtidal, lower Zone 2 and Zone 3. Shell narrow, elongate, length about six times width. Hinge at end of valve, two cardinal teeth on one valve. Glossy olive to dark brown. To 10 inches.

5. **Little Green Jackknife Clam** *Solen viridis.* Buried in sand, intertidal to subtidal, lower Zone 2 and Zone 3. Shell smaller and straighter than that of common jackknife clam. White with light green to brown periostracum. Hinge at end of valve, one cardinal tooth. To 2 inches.

6. **Hard Clam** (Quahog, littleneck, cherrystone) *Mercenaria mercenaria.* Buried in muddy-sand, sand, intertidal to subtidal, lower Zone 2 and Zone 3. Shell thick, strong, beaks angled forward. Outside gray, sculptured with concentric rings; inside white with purple stains. Hinge with three cardinal teeth. To 4 inches.

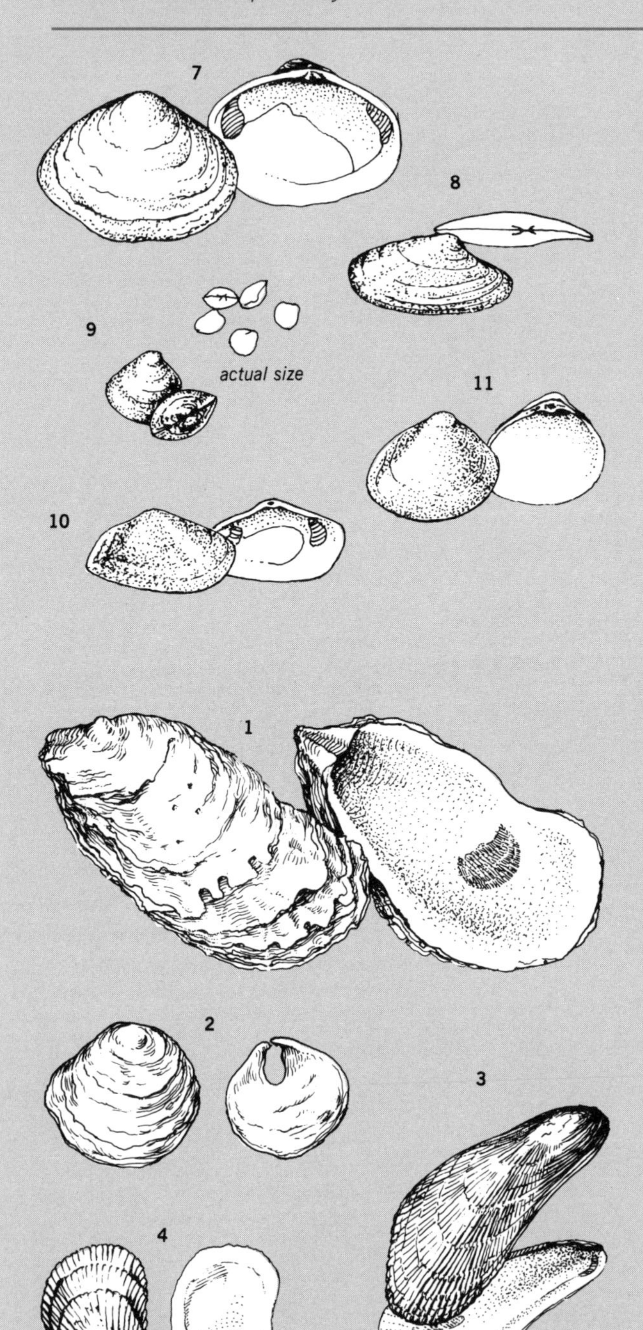

7. Baltic Macoma Clam *Macoma balthica.* Buried in mud, sandy-mud, intertidal to subtidal, Zones 2 and 3. Shell thin, variably oval shaped. Dull white, sometime pinkish. No lateral teeth; gray periostracum thin and flaky. To 1½ inches.

8. Narrowed Macoma Clam *Macoma tenta.* Buried in mud, sandy-mud, sand, subtidal, lower Zone 2 and Zone 3. Shell thin, fragile, elongated, smooth. White to yellowish, slightly iridescent. Narrowed posterior end slightly twisted to left; no lateral teeth. To ¾ inch.

9. Gem Clam (Amethyst clam) *Gemma gemma.* Buried in sand, intertidal to subtidal, Zones 2 and 3. Minute shell, triangular, thin, glossy. White, tan to purplish. Inner margin of shell finely crenulated. Hinge with well-developed teeth. To ⅛ inch.

10. Northern Dwarf Tellin *Tellina agilis.* Buried in sandy-mud, mud, sand, subtidal, Zones 2 and 3. Very small shell, thin, fragile, elongate, glossy, iridescent. White to pink. Large rounded pallial sinus. Hinge small with tiny teeth. To ½ inch.

11. Little Surf Clam (Coot clam) *Mulinia lateralis.* Buried in mud, sand, subtidal, lower Zone 2 and Zone 3. Shell smooth, sloping into a posterior ridge. White to yellowish. Hinge with cuplike indentation, smooth lateral teeth. To ¾ inch.

THE OYSTER AND JINGLE SHELL, MUSSELS, AND ANGEL WINGS

1. American Oyster *Crassostrea virginica.* On hard bottoms, sandy-mud, mud, sand, intertidal to subtidal, Zones 2 and 3. Shell irregularly shaped, thick, rough surface. Grayish, interior white with purple muscle scar. To 10 inches.

2. Jingle Shell (Atlantic jingle shell) *Anomia simplex.* Attached to shells, rocks, and other hard substrates, intertidal to subtidal, lower Zone 2 and Zone 3. Shells irregularly shaped, upper valve cupped, thick. Yellow, orange, or silvery black. Bottom valve flat, more fragile with distinct hole, whitish. To 1½ inches.

3. Atlantic Ribbed Mussel *Geukensia demissa.* Buried in mud, among marsh plants, intertidal, Zones 2 and 3. Shell thin, slightly curved, beak not quite at end as in other mussels; surface with numerous radiating ribs, no hinge teeth. Dark green to yellowish brown, interior iridescent bluish, slightly purple at edge. To 4 inches.

4. Hooked Mussel (Bent mussel, curved mussel) *Ischadium recurvum.* Attached to shells, rocks, and other hard substrates, intertidal to subtidal, Zones 1, 2, and 3. Shell sharply bent, beak at end of shell, surface with strong radiating ribs. Hinge with three or four small teeth. Black, dark gray to brown, interior rosy-brown to purplish. To 2 inches.

5. Blue Mussel *Mytilus edulis.* Attached to shells, rocks, and other hard substrates, intertidal to subtidal, Zone 3. Shell smooth, beak at end of shell, long external ligament, hinge with fine teeth. Glossy blue-black or brownish, interior light purple with distinct dark muscle scar. To 4 inches.

6. Dark Falsemussel *Mytilopsis leucophaeata.* Attached to rocks, other hard substrates, intertidal to subtidal, Zones 1 and 2. Small smooth-shelled mussel, beak at end of shell, shelf under hinge. Brown to tan, interior bluish tan. To ¾ inch.

7. Atlantic Mud-piddock (Fallen angel wing) *Barnea truncata.* Bored into clay, mud, intertidal to subtidal, lower Zone 2 and Zone 3. Shell looks like angel wing but smaller and more fragile, one end squared and smooth, other end with radiating ribs. Interior with thin, narrow extension under hinge. Chalky white. To 2 inches.

8. False Angel Wing *Petricola pholadiformis.* Bored into stiff mud or clay, intertidal, lower Zone 2 and Zone 3. Shell looks like angel wing but smaller and more fragile, elongate, radiating ribs more distinct toward one end. Hinge small with tiny teeth. Chalky white. To 2 inches.

9. Angel Wing *Cyrtopleura costata.* Bored into mud or clay, intertidal, lower Zone 2 and Zone 3. Shell large, thin, somewhat fragile; strong elongate radiating ribs over most of shell. Hinge flaring with spoon-shaped extension into interior. Whitish with thin gray periostracum. To 7 inches.

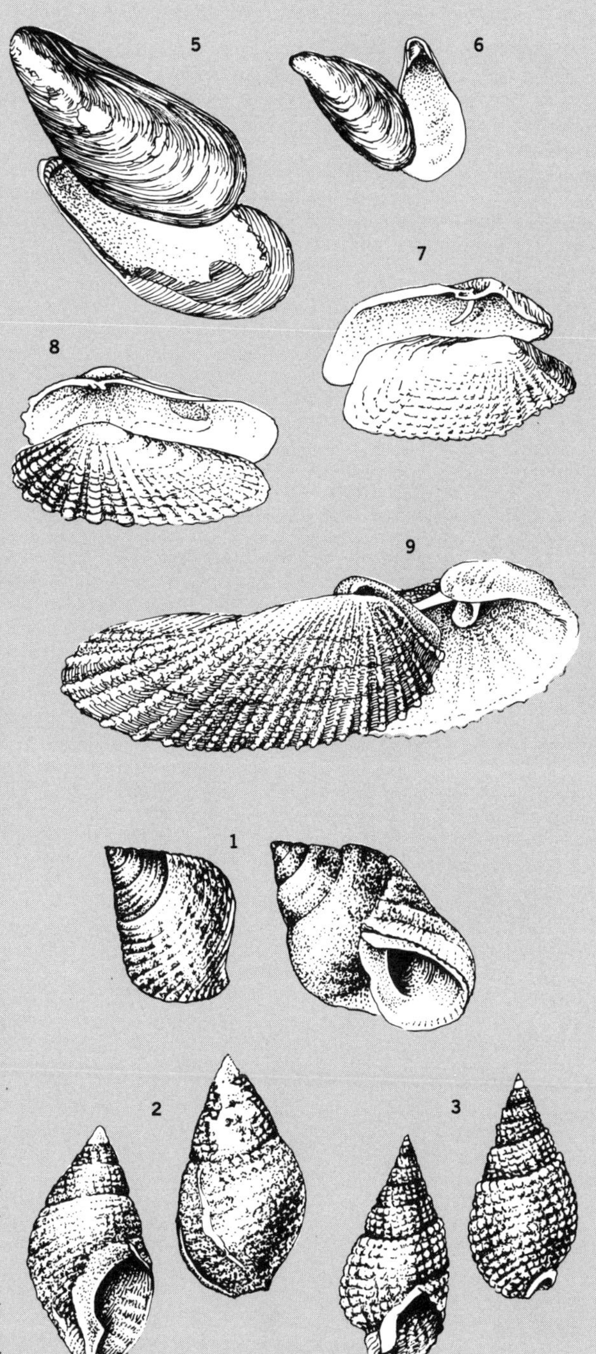

SNAILS OF THE BAY

1. Marsh Periwinkle *Littorina irrorata.* On and among marsh plants, rocks, intertidal, Zones 2 and 3. Shell robust with many shallow spiral grooves, sutures indented. Grayish white to yellow-tan with short reddish brown dashes along grooves. Body whorl of older periwinkles with shallow ribs. To 1 inch.

2. Eastern Mudsnail (Eroded basket shell) *Ilyanassa obsoleta.* On mud flats, muddy-sand, seagrass meadows, intertidal to subtidal, lower Zone 2 and Zone 3. Dark grayish-brown, weakly sculptured with spire generally eroded. Aperture and inner lip dark, outer lip thin. To 1 inch.

3. Threeline Mudsnail *Ilyanassa trivittata.* Over sand, in seagrass meadows, primarily subtidal, lower Zone 2 and Zone 3. Whorls sharply beaded, sutures channeled. Light gray to yellow gray, inner lip glazed with white. To ¾ inch.

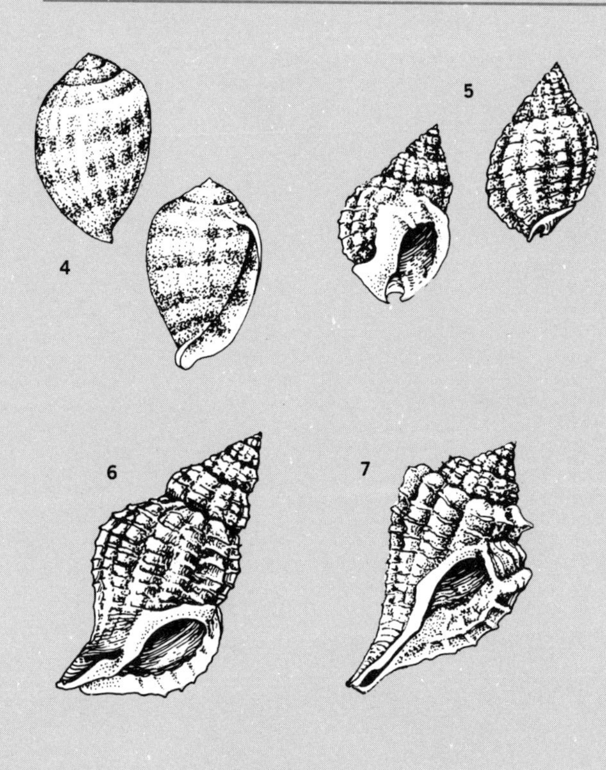

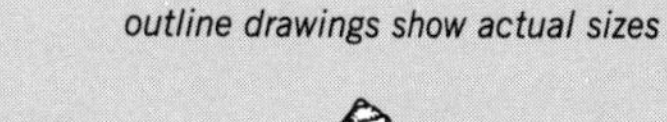
outline drawings show actual sizes

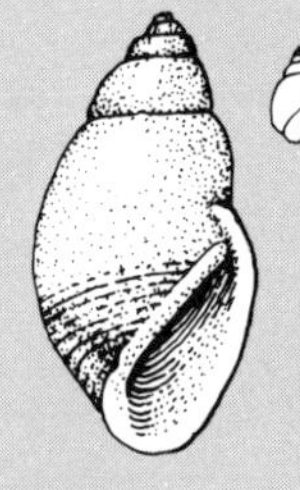

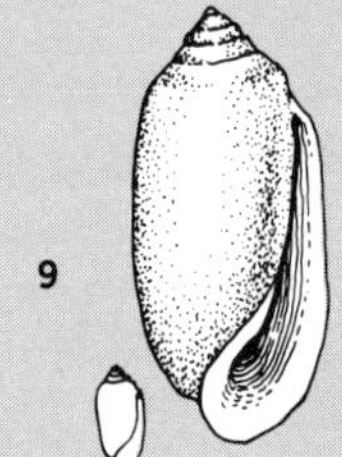

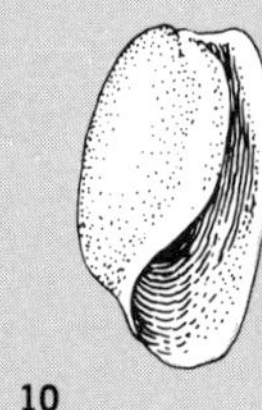

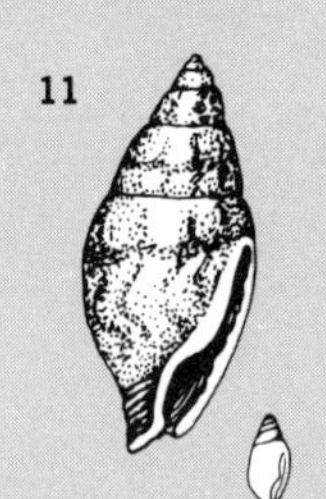

4. Saltmarsh Snail *Melampus bidentatus.* Marshes, intertidal, Zones 2 and 3. Shaped like a top, thin shelled, and smooth. Light brown, usually with dark brown bands. To ½ inch.

5. Nassa Mudsnail (Mottled dog whelk) *Nassarius vibex.* On mud, sandy-mud flats, in seagrass meadows, lower Zone 2 and Zone 3. Whorls coarsely beaded, sutures shallow, inner lip wide and flaring and glazed with yellow-white. Gray sometimes mottled with brown. To ½ inch.

6. Atlantic Oyster Drill *Urosalpinx cinerea.* Oyster bars, seagrass meadows, intertidal to subtidal, lower Zone 2 and Zone 3. Shell ribbed with pointed spire, outer lip flared with two to six small teeth. Gray to tan, some with brown spiral bands. To 2 inches.

7. Thick-lipped Oyster Drill *Eupleura caudata.* Oyster bars, seagrass meadows, intertidal to subtidal, lower Zone 2 and Zone 3. Similar to common oyster drill but anterior canal long, tubular, almost closed; outer lip thick with no flare and about six distinct teeth. To 1 inch.

8. Pitted baby-bubble *Rictaxis punctostriatus.* Sand, intertidal to subtidal, Zones 2 and 3. Spiral, globular shell. Shiny white, lower half of body whorl with rows of tiny dark dots. To ¼ inch.

9. Barrel Bubble *Acteocina canaliculata.* Mud, intertidal to subtidal, Zones 2 and 3. Glossy, smooth, cylindrical shell with shallow spire, often eroded. White to cream, sometimes with rust stains. To ¼ inch.

10. Solitary Bubble *Haminoea solitaria.* Mud, sand, intertidal to subtidal, Zones 2 and 3. Shell fragile, rounded, spire barely noticeable in depression at top edge of aperture. Glossy blue-white to amber. To ½ inch.

11. Lunar Dove Shell (Crescent mitrella) *Mitrella lunata.* Seagrass meadows, mud, subtidal, lower Zone 2 and Zone 3. Shell glossy, smooth, translucent, marked with zigzag stripes of brown to yellow. To ¼ inch.

12. Seaweed Snails *Hydrobia* spp. Weed beds, mud flats, intertidal to subtidal, Zones 1, 2, and 3. Minute spiral shell, glossy, translucent brown, apex blunt. To ⅕ inch.

13. Spindle-shaped Turret Snail *Mangelia plicosa.* Bamboo worm communities, seagrass meadows, oyster beds, subtidal, Zone 3. Minute spindle-shaped shell, sharply etched ribs and spiral cords giving a crisscross pattern, a distinct notch at upper edge of outer lip. To ⅛ inch.

14. Grass Cerith (Variable bittium) *Bittium varium.* Seagrass meadows, subtidal, lower Zone 2 and Zone 3. Turreted shell, body whorl relatively small, with thick rib on back. Shell with crosshatched pattern of ribs and spiral lines, aperture wide and rounded. Brown to gray. To ⅛ inch.

15. Black-lined Triphora *Triphora nigrocincta.* Eelgrass meadows under stones and shells, subtidal, lower Zone 2 and Zone 3. Spiral fusiform shell, sinistral. Dark brown with glossy beads and black bands just below suture. To ¼ inch.

16. Interrupted Turbonille *Turbonilla interrupta.* Sand, mud, bamboo worm cases, intertidal to subtidal, Zone 3. Shell a slender spire, shallow ribbing, waxy yellow. To ⅓ inch.

17. Impressed Odostome *Boonea impressa.* Predator on oysters and other mollusks; oyster bars, seagrass meadows, subtidal, lower Zone 2 and Zone 3. Elongated conical shell, milky white, whorls flattened and indented. To ¼ inch.

18. Two-sutured Odostome *Boonea bisuturalis.* Predator on oysters and other mollusks, worms, oyster bars; mud bottom on worm cases, subtidal, Zone 3. Elongate conical shell, milky white, whorls smooth with second incised line below suture. To ¼ inch.

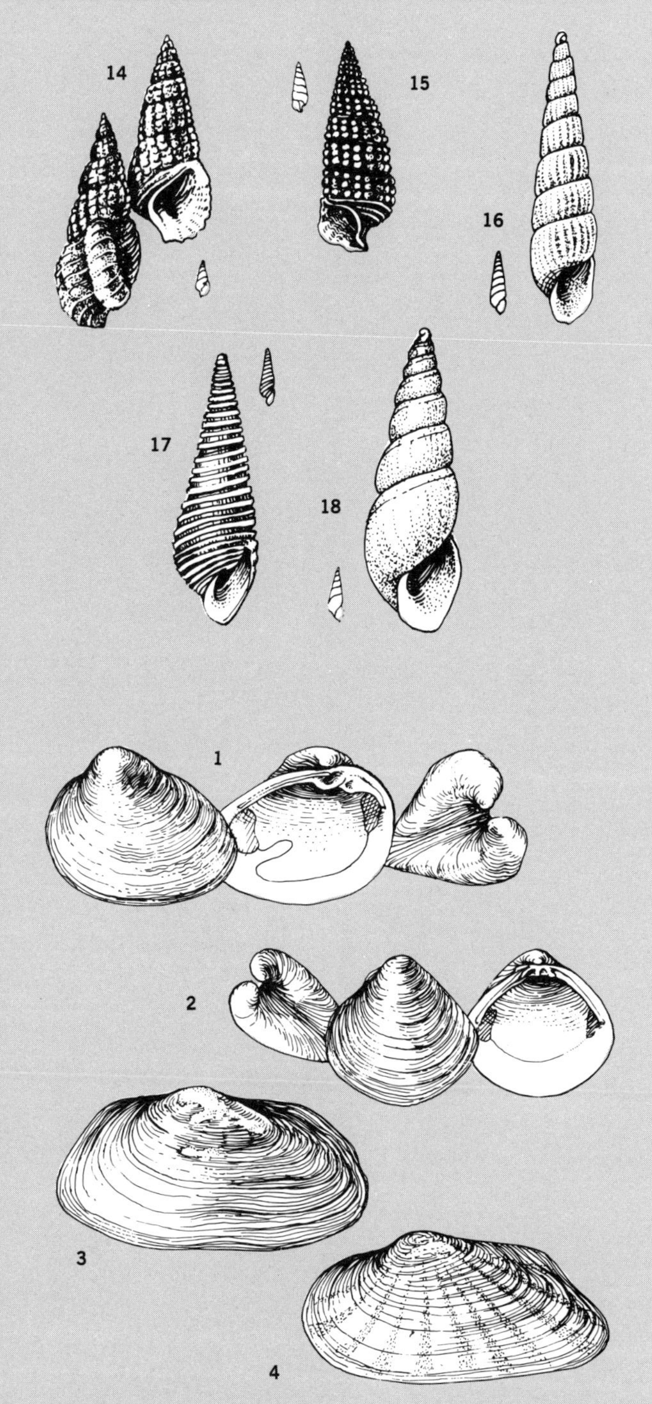

MOLLUSKS OF FRESH AND BRACKISH WATERS OF THE BAY

1. Brackish-Water Clam (Wedge rangia) *Rangia cuneata.* Mud, sand, marshes, intertidal to subtidal, Zones 1 and 2. Shell strong, thick, heart-shaped in cross section; beaks bulbous and pointed inward and forward; pallial sinus small, deep, and distinct. Hinge with distinct lateral and cardinal teeth, spoon-shaped chondrophore. Yellow to brown, interior glossy white. To 2½ inches.

2. Asian Clam *Corbicula fluminea.* Mud, sand, gravel, intertidal to subtidal, Zone 1. Shell triangular, heart-shaped in cross section; beaks distinct; no pallial sinus; hinge with three distinct cardinal teeth, two lateral teeth. Yellow to brown, interior glossy blue-gray. To 2 inches.

3. Freshwater Mussels (Freshwater clams) *Anodonta* spp. Mud, subtidal, Zone 1. Shell thin, generally shiny, species usually elliptical but variable in shape, with concentric rings, no teeth on hinge. Brown to blackish; beak often eroded showing chalky white, interior pearlescent white. To 5 inches or more.

4. Freshwater Mussels (Freshwater clams) *Lampsilis* spp. Mud, sand, subtidal, Zone 1. Shell often thick, oval to elliptical but variable in shape, with concentric rings, hinge with distinct teeth, two false cardinal teeth and one or two long lateral teeth. Yellowish to dark brown, often with dark radiating stripes; beak often eroded, interior shiny white. To 5 inches or more.

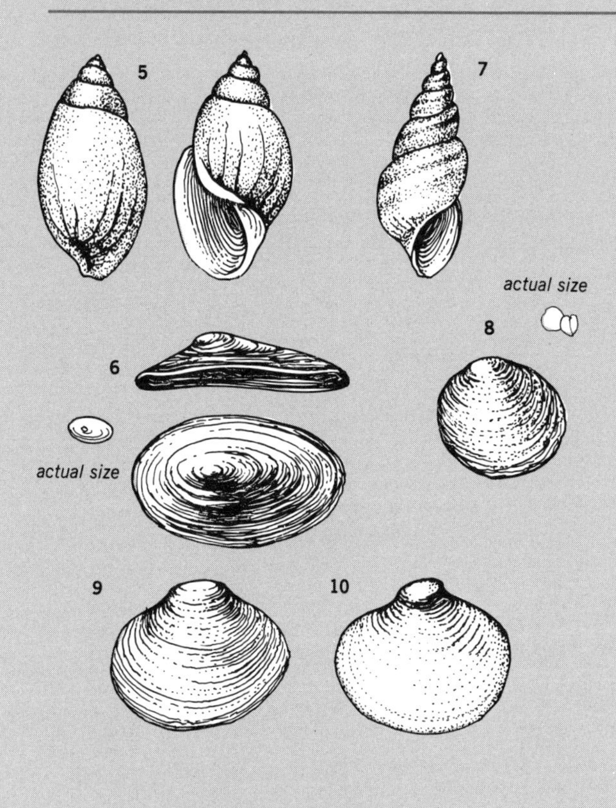

5. Pouch Snail (Tadpole snail) *Physa gyrina.* Mud, intertidal to subtidal, Zone 1. spiral, sinistral shell, with coarse, shallow sculpturing. To 1 inch.

6. Coolie Hat Snails (Freshwater limpet) *Ferrissia* spp. On shells, stones, plants, intertidal to subtidal, Zone 1. Tiny cup-shaped shells with peaked apex. Species variably colored brown to green. To 1⁄4 inch.

7. Hornshell Snail *Goniobasis virginica.* Weed, beds, intertidal to subtidal, Zone 1. Elongated spiral shell, thin, smooth but not glossy. Brown to olive, sometimes with reddish bands on whorls, aperture bluish. To 1 inch.

8. Pill Clams *Pisidium* spp. Weed beds on plants, on rocks, subtidal, Zone 1. Minute rounded, pea-like clams, shell thin, smooth, fragile; beaks toward one side. Light to dark brown. To 1⁄8 inch.

9. Short-siphoned Fingernail Clams *Sphaerium* spp. On plant stems, weed beds, also buried in mud, subtidal, Zone 1. Small clams, shells firm, oval; beak central; two small cardinal teeth on left valve, one on right. Yellowish to brown. To 1⁄3 inch.

10. Long-siphoned Fingernail Clams *Musculium* spp. Weed beds on leaves, also buried in mud, subtidal, Zone 1. Small clams, shell thin, polished, almost transparent; beak prominent, often raised into a cap; cardinal teeth minute or absent. Yellowish to light brown. To 1⁄2 inch.

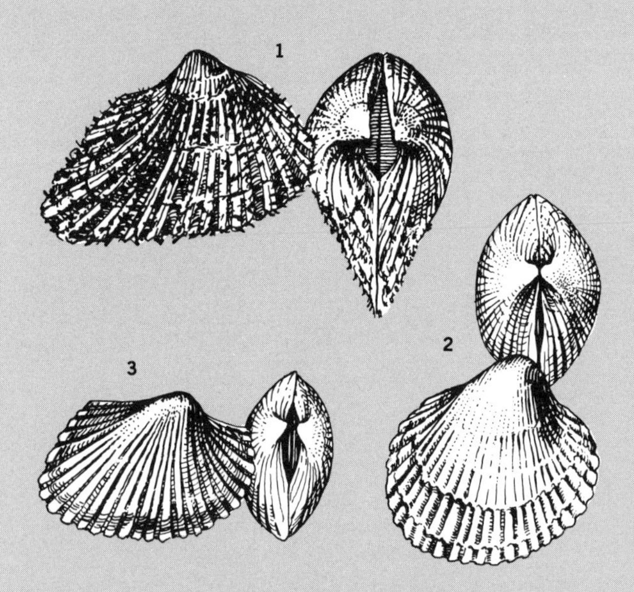

MOLLUSKS OF THE LOWER BAY

1. Ponderous Ark *Noetia ponderosa.* Sand, muddy-sand, subtidal, Zone 3. Shell thick, somewhat truncate with squared raised ribs split in center with incised line; deep black periostracum thick and velvety but often eroded over beaks. Beaks large, pointed backward and separated by broad transversely striated ligament. To 2½ inches.

2. Blood Ark *Anadara ovalis.* Mud, sand, subtidal, Zone 3. Shell thick, more rounded, ribs square; periostracum black-brown and hairy, often eroded. Beaks pointed toward each other, closely aligned; ligament narrow, extending only in front of beaks. To 2 inches.

3. Transverse Ark *Anadara transversa.* Sand, muddy-sand, subtidal, Zone 3. Smallest of three arks, with ribs smaller and more numerous. Shell thick, truncate, left valve overlaps right valve; periostracum gray-brown, often eroded. Beaks smallish, pointed toward each other; ligament relatively strong, extending on both sides of beaks. To 1½ inches.

4. File Yoldia *Yoldia limatula.* Mud, sand, subtidal, Zone 3. Shell elongate, thin and fragile, gaping at both ends, posterior end long and narrow. Glossy greenish tan to brown, about 20 coarse hinge teeth on each side of beak. To 2½ inches.

5. Convex Slipper Shell *Crepidula convexa.* On eelgrass blades, other shells and hard substrates, intertidal to subtidal, lower Zone 2 and Zone 3. Shell highly arched with hooked apex, interior platform about one-third of aperture, edge somewhat curved. Color dark reddish brown to tan. Shells on eelgrass smaller, thinner shelled, and more elongated than those on other substrates. To 1½ inch (⅓ inch on eelgrass).

6. Common Atlantic Slipper Shell *Crepidula fornicata.* On shark eye snails, whelks, oysters, other hard substrates, intertidal to subtidal, Zone 3. Shell usually arched, but variable, apex turned to one side, interior platform about one-half of aperture, edge somewhat wavy. Color dirty white to tan, often with fine spotting. To 1½ inches.

7. Flat Slipper Shell (White slipper shell) *Crepidula plana.* On shells and other hard substrates, subtidal, Zone 3. Shell flat or molded to shape of shell to which it was attached, convex to concave. Color milky white, inside and out. To 1 inch.

8. Knobbed Whelk *Busycon carica.* Sand, sandy-mud, intertidal to subtidal, Zone 3. Largest shell in Chesapeake Bay. Shell strong, body whorl large in relation to spire, anterior canal elongated, aperture wide with interior orange-yellow to brick red in color. Knobs on shoulder of body whorl and spire. To 9 inches.

9. Channeled Whelk *Busycotypus canaliculatus.* Sand, sandy-mud, intertidal to subtidal, Zone 3. Similar in shape to knobbed whelk but with grooved sutures between whorls and a heavy, thick, fuzzy, gray periostracum. Aperture large with interior brownish color. To 7 inches.

10. Shark Eye (Atlantic moon snail) *Neverita duplicata.* Sand, sandy-mud, intertidal to subtidal, Zone 3. Shell globular, smooth, with low spire flattened, aperture wide and open, deep umbilicus covered with a purple-brown callus flap at side of parietal wall. Color gray to tan, covered with horny, light brown periostracum. To 3 inches.

11. Atlantic Surf Clam (Ocean clam) *Spisula solidissima.* Shells only on Chesapeake Bay beaches near mouth of Zone 3; live surf clams offshore in ocean. Shell large, thick, strong, oval with fine concentric lines. Chondrophore large and spoon-shaped; pallial sinus turned slightly upward. Periostracum thin, yellowish. To 7 inches.

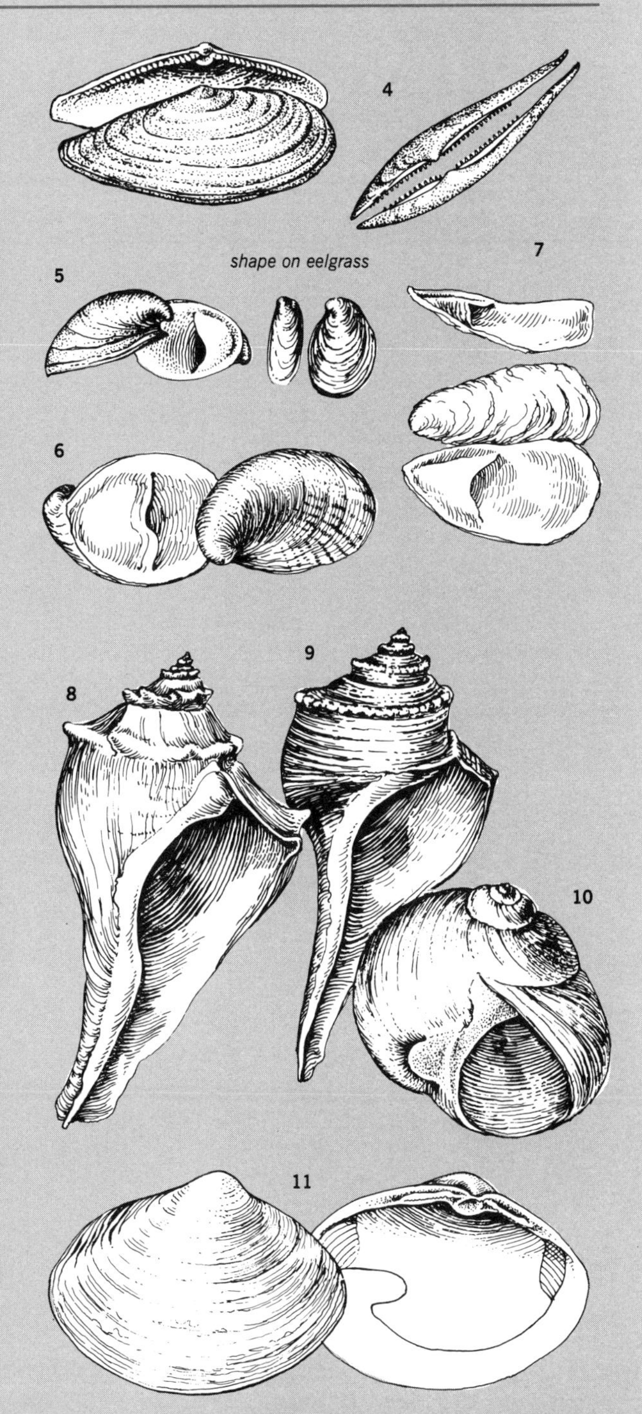

INTERTIDAL FLATS

An open stretch of intertidal muds appears inhospitable if not downright ugly to most people, surely to be avoided and certainly barren of life. Yet the intertidal flat harbors many more types and numbers of plants and animals than a sand beach. Once aware of this, the inveterate Bay-watcher will begin to observe and learn about a vital and surprisingly interesting habitat of the Chesapeake Bay.

Many intertidal shorelines are soft and mushy and almost impossible to traverse without sinking ankle-deep in mud; but others are firm, and you can venture easily across them. You will be amply rewarded. The type of flat depends on the kinds of sediments deposited there; soft-bottom flats contain a high percentage of very fine silt and clay particles, whereas hard-bottom flats are composed mostly of sand particles. Marine animals are sensitive to the size of sediment particles; consequently, certain animals common to soft muds are never found in sandy-mud areas, and vice versa. However, the composition of mud flats in an estuary such as the Chesapeake is highly variable, and many estuarine organisms have adapted to a wide range of sediment types.

The intertidal flat, like the beach, is a rigorous environment for marine plants and animals, as they are intermittently exposed to the heat of the sun with each tidal cycle and to the drying action of air and wind. Distinct intertidal zones exist here as on the sand beach. The high-tide zone is occupied by semiterrestrial crustaceans that can live out of water for long periods. The variety and numbers of intertidal plants and animals gradually increase toward the lowest intertidal zone, which is exposed only during the lowest tides. Here you will find not only typical intertidal species but also some essentially subtidal ones, able to survive out of water for short periods of time.

The intertidal mud or sand flat habitat is continuous with many other habitats. Landward, it may be bordered by a beach, marsh, bulkhead, or stretch of riprap. Beyond the

3. INTERTIDAL FLATS

water's edge there may be a rich stand of aquatic plants or an oyster bar. Pier pilings will be covered with seaweeds, barnacles, sponges, and a myriad of other fauna. Free-moving worms, crabs, and snails from these other habitats will often move onto the mud flats in a quest for food. The buried animals of the flats feed on the rich supply of planktonic food borne by the flooding tide and are, in turn, fed upon by wading birds which are well adapted for searching out the animals of the intertidal zone.

Rooted aquatic plants are not characteristic of intertidal flats, but when a strong prevailing wind pushes the ebbing waters out below the normal tide mark, stranded

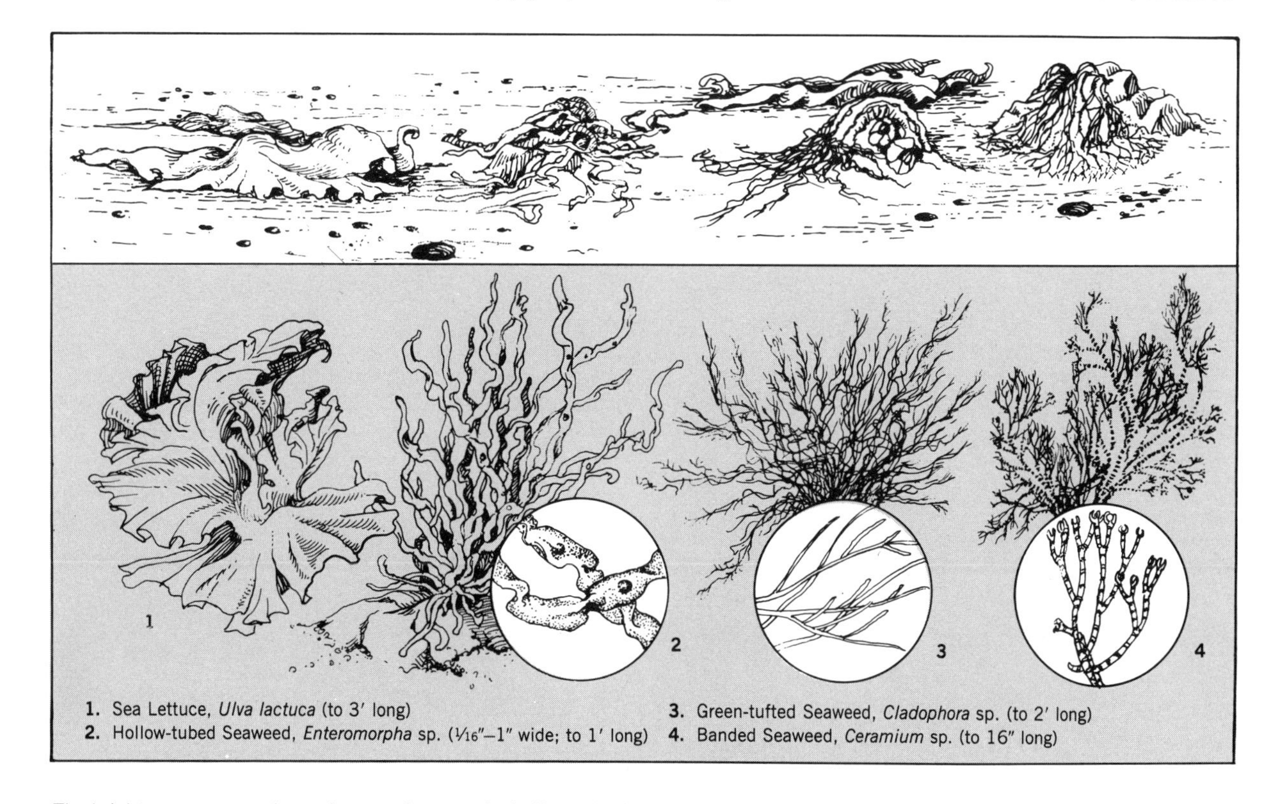

1. Sea Lettuce, *Ulva lactuca* (to 3′ long)
2. Hollow-tubed Seaweed, *Enteromorpha* sp. (1⁄16″–1″ wide; to 1′ long)
3. Green-tufted Seaweed, *Cladophora* sp. (to 2′ long)
4. Banded Seaweed, *Ceramium* sp. (to 16″ long)

The bright green seaweeds, such as sea lettuce, the hollow-tubed seaweeds, and the green-tufted seaweeds, are the easiest to spot on the intertidal flat. The reddish traceries of banded seaweeds are also distinct, but their natural form is not evident until the tide comes in and they float free.

plant beds may be exposed. However, other forms of plant life thrive on the flats, much of it microscopic algae visible only as a brown stain or a bright green tint. Bacteria and algae are highly productive on flats and form thin sheets covering shells and mud.

PLANT LIFE ON THE FLATS

Seaweeds wrenched from their original holdfast float in with the tides and are often found on intertidal flats in windrows or matted clumps. Individual plants often grow on scattered shells, on pebbles, or on other hard-surfaced debris. Seaweeds are marine algal plants related to the microscopic planktonic algae (phytoplankton) but are much larger and attach to almost any firm substrate. Some seaweeds, such as the ocean kelps, grow to 15 feet or more, but seaweeds of the Chesapeake Bay are generally only a few inches tall, although aggregations of small plants can carpet broad areas. Seaweeds grow on piers and rocks, in the shallows of coves and bays, and are attached to fronds of eelgrass or widgeon grass. They need sunlight to grow, so as the turbid waters deepen offshore, the plants thin out and eventually disappear.

There are many different kinds of seaweeds in the Chesapeake, but only a few are distinct enough to be easily identified. Most closely related seaweeds are so similar in appearance that even algologists (those who study algae) have difficulty identifying them to species. Seaweeds stranded on the intertidal shore lie limp and formless on the muds or sands. They should be placed into a jar of water, where their true beauty and lacy delicacy can be appreciated as they unfold. However, your first introduction to seaweeds might be through your nose rather than your eyes. Huge windrows of rotting sea lettuce, *Ulva lactuca,* are common along the shore and accumulate in such masses that they create a nuisance. Sea lettuce is brilliant green and broad-leaved with delicate, curly edges. It is quite beautiful when seen among the fauna of a pier piling or a rock jetty, gracefully undulating in the currents just below the surface. Probably the next most commonly encountered species are the hollow-tubed seaweeds, *Enteromor-*

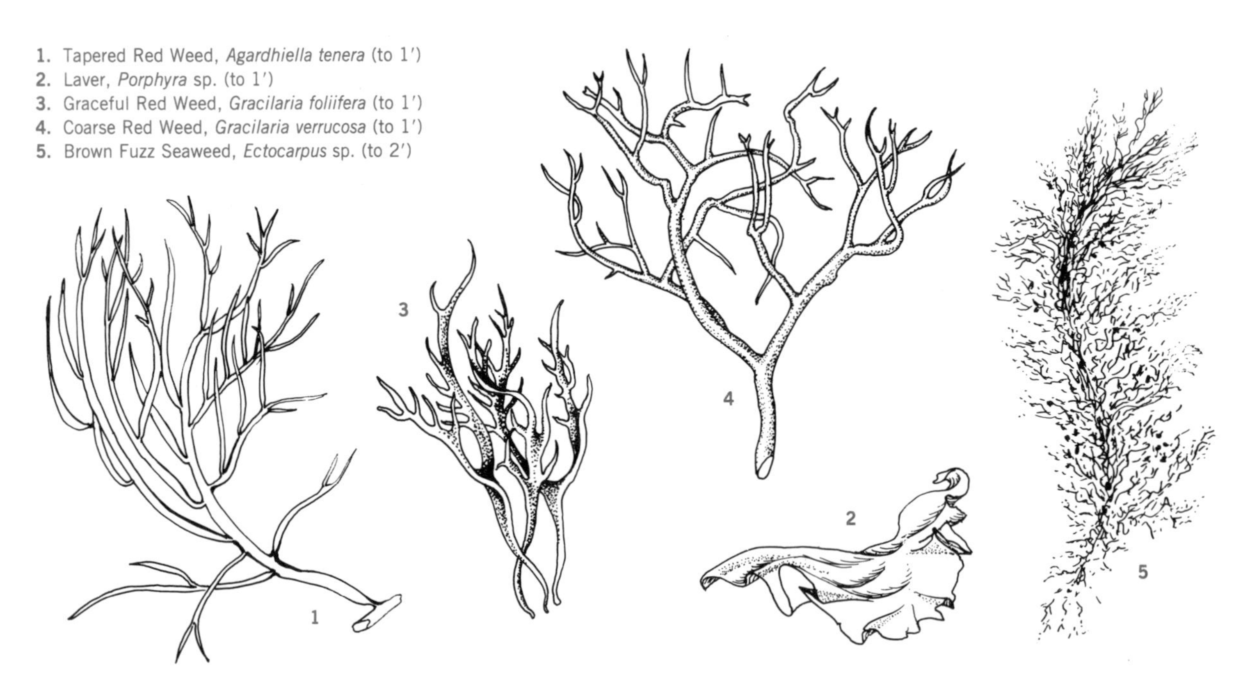

Red weeds and laver seaweeds are found only in the higher salinities of the lower Chesapeake Bay. Brown fuzz seaweeds are more widespread, particularly during colder months.

pha spp., plants as brightly green as sea lettuce. Small clumps on intertidal flats are often seen attached to the tiniest fragments of a shell or to a small pebble. The hollow strands are curved or crinkled, tubular or flattened, and some species have tiny air bubbles within the strands. Raise the strands to the light and the bubbles will glisten.

The silky, filamentous, green-tufted seaweeds, *Cladophora* spp., are often draped and flattened over a rock, shell, or stick on the intertidal flat. If a bit is picked up and dropped into the adjacent shallows it will turn into a soft greenish mist spreading through the water. The delicate filaments are intricately branched and intertwining.

Many reddish seaweeds also occur in the Chesapeake. Intertidally, the most common species are probably the banded seaweeds, *Ceramium* spp. These are soft and bushy seaweeds that appear to be nondescript reddish brown clumps when found stranded on a flat. They have an intricately branching and delicate shape when floating in water. This group can be identified by looking very closely (a hand lens can be helpful) at the tips of the branches, which have tiny "pincer" ends pointing inward and delicate banding on each branch and stem. Species are variably colored, usually in shades of red.

In the higher salinities of the lower-Bay region, the red weeds are bushy and coarse, with thick, fleshy branches. Unlike so many other seaweeds, which are almost impossible for the layman to identify, red weeds are worth the attempt; if successful, you should be greatly satisfied. The tapered red weed, *Agardhiella tenera,* is a dark red species with rounded branches tapered at the tips and at the base, where each joins the stem. Graceful red weed, *Gracilaria foliifera,* has flattened branches and is variably colored, red to purple or yellowish. Coarse red weed, *Gracilaria verrucosa,* is also variably colored, but has rounded branches that are not tapered at the base. Laver seaweeds, *Porphyra* spp., other red seaweeds of the lower Bay, are tissue-thin reddish versions of sea lettuce. Laver seaweeds are more abundant in winter months than in summertime.

In winter, when most seaweeds have disappeared, especially the green seaweeds, the brown fuzz seaweeds, *Ectocarpus* spp., appear as long, entangled strands of hairlike filaments strung along the intertidal flat and often attached at one end to a bit of rock or shell. When expanded under water, they look like brown fuzz, lacy and soft, much like the green-tufted seaweeds. The plants of the intertidal flats are varied in form and color; however, the animal species of this habitat are even more diverse.

Three general types of invertebrate animals inhabit the intertidal flats: burrowers in the substrate, called infauna; immobile sessile (attached) animals on the surface; and motile animals moving readily over the surface, collectively called epifauna. The soft, silty surface of an intertidal habitat offers little succor for most sessile animals. In the Chesapeake Bay, oysters and mussels may be seen in certain areas, especially during the lowest tides, but they are more characteristic of subtidal areas.

ROAMERS OVER THE FLATS

The epifaunal motile animals gliding and scooting over the surface muds or sands are primarily crustaceans and snails that prey on the rich supply of buried infauna. Many foragers, such as blue crabs, small fishes, and shrimp, come in with the tides to feed on surface detritus or to prey on intertidal burrowers, but they leave the flats with the receding tides and are more properly at home in shallow-water communities.

Wanderers from the Land to the Edge of the Sea

Fiddler Crabs. Fiddler crabs will be found in the muddy marshlands bordering the Bay and its tributaries. These semiterrestrial crabs are unmistakable in appearance, since the male sports one remarkably large claw, with which he wards off intruders or waves in a fiddling motion to attract a mate. Females do not have the distinctive fiddle claw so necessary for identifying these crabs to species, but there will usually be a male close by to provide the essential clues for identification. The large claws of the males are virtually useless in feeding, which gives the females an advantage as a drove of fiddlers moves over the muds in search of bits of algae or decaying marsh plants. Fiddler crabs are active during daylight hours, busily digging and enlarging their burrows. They move onto the flats when the tide is low. At night they retreat to their burrows, carefully plugging the entrances with mud or sand brought up from below, and there they remain protected from predators. In winter, fiddler crabs remain deep in their covered burrows below the frost line, emerging by the thousands in spring to resume their frenzied activities.

Three species of fiddler crabs can be found in the Bay, each slightly different in appearance and each preferring a slightly different habitat. Perhaps the most widespread is the red-jointed fiddler crab, *Uca minax,* which is also

The red-jointed fiddler crab, *Uca minax,* largest of the Chesapeake Bay fiddler crabs, courts the small-clawed female with his waving fiddle claw.

known as the brackish-water fiddler because its greatest populations occur in low- to mid-salinity waters. It is the largest of the three species, almost as large as its close relative the ghost crab, and can be identified most easily by the bright red joints of its "fiddle" claw. This crab prefers muddy areas, where it digs its burrow above the high-tide line, often constructing a ledge of mud to shade the entrance. The marsh or mud fiddler crab, *Uca pugnax,* is also widespread in the Bay, although not generally as tolerant as the red-jointed fiddler of low salinities. It is the smallest Bay fiddler, its body generally less than an inch wide. The sand fiddler, *Uca pugilator,* is also a fairly small crab, and as its common name implies, it prefers sandy areas. In fact, it will not survive if placed in silty mud. Apparently its feeding mechanisms are adapted for working over a specific size range of sediment particles. Sand fiddlers are a lighter color than the other two species, but to be absolutely sure of your identification capture one, preferably holding the tips of the large claw together, and scrutinize the underside of the fiddler claw. A sand fiddler's will be smooth. If there is a row of tubercles present, look for red joints on the claw and an indented groove on the back behind the eye—if present, the specimen is a red-jointed fiddler; otherwise, it is a marsh fiddler.

Little Square Crabs. You might catch sight of a small, scurrying, square-backed crab in many of the same locales as fiddler crabs. At first glance the square-backed crab might be mistaken for a female fiddler crab. If the eyes are at the corners of its square body, it will be one of two species of small Chesapeake Bay shore crabs: the marsh crab, *Sesarma reticulatum,* or the wharf crab, *Sesarma cinereum.* The marsh crab has a notched margin at the front angle of the carapace; the wharf crab's carapace is smooth. These little crabs are most likely to be spotted on intertidal flats bordering marshlands or within the marsh-

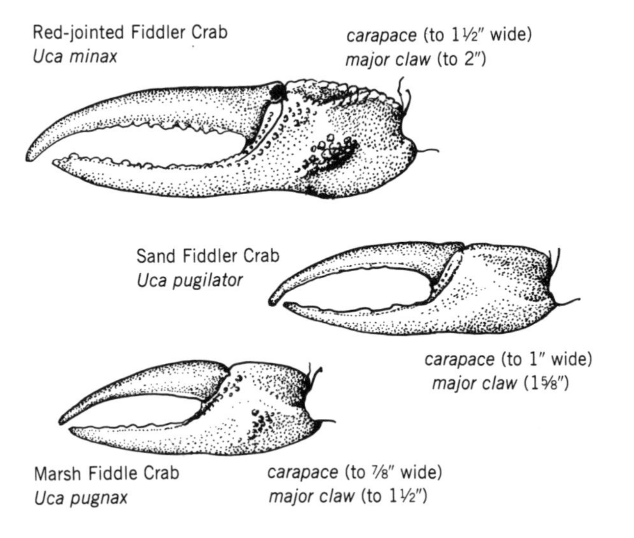

The males of the three species of fiddler crabs of the Chesapeake Bay are identifiable by examining their large claw.

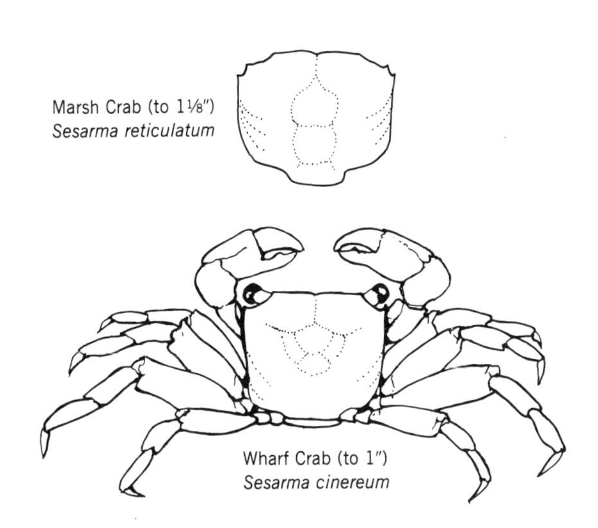

es. Wharf crabs are particularly adventurous wanderers and often scamper over piers, where they will hide in any nook or cranny.

Freshwater Crayfish. Crayfish are little known along the shores of the Chesapeake and when one is captured it becomes an object of great interest; some people wonder where the lobster came from, and others, while properly identifying the beast, are certain that no crayfish occur naturally in the Bay area. There are, indeed, crayfish in the Bay area.

Crayfish—freshwater crustaceans that look very much like diminutive lobsters—are more populous inhabitants of the coastal plains, the rolling piedmont, and the mountainous areas of Western Maryland and Virginia, but two species are common in Zone 1 of the upper Bay and tributaries.

Crayfish do not have a complicated life cycle such as that seen in crabs or marine mollusks. The males and females generally mate in the autumn and the spring. The female subsequently deposits her eggs in a sticky mass on the underside of her long abdomen; the eggs develop and hatch into small replicas of the parents, and the young reach sexual maturity in about one year.

Crayfish are opportunistic feeders, obtaining nourishment from algae, rooted plants, and recently dead organisms; they are occasionally able to capture small fish and other crayfish. The first pair of a crayfish's five pairs of walking legs have very large claws or pincers. The second and third pairs of walking legs also terminate in pincers; however, they are much smaller than the first. Their powerful and threatening claws are used more for defense and in the mating ritual than as offensive weapons.

The coastal plains river crayfish, *Orconectes limosus,* is a species that prefers slow-moving water and soft, silt bottoms and vegetated areas. This mottled, olive-brown crayfish may sometimes be found in the brackish waters of rivers and the upper Bay, particularly after high-river flows and freshets in the spring.

The burrowing crayfish, *Cambarus diogenes,* is a widely distributed species and is found throughout the Bay region, particularly in moist areas along streams and around ponds and lakes. This species constructs burrows, sometimes with side chambers, that may extend to a depth of three feet or more. The material cast up from the burrow is used to build chimneys, which cap the entrance to the burrows. The chimneys are quite obvious and easy to recognize, composed of pellets of soil which harden into a brittle clay and sometimes exceeding a height of six inches.

The burrowing crayfish is a solitary, secretive species

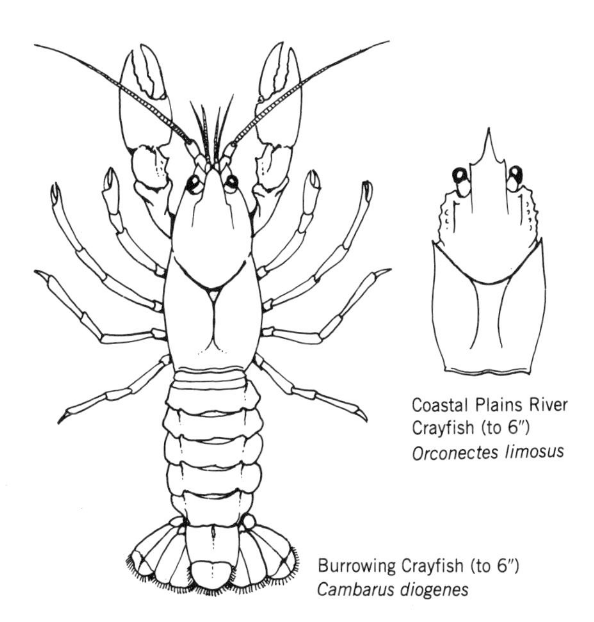

Crayfish species are similar in appearance and often difficult to identify. The shape and sculpturing of the carapace are good characteristics for separating the two crayfish common in tidal freshwater regions of the Chesapeake Bay.

The burrowing crayfish, *Cambarus diogenes,* spends time outside his burrow as well as within. The entrance to the burrow is the hole at the top of the mud pellet chimney.

and, unlike the coastal plains river crayfish, is active only at night. Like most crayfish, its overall color is olive-brown; however, it is often tinged with bright orange-red markings on the claws and carapace.

Wanderers from the Sea to the Edge of the Land

Hermit Crabs. On a shallow intertidal flat, if you spot a snail shell rapidly scrambling about you can be sure it is occupied by a hermit crab. These crabs, which carry their homes with them, are adapted to fitting their spirally coiled, soft, unprotected abdomens into a snail shell. Since almost all snails have a right-handed spiral—that is, the snail's aperture is on the right when the apex is pointed up and the aperture faces the observer—the hermit crab's abdomen is asymmetrical to conform to the inner spiraling of the snail shell. A hermit crab that attempts to occupy the occasional left-handed shell must have the same problem as those of us who attempt to fit a right-handed glove on our left hand. The last pair of the crab's abdominal appendages is modified so that the crab is securely held in the shell. It is almost impossible to remove a hermit crab from its housing without literally tearing the animal apart.

As hermit crabs grow they require increasingly larger snail shells and will actively search for a shell of the appropriate size. Sometimes that shell is occupied by another hermit crab, which may be evicted by the more aggressive and larger crab looking for a new home.

An encrusting organism, the colonial hydroid, *Hydractinia echinata,* appropriately called snail fur, is often found on snail shells occupied by hermit crabs. This interesting species may completely cover the shell giving the surface a pink, furry texture (see Chap. 4 for discussion on hydroids).

Hermit crabs live their entire lives within a shell except during their free-swimming larval stages. They develop in generally the same manner as do other crabs. The eggs hatch as tiny larval forms called zoea and develop through several molts into megalopae, which, although quite small, are recognizable as crustaceans. The megalopa then molts into its final form, the juvenile crab, which continues to grow, molting into the adult crab.

The omnivorous hermit crabs feed on anything avail-

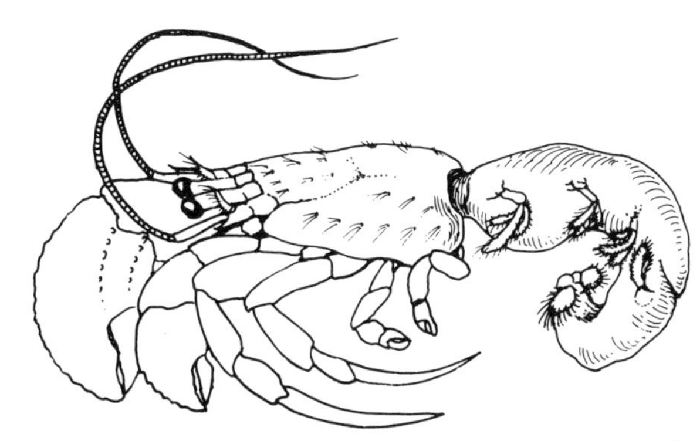

A broad-clawed hermit crab removed from its shell home. Its soft body curls to the right to fit the spiral coil of the shell.

able, including organic matter, algae, dead animals—and sometimes one another.

Three species of hermit crabs commonly occur in the Cheasapeake from Tangier Sound south to the mouth of the Bay. The smallest species, the banded hermit crab, *Pagurus annulipes,* can be recognized by its hairy chelipeds, or claws, and the brown banding around each jointed section of the walking legs. These crabs occupy small shells of eastern mudsnails, oyster drills, and periwinkles. They are common in Zone 3 from the low-tide mark, on a variety of bottom types, to deep water.

The long-clawed hermit crab, *Pagurus longicarpus,* is also a small hermit crab and has, as the name suggests, long claws, which are virtually hairless. The hand of the claw is white with a gray to brown stripe. This is a very common species in shallow water, ranging from lower Zone 2 to Zone 3.

The broad-clawed hermit crab, *Pagurus pollicaris,* is the largest hermit crab in the Bay, with a heavy, broad right cheliped distinguished by tubercles, or wartlike projections, on the wrist. The overall color varies from tan to brownish red. This large species occupies shark eye snails and whelk shells and is found in the deeper waters of the flats to open Bay waters in Zone 3.

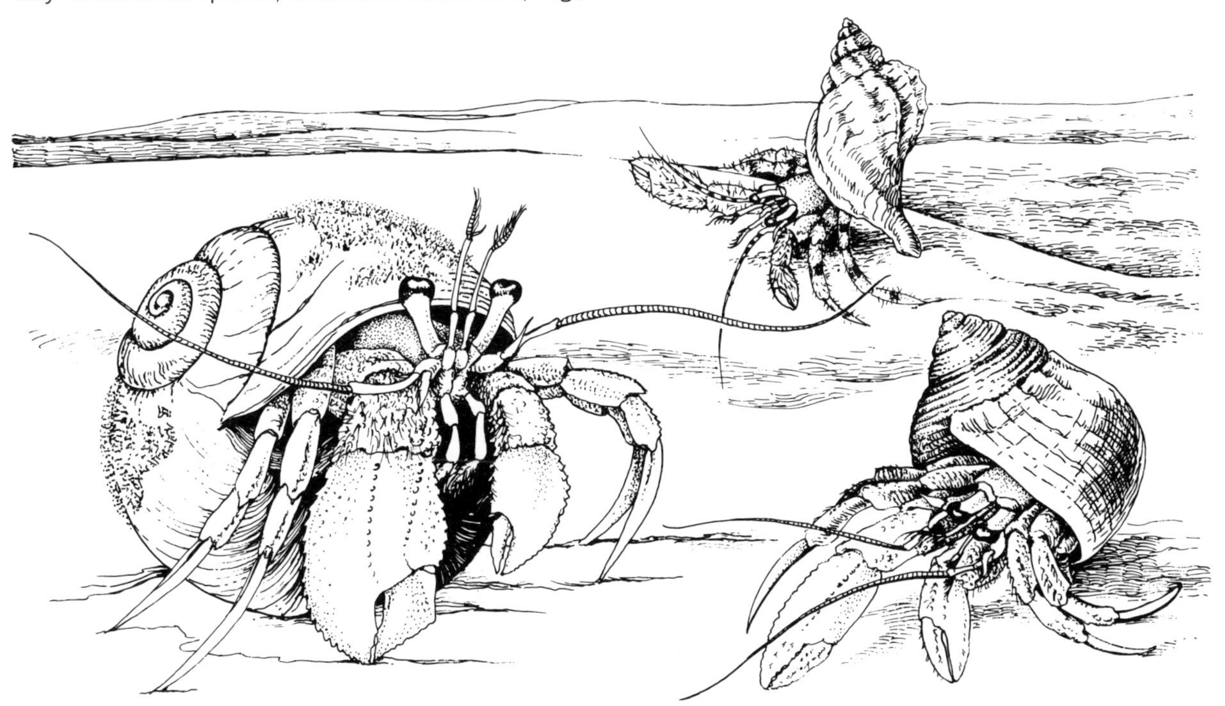

The intertidal flat is home to the broad-clawed hermit crab, *Pagurus pollicaris* (to 4 inches), in a shark eye shell covered with snail fur; the banded hermit crab, *Pagurus annulipes* (to 1 inch), in an Atlantic oyster drill shell; and the long-clawed hermit crab, *Pagurus longicarpus* (to 1½ inches), in a marsh periwinkle shell.

Eastern mudsnails, *Ilyanassa obsoleta* (to 1 inch), scavenging over a dead blue crab on the intertidal flat.

Snails. The gastropods include the familiar snail, which has a single coiled shell. Snails are widely distributed on land and in fresh, estuarine, and marine waters. Some gastropod groups are separated according to the evolution of breathing mechanisms. One group, the pulmonates, have a lung and utilize atmospheric oxygen; the prosobranchs possess gills and extract dissolved oxygen from the water. Most Bay snails, with few exceptions, are prosobranchs. When prosobranch snails are stranded on a flat during low tide they retract their bodies into the shell and seal the opening with their leatherlike operculum to prevent drying out.

Gastropods have various methods of reproduction. Some of the primitive marine species, such as the saltmarsh snail, produce very yolky eggs and free-swimming larvae. Periwinkles deposit into the water single egg capsules, which hatch into larvae. Whelks deposit eggs in the rubbery capsules connected in long chains which are found so often on the beach. Larval forms serve an important purpose—finding new settling sites. They settle in enormous numbers in good years and, thus, depending on seasonal conditions, are subject to great fluctuations. Those species that deposit their eggs in individual capsules and in which the embryos develop directly into small snails rather than larvae generally lay fewer eggs but have a higher rate of survival.

The nassa mudsnail, *Nassarius vibex,* threeline mudsnail, *Ilyanassa trivittata,* and eastern mudsnail, *Ilyanassa obsoleta,* are all common and abundant mud snails of the intertidal flats. All three species are widely distributed from Nova Scotia to the Caribbean. They live on bottoms ranging from coarse sand to silt and are limited to the middle and lower end of the Chesapeake Bay (lower Zone 2 and Zone 3). The Guide to Common Shells in Chapter 2 will help identify the snails and other mollusks living in the intertidal flats and other habitats of the Bay.

The most common species of mud snail in the Bay, the eastern mudsnail, is a grayish brown, rather ordinary-looking snail three-quarters to an inch long with a spire that is usually badly eroded. Although the shell is weakly sculptured and has a granular appearance, it is distinguished by a covering of mud, debris, and algal growth.

The eastern mudsnail feeds primarily on bottom sediments rich in tiny algal forms such as diatoms. However, it is not uncommon to come upon hundreds of these snails feeding on a dead clam, crab, or fish. They are quick-

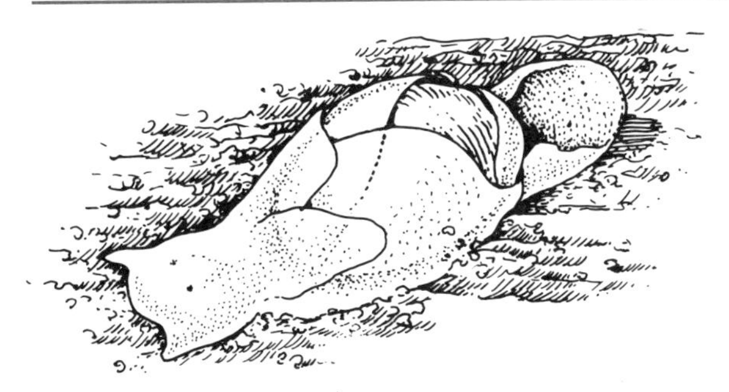

The minute solitary bubble snail, *Haminoea solitaria* (less than ½ inch), roams over a mud flat in search of live prey, its fragile shell barely visible under its enveloping mantle.

moving and agile and have a remarkable ability to detect, almost instantaneously, dead organisms that wash up on the flat. During the winter months eastern mudsnails aggregate in large numbers in eelgrass beds, in water several feet deep, returning to the intertidal flats the following spring. Hermit crabs in the lower Bay are often found occupying empty mudsnail shells.

The solitary bubble snail, *Haminoea solitaria,* the pitted baby-bubble snail, *Rictaxis punctostriatus,* and the barrel bubble snail, *Acteocina canaliculata,* are very small snails, less than half an inch long, with large bodies. The shells are thin and fragile and are often completely enveloped by the mantle. The shell of the solitary bubble snail varies in color from glossy blue-white to amber. The living animal within gives the shell a gray color. The spire rests in a depression that gives the shell a flat-topped appearance. The shell of the baby-bubble snail is shiny white with a low but prominent spire, and the shell of the barrel bubble snail is cylindrical and shiny-white, with a shallow spire. Bubble shell snails are found throughout Zones 2 and 3 on the intertidal flats and in deeper waters, particularly in vegetated areas. In contrast to the eastern mudsnails, the bubble shell snails are carnivorous and seek live prey rather than feeding on algae or occasional carrion.

Shark eyes are large, solid, globular snails found in sandy areas. The shark eye, *Neverita duplicata,* found in the lower Bay in high-salinity waters, is a highly predacious hunter, drilling holes in the shells of clams and other mollusks with its sharp teeth and feeding on the flesh inside with its extensible proboscis. These characteristic holes can frequently be seen near the beak in dead clam shells.

Shark eyes plow through the sand with an enormous foot and leave a telltale furrow, which can be used by the careful observer as a trail marker. This is the species that forms sand collars (see Chap. 2), the distinctive egg cases, some six inches in diameter, that are commonly cast up on the beach.

Whelks are large snails with massive shells familiar to even the most casual beachcomber or occasional frequenter of shell stores. The knobbed whelk, *Busycon carica,* and the channeled whelk, *Busycotypus canaliculatus,* are aggressive carnivores that prey on other mollusks, particularly clams. These large, abundant gastropods are found only in the lower Bay in the high-salinity waters of Zone 3, from the shallow tidal flats to depths of about 50 feet.

The knobbed whelk, the largest snail found north of Cape Hatteras, often attains the impressive length of 9 to 10 inches. The shell is characterized by a series of rounded knobs on the shoulder of the body whorl, while the inner lining of the shell may be yellow-orange to brick-red. The animals of both the knobbed and channeled whelks have large gray, fleshy bodies with a broad foot.

The channeled whelk does not quite reach the size of its close relative, but it may grow to eight or nine inches. Its shell has a groove, or channel, at the suture, the furrow separating each whorl. The aperture is yellowish to tan-colored. The channeled whelk is often found in crab traps and, in New England, in lobster pots. It feeds on bait found in traps and pots, probably to a greater extent than the knobbed whelk. A voracious predator, the knobbed whelk pries and rasps a gap between the valves of its prey by exerting inexorable force with its muscular foot. When a small opening is created, it uses its shell as a wedge to maintain the opening and then consumes the soft body of the clam.

Whelks are harvested along the Atlantic coast and are marketed primarily in the Northeast, where they are savored. A popular Italian dish, scungili, is prepared from the rubbery foot of the whelk. Hermit crabs, particularly the large, flat-clawed hermit crab, often occupy empty whelk shells along with slipper shells, which colonize the aperture.

The common Atlantic slipper shell, *Crepidula fornicata,* and its relatives the convex slipper shell, *Crepidula convexa,* and the flat slipper shell, *Crepidula plana,* are also known as slipper limpets and boat shells. These species are abundant along the Atlantic coast and in Zone 3 in the lower Chesapeake. They can be found in impressive numbers washed up on the beach after a storm. Look for them attached to large dead whelk shells and other mollusks and particularly on the stranded remains of horseshoe crabs.

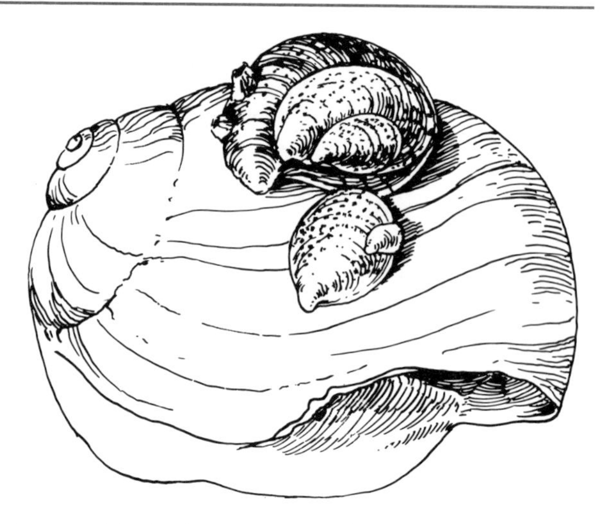

Common Atlantic slipper shells, *Crepidula fornicata* (to 1½ inches) pile one atop another in a communal stack over a shark eye shell. The smaller and younger slipper shells on top are males; the larger ones on the bottom, females. The males will eventually develop into females.

These shallow-water snails are cup-shaped, and the underside of the shell is reinforced by a platform or deck. The shells are variable in shape, from somewhat flattened to a high-arch form. The surface can be corrugated, ribbed, nodular, or smooth, depending on whether the slipper shell has attached to a scallop, a bottle, or the interior of a whelk shell. In other words, their highly plastic shells are influenced by their place of attachment. The slipper shell's reproductive cycle is unusual in that this snail exhibits sex reversal. All the young are males; when fully grown, they develop into females. Female characteristics are controlled by a waterborne hormone. The males are smaller than the females. Generally, the male moves to the vicinity of a female of the same species and may even attach itself to the female. Where there are soft bottoms with few hard objects on which to fasten, common slipper shells will often attach to one another in a continuing communal stack, the older and larger snails, the females, on the bottom and newly arrived males on the top. The eggs are laid in thin-walled capsules and are brooded by the female.

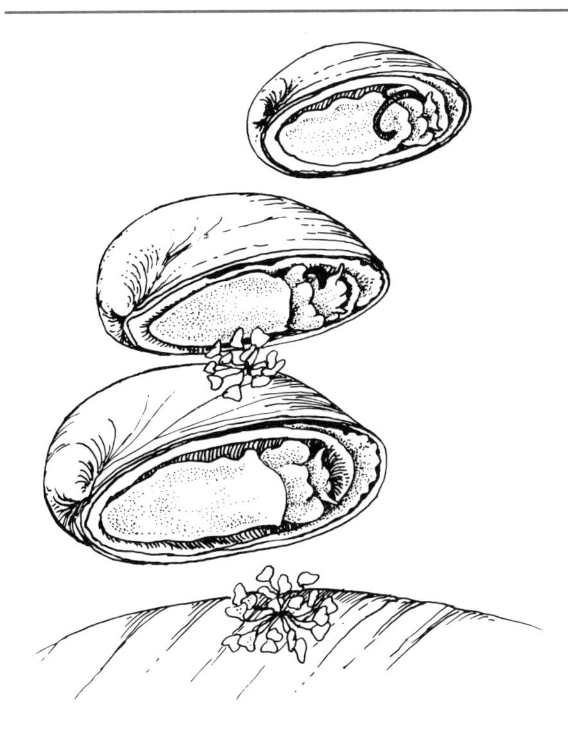

Common Atlantic female slipper shells lay bundles of egg-filled capsules attached by strands to the surface below them and brood the eggs under their foot. The dark curled penis of the male on top is slipped under the edge of the shell of the underlying female to fertilize her eggs. During the period of transition from male to female, the middle slipper shell retains a residual penis.

Slipper shells are algal feeders and incorporate their food in mucus within the mantle. They are often found in shallow water on shells occupied by hermit crabs. The convex slipper shell is abundant in eelgrass beds.

"Bugs" of the Flats

Amphipods are shrimplike—some refer to them as flealike—crustaceans that are distributed worldwide and, like crayfish, are adapted to living in freshwater lakes and rivers and the subterranean waters of caves. They are most abundant, however, in estuaries and in the oceans, and they may be found from the high-tide line on beaches to the profundal depths of the open seas. There are an estimated 5,500 species of amphipods in the world, and only an expert in this field—of which there are few—can identify them to the species level. Amphipods are, in general, abundant and ubiquitous, and they form the major food of many fish species. The amphipods that are usually seen in the intertidal zone belong to the family Gammaridae. They are compressed or somewhat flattened from side to side and include the beach flea types (discussed in Chap. 2) and the scuds, *Gammarus* spp. Two quite typical species of intertidal flats in Zones 2 and 3 are the saltmarsh flea, *Orchestia grillus,* which literally means "leaping-sea-fish-like-a-cricket," and the spine-backed scud, *Gammarus mucronatus.* The larger males of both species can frequently be seen carrying the smaller females around in what is considered to be prenuptial courtship behavior. The fertilized eggs are carried in the female's brood pouch and undergo development into juvenile forms that closely resemble the adult. There are no complicated larval forms as in some other invertebrate species.

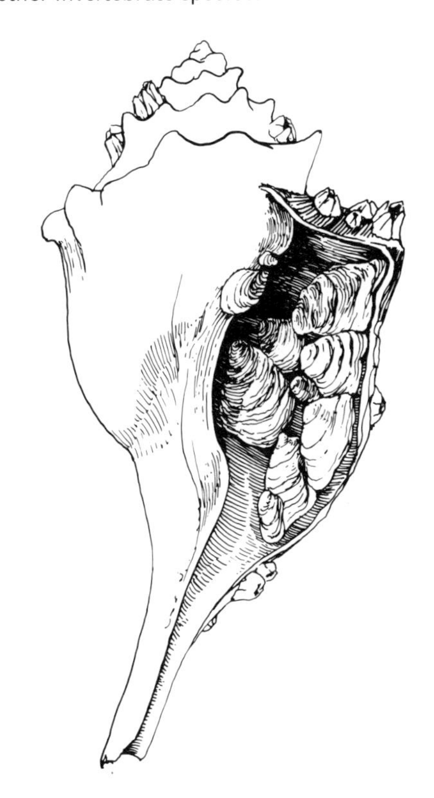

Flat slipper shells, *Crepidula plana* (to 1 inch), cover the inner surface of an empty knobbed whelk shell, *Busycon carica.*

Saltmarsh fleas are aptly named, as they are very agile and, when disturbed, can flex their abdomens and spring into the air. When undisturbed they walk upright, as do most other crustaceans. The spine-backed scuds, on the other hand, are more closely tied to a watery environment and do not have the ability to jump. They swim dorsal- or topside up, with alternating beats of their swimming legs. When grazing on the bottom or stranded at the edge of the water they sidle along on their sides, pushing their bodies with their legs.

The saltmarsh flea, spine-backed scud, and other scuds are found in protected bays, estuaries, and salt marshes. Saltmarsh fleas and beach fleas can be found in the shallow waters of intertidal flats to above the high-tide mark in drift materials cast up by tides and wind. They are found in large numbers under debris and among the roots of marsh plants, where they maintain nesting areas and runways. Saltmarsh fleas often climb the stems of marsh grasses to avoid the flooding tides. They grow to a maximum of about three-quarters of an inch, with the male being slightly larger than the female, and are olive to brownish red in color.

Scuds abound in shallow waters of intertidal flats and are plentiful in masses of green algae and underneath stones at the water's edge, but they do not burrow. The spine-backed scud is green to olive-green with red or brown patches and is about half an inch long. Other scuds of the flats and shallows are similar in appearance but variable in hue; the body color of amphipods is often reflective of its habitat or food habits.

THE WORLD BENEATH THE INTERTIDAL MUDS

By far the most populous and varied dwellers of the intertidal habitat are the unseen scores of animals that live below the surface muds. Numerous kinds of worms, bivalve mollusks, burrowing amphipods, and shrimps occupy this buried world. Some creatures occasionally emerge from their tunnels and burrows to wander over the surface, but most are permanent dwellers within the muds. Some tunnel slowly through the sediments, whereas others are encased in various types of tubes or burrows. Many raise their foreparts out of the mud to feed on bacteria and microal-

Amphipods grazing on the bottom. The larger species, the saltmarsh flea, *Orchestia grillus,* springs into the air, while the smaller scuds, *Gammarus* spp., propel themselves along on their flattened sides.

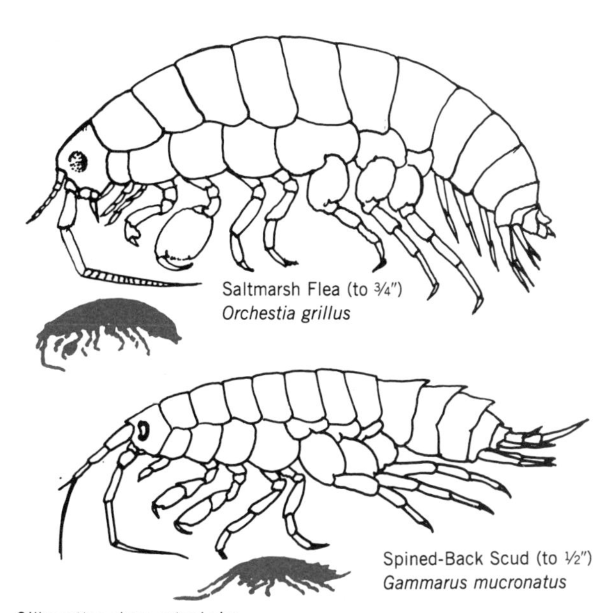

Silhouettes show actual size.

gae and planktonic organisms carried in with the tides and then retreat into their protective havens as the waters ebb. The animals themselves may be hidden, but the signs of buried life are everywhere. Small holes pepper the surface, and little mounds of fecal mud pellets give witness to their presence.

Worms, Worms, and More Worms

Worms are not everyone's favorite subject, but in the Chesapeake system they are so variable, in color, shape, and form, and so abundant and important to the ecology of the Bay that we must devote considerable attention to them. Worms provide food for a host of other creatures. Crabs, fishes, and birds of the Bay are major predators. Burrowing worms work up the bottom constantly, serving much the same function as earthworms on land. There are so many species of worms in the Bay that only the commoner or the most distinctive are covered in this book.

Intertidal flats harbor diverse groups and are the best places to start worm hunting. Many of the same species may be found in other habitats—at the base of aquatic plants, crawling between barnacles, sponges, and hydroids, or living in the pockets of mud between oysters. Some worms are adapted to living in the muds of the deepest parts of the Bay, where dissolved oxygen levels are low or absent and where the soft channel silts are too unstable for many other bottom-dwellers.

Ribbon Worms. Ribbon worms, as they are aptly named, are common in intertidal flats of the Chesapeake Bay. Most of them burrow deep into the soft sands and clays, but they may also be discovered curled under rocks and stones, or even swimming. Ribbon worms are fragile and fragment easily when handled, so care must be taken not to end up with a handful of bits and pieces. Over 20 species of ribbon worms, known to scientists as nemertean worms, occur in the Bay, particularly in the intertidal flats or in eelgrass or other aquatic plant beds.

The milky ribbon worm, *Cerebratulus lacteus,* is a large, flat, pale-white to yellow-pink worm that may attain a length of three or four feet when fully extended. It twists and turns when lifted and will often knot itself into a convoluted mass. An identifying slitlike mouth can be seen on the underside of its grooved head. At times it will evert its very long proboscis through a pore of the tip of its head. Although normally pale, breeding milky ribbon worms are a dark-reddish color when they swarm to the surface in late spring and summer. It is quite startling suddenly to sight a huge swimming ribbon worm! The milky ribbon worm prefers a fine sandy bottom and is common only in the lower half of the Bay and in the higher-salinity portions of the tributaries.

The bright red ribbon worm, *Micrura leidyi,* is widely distributed in both sandy and muddy intertidal flats, from fairly low-salinity waters to the mouth of the Bay. It does not grow quite as long as the milky ribbon worm but can reach 12 inches, still a formidable length for a worm. It is thin, rounded to flat, and has a small round mouth rather than a slit.

The sharp-headed ribbon worm, *Zygeupolia rubens,* is common in sandy flats of the lower Bay. Its long, pointed head distinguishes this small, three-inch nemertean from other ribbon worms. It is usually rosy-colored with a whitish head.

A curious little inch and one-half long nemertean, the leech ribbon worm, *Malacobdella grossa,* lives within the mantle cavity of such bivalve mollusks as hard clams and

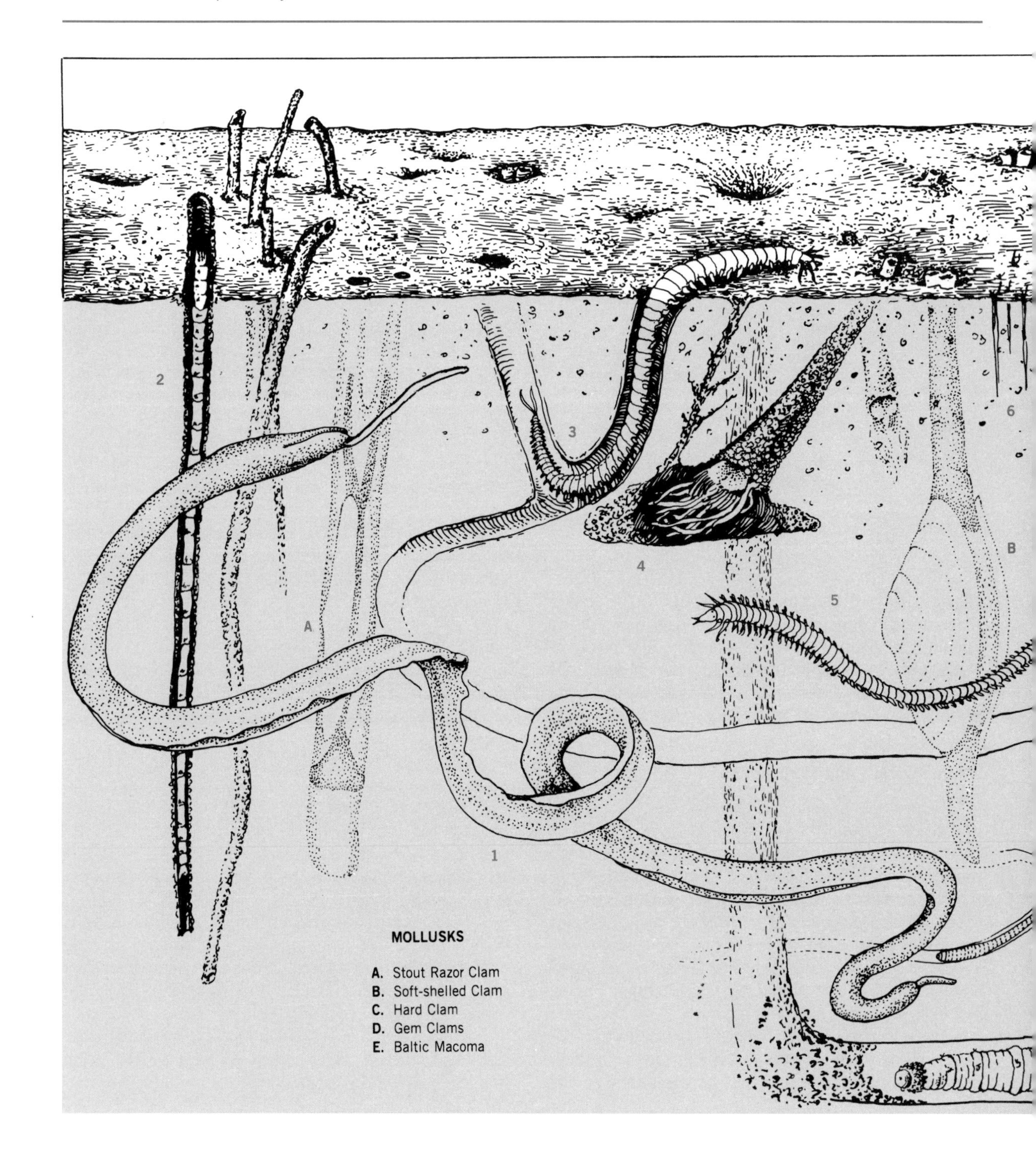
2
3
4
5
6
A
B
1
MOLLUSKS
A. Stout Razor Clam
B. Soft-shelled Clam
C. Hard Clam
D. Gem Clams
E. Baltic Macoma

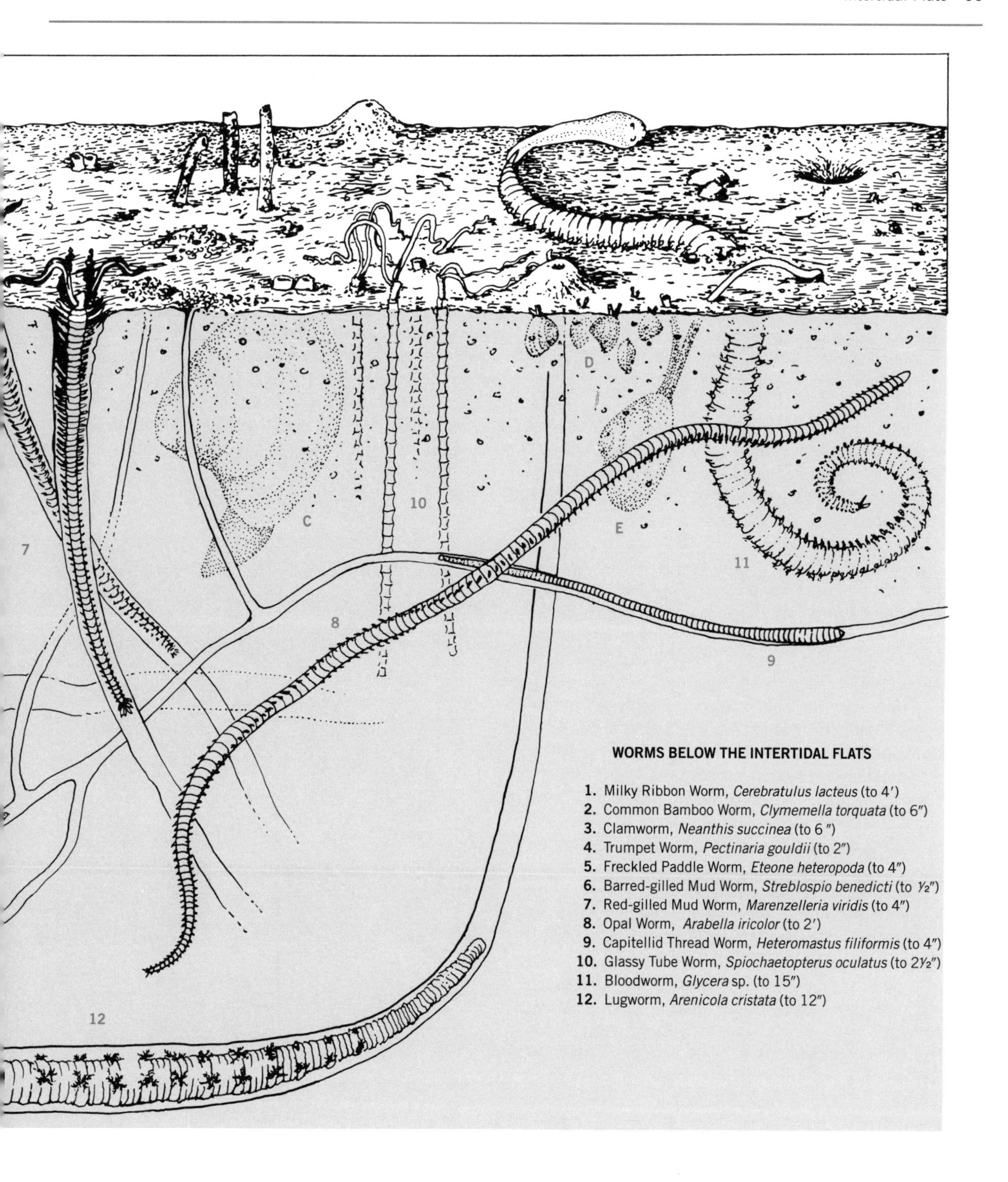
C
D
E
7
8
9
10
11
12
WORMS BELOW THE INTERTIDAL FLATS
1. Milky Ribbon Worm, Cerebratulus lacteus (to 4′)
2. Common Bamboo Worm, Clymemella torquata (to 6″)
3. Clamworm, Neanthis succinea (to 6 ″)
4. Trumpet Worm, Pectinaria gouldii (to 2″)
5. Freckled Paddle Worm, Eteone heteropoda (to 4″)
6. Barred-gilled Mud Worm, Streblospio benedicti (to ½″)
7. Red-gilled Mud Worm, Marenzelleria viridis (to 4″)
8. Opal Worm, Arabella iricolor (to 2′)
9. Capitellid Thread Worm, Heteromastus filiformis (to 4″)
10. Glassy Tube Worm, Spiochaetopterus oculatus (to 2½″)
11. Bloodworm, Glycera sp. (to 15″)
12. Lugworm, Arenicola cristata (to 12″)

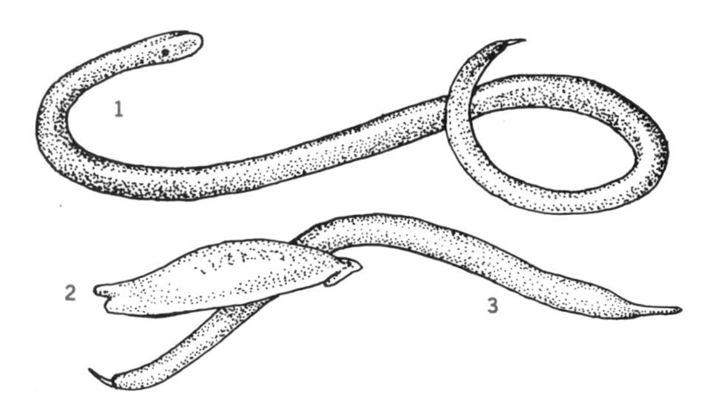

1. Red Ribbon Worm, *Micrura leidyi* (to 12")
2. Leech Ribbon Worm, *Malacobdella grossa* (to 3")
3. Sharp-headed Ribbon Worm, *Zygeupolia rubens* (to ½")

soft-shelled clams. Although it looks like a leech with its suckerlike end, it lives commensally, primarily in the saltier waters of the lower Bay, finding shelter and consuming some of the food collected by the host rather than living off its host as a parasite.

Bristle Worms—The Omnipresent Polychaetes. Polychaete annelid worms, commonly known as bristle worms because of their hairy appearance, are the most diverse group of worms in the Chesapeake Bay. The bristled, paddle-shaped appendages extending from each body segment identify them as bristle worms. Species of bristle worms are identified by the number and kind of head appendages and the number and type of bristles, or paddles, along their sides, eyespots, proboscises, palps (fleshy bulbs extruding from the head region), gills, and other structures. The relative size and the color of the worm will often suffice to distinguish it. Most of the worms described here have characteristics of form or habit which make it easy for the amateur to identify them. Two general types of bristle worms live in the intertidal flat habitat. The first group are predators that wander over the surface or burrow through the substrates in search of food. They move through the mud and sand by means of peristaltic contractions. Some wandering worms construct tubes within the bottom from which they leave and to which they return, while others burrow freely through the substrate particles without ever constructing permanent "homes." Little piles of mud at the edge of a hole, conical-shaped depressions, or tough, debris-coated encasements sticking out of the bottom of intertidal flats are all signs of bristle worms. These errant worms, as they are called, are widely distributed in many habitats of the Bay, and most are found subtidally as often as intertidally. The second type do not move freely about but live more or less permanently within tubes or other discrete burrows below the surface.

Burrowing Wanderers. Common errant worms belong to various groups referred to as clamworms, thread worms, bloodworms, red-lined worms, and paddle worms.

Clamworms are probably the most abundant and widespread polychaete worms in the Chesapeake Bay, extending from the tidal freshwaters of rivers and streams down to the mouth and occupying every intertidal and shallow subtidal habitat. They bury themselves in intertidal muds, crawl among the recesses of surface debris or shells, and graze over the blades of eelgrass. There are seven species in the Bay, but a single species, the common clam-

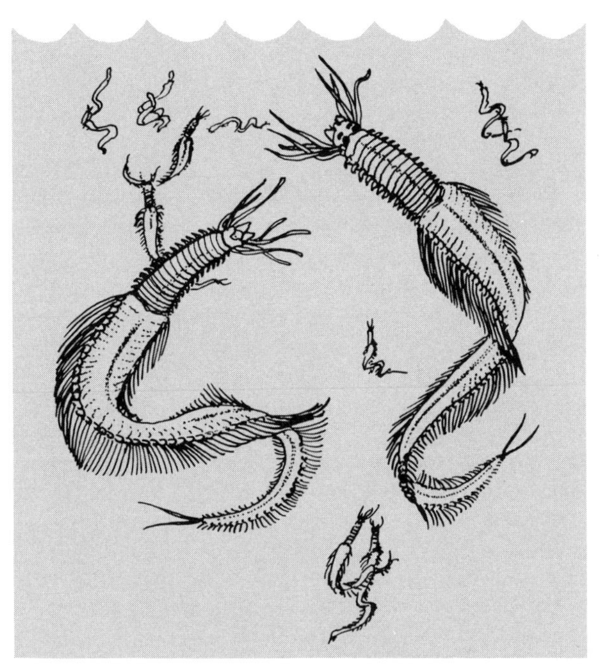

A swarm of the sexually transformed heteronereises of common clamworms, *Neanthis succinea,* rising to the surface on a moonless spring night.

worm, *Neanthis succinea,* is the one most likely to be encountered.

Common clamworms may be fairly large (five to six inches), but smaller sizes are more common. The anterior portion of the body is a brownish bronze and has a slightly different shape from the rest of the body, which is reddish. A pulsating bright red blood streak down the middle of the back can be seen just under the skin. The back is smoothly rounded and the parapodia are relatively large and distinct. The head has four tiny eyes, four pairs of tentacles, and a pair of fleshy protuberances called palps. Don't be surprised if a clamworm you are examining suddenly ejects a large, club-shaped sack (the proboscis) with two light, amber-colored hooks on the end. These worms are active and voracious predators whose proboscis is used to grasp any soft-bodied prey, such as other worms, bits of dead fish, or even algae. As the proboscis retracts the beaks close over and draw the prey into the worm's mouth.

Clamworms bury in bottom muds and as they move through the substrate, they exude a mucous that hardens into a sheath. Sand grains adhere to the outside of the sheath and form a strong, flexible tube. Clamworms can go in and out of their tubes quickly, and they often leave their tubes to roam over the surface. They are easy prey and favored food for bottom-feeding fishes and crabs. During the breeding season clamworms metamorphose into special sexual forms called heteronereises, which have enlarged parapodia especially adapted for swimming. During the dark of the moon, in spring or summer, particularly in May, heteronereises swarm to the surface in frenzied mating dances. They are attracted to light, and if you walk out on a pier during this time and shine a flashlight onto the water, chances are you will catch sight of hundreds of worms swimming rapidly in small circles just below the surface. They will appear for several successive nights; finally, they will release their eggs and sperm and die. Meanwhile, tiny planktonic larvae develop from the eggs and eventually become recognizable clamworms. They then descend to the bottom and resume the benthic existence of their parents.

Thread worms are long, slim, iridescent worms that burrow in sand and sandy muds. They often occur in dense populations with their burrows intricately intertwined. Thread worms live in lower Zone 2 and Zone 3. The opal worm, *Arabella iricolor,* is a beautiful bright green iridescent thread worm that grows to two feet long. It has no appendages on its conical head, and its tiny parapodia break easily if handled. Other species of thread worms look

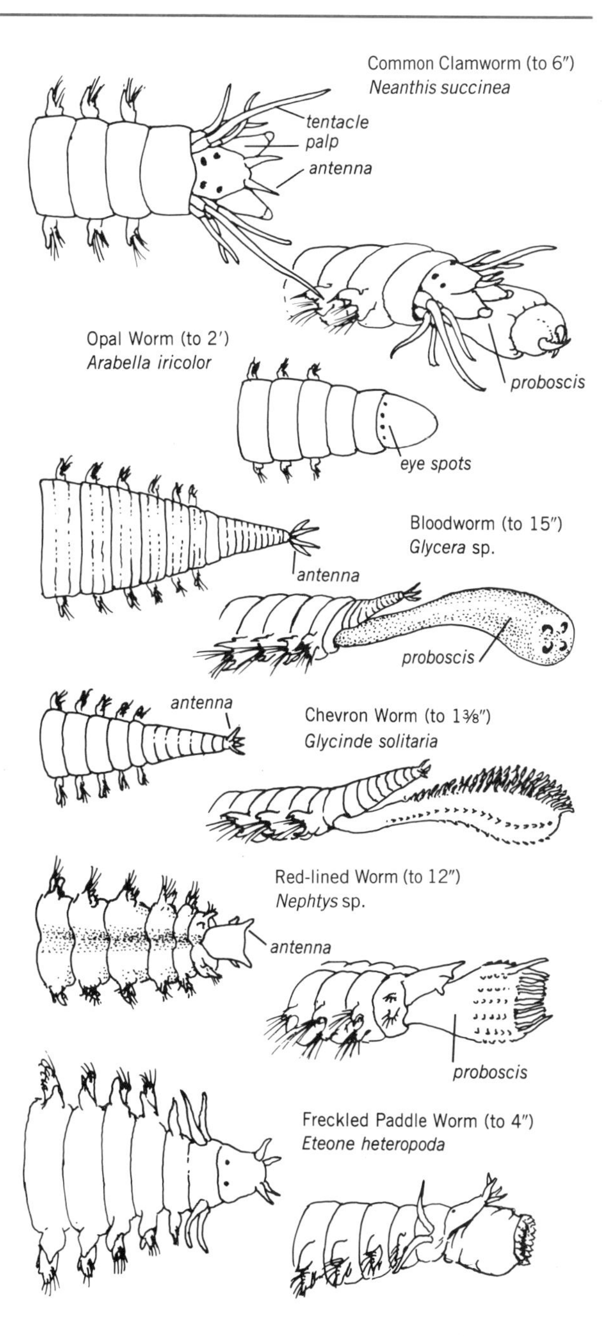

Common wandering bristle worms of the Chesapeake Bay. All but the opal worm have a large evertible proboscis, shown retracted and extended.

very much like the opal worm but are reddish rather than greenish.

Bloodworms are known to many a fisherman who has used them as a favored bait. Two species of bloodworms, *Glycera* spp., are ubiquitous in the Bay (Zones 2 and 3), common not only in intertidal flats but in eelgrass beds, oyster bars, and marshes. They are robust worms, pale purplish red and somewhat iridescent and generally much smaller in the Chesapeake Bay than in New England waters, where they grow to 15 inches. As in clamworms, the red blood can be seen streaming through a vessel under the skin. They are smooth-backed, with small bristles along the whole length of the body, and are tapered at both ends with a barely noticeable pointed head. They too can eject a large proboscis equipped with four black beaks at its tip. Beware, as they can give you quite a nip. Bloodworms also arise from the bottom and swim upward in swirling spirals on dark summer nights. If caught unawares on the intertidal flat, they will quickly corkscrew themselves into the bottom.

Small bloodworms may easily be confused with the chevron worm, *Glycinde solitaria,* until the latter projects its huge bulbous proboscis, with a fuzzy surface on top and a series of small hooks below. These small polychaetes (only about one and three-eights inches long) are widely distributed in the Bay and are found throughout Zone 2 and Zone 3 in both sand and mud bottoms.

Red-lined worms, *Nephtys* spp., are similar to clamworms and bloodworms in size, general appearance, and habits. They are predatory burrowers and wanderers over the surface throughout lower Zone 2 and Zone 3. Red-lined worms are grayish, with the bright red line of the dorsal blood vessel sharply marked along the mid-back. The back is indented rather than smoothly rounded as in clamworms and bloodworms. Their shovel-shaped head also helps to identify them. The end of their club-shaped proboscis is covered with soft papillae.

Paddle worms are still another group of errant polychaete worms. These are raggedy-looking worms sometimes even called rag worms. A common Chesapeake Bay paddle worm, the freckled paddle worm, *Eteone heteropoda,* is three to four inches long. It is a pale worm, freckled with brown, with a small, triangular head that has two eye spots and four small tentacles at its tip. The first body segment has two more pairs of tentacles. The large, bulbous proboscis ends in small, soft papillae rather than in hooks.

Tube-building Worms. Tube-dwelling bristle worms generally differ from wandering bristle worms in having smaller parapodia and differently shaped body parts (an anterior thorax region and a posterior abdomen). They are homebodies, although many species move freely within their tubes. Unlike the predatory wanderers, they must wait for food to come to them, so most species have specialized structures, such as long tentacular palps or other appendages, to aid them in food gathering. The variety of species is impressive; some are inconspicuous and others obvious, some tiny and others large. They proliferate in all the intertidal and subtidal bottom sands and muds of the entire Bay system. Many species are also abundant among the roots of eelgrass and other aquatic plants.

The capitellid thread worm, *Heteromastus filiformis,* abounds in the sandy-mud bottoms throughout Zones 2 and 3. A thin worm, it resembles a smooth earthworm with a somewhat swollen thorax and vestigial parapodia. Capitellid thread worms build intricate networks of mucus-lined tubes, constructed as they literally eat their way through the substrate. They ingest mud and digest any available organic material. Fecal mud pellets are produced which are then moved up a vertical tube and placed at the entrance of their burrows.

Mud worms are infaunal species widely distributed from tidal freshwaters to the mouth of the Bay. One of the larger mud worms, the red-gilled mud worm, *Marenzelleria viridis,* which grows up to four inches or so, is typical of the mud-worm group. This species is widespread in intertidal sandy areas of all zones. It burrows vertically in a mucus-and-mud-covered tube, protruding its head above the surface to gather food particles with long, probing palps. It has bright red gills and prominent parapodia extending from each segment of the anterior part of its body. This creature spawns near the surface on an ebbing tide during late winter nights.

The tiny barred-gilled mud worm, *Streblospio benedicti,* is found in all types of bottoms throughout the Bay. It is only one-half inch long, or less, and will rarely be noticed. The larger, fringe-gilled mud worm, *Paraprionospio pinnata,* is also common throughout the Bay region.

The glassy tube worm, *Spiochaetopterus oculatus,* lives in the bottom in vertical tubes, not in muddy tubes as do most other worms, but in transparent tubes ringed like bamboo. The glassy tubes and the worm inside are very small, up to about two and one-half inches long but only a

COMMON TUBE-DWELLING WORMS OF THE CHESAPEAKE BAY

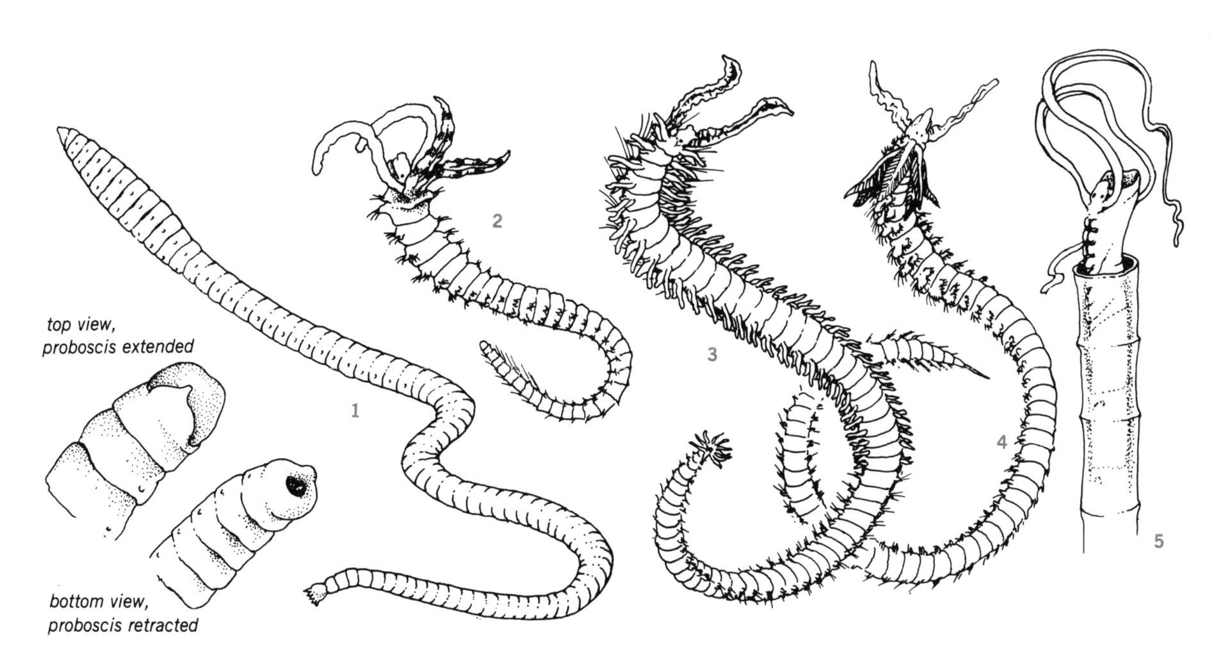

1. Capitellid Thread Worm, *Heteromastus filiformis* (to 4″)
2. Barred-gilled Mud Worm, *Streblospio benedicti* (to ½″)
3. Red-gilled Mud Worm, *Marenzelleria viridis* (to 4″)
4. Fringed-gilled Mud Worm, *Paraprionospio pinnata* (to 4″)
5. Glassy Tube Worm, *Spiochaetopterus oculatus* (to 2½″)

fraction of an inch in diameter. Like the mud worms, it has two long, curling palps on its head which are used for food gathering.

One of the most interesting intertidal worms is the common bamboo worm, *Clymenella torquata.* Dense beds of common bamboo worms can be sighted at the low-tide line of quiet, protected sandy-mud flats. The muddy surface will be textured with neat, mud-encrusted tubes projecting a short distance above the surface. Bamboo worms are appropriately named because they look like sticks of brick-red bamboo. Their elongated body segments account for this appearance. They live head down with the tip of the tail terminating in a funnel, edged with small papillae that can close off the top of the tube. At the bottom of the tube, their heads are blunt and are equipped with an eversible proboscis. Bamboo worm colonies often attract a bevy of other animals. The nassa mudsnail and the tiny spindle-shaped turret snail, *Mangelia plicosa,* may often be collected in these colonies as they scavenge for food. A tiny amphipod, the bamboo worm amphipod, *Listriella clymenellae,* just a fraction of an inch long, lives commensally with the bamboo worm. Parasitic snails common in seagrass meadows also inhabit bamboo-worm tubes and feed on their host by inserting long siphons into the worm's flesh.

A small "ice-cream cone" of sand particles neatly cemented together, with coarse particles at the large end and finer grains at the other, signals the home of the trumpet worm, *Pectinaria gouldii.* The trumpet worm lives buried head down at the low-water line of intertidal flats, as well as subtidally in shallows and in deeper waters. The worm inside is a strange-looking creature, relatively small (two

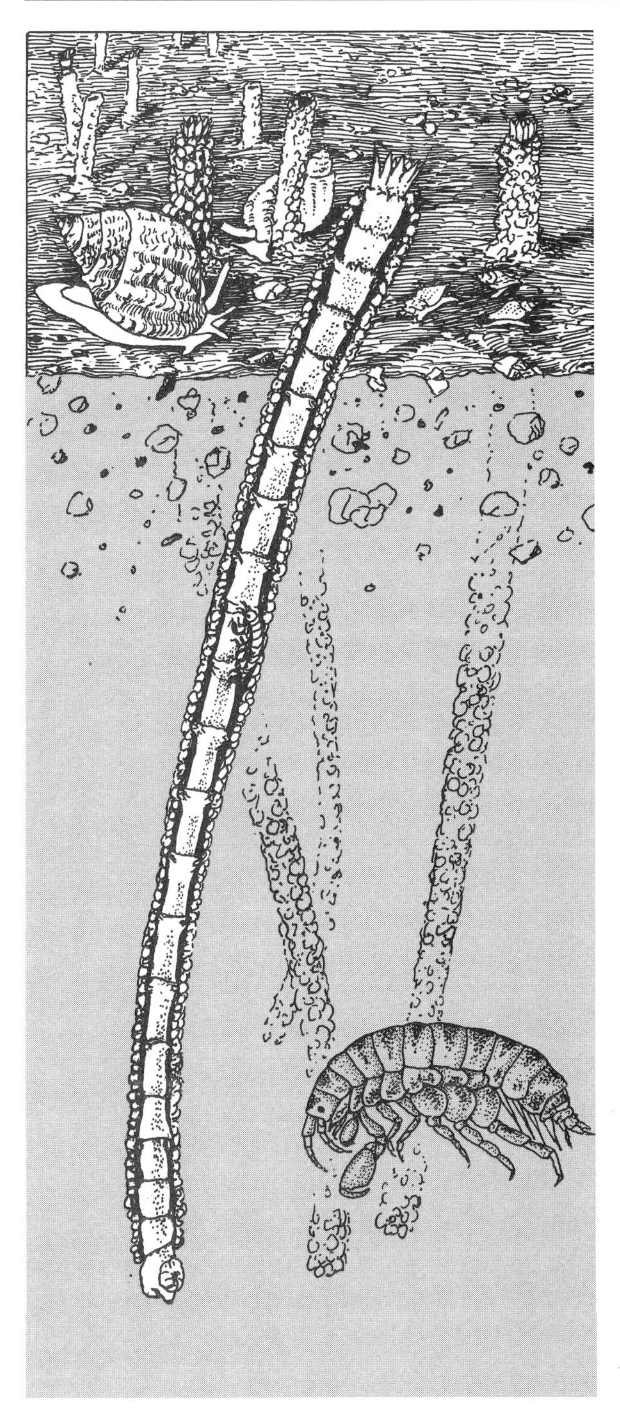

inches or less) with a large, truncated head and a fan of glistening golden bristles, which it uses for digging burrows in the bottom. Its cluster of long, flesh-colored tentacles extends out into the open spaces of the burrow. In life, only the small end of the tube projects above the surface, but when empty, these slightly curved trumpet shells can often be found strewn on intertidal flats or beaches.

The parchment worm, *Chaetopterus variopedatus,* builds a U-shaped, parchmentlike tube in the bottom which can be up to a foot long. Each end projects a little above the surface. These tube openings can be seen dotting the bottom of the very low intertidal and shallow-water zones, mostly in the higher-salinity regions of the Bay. The worm inside is a very curious creature—pale, brightly luminescent in the dark, but oddly shaped, with various appendages along different parts of the body. Paddle-shaped body parts in the midsection of the parchment worm keep water currents moving into the front opening of the tube and out the other opening. Feeding is unique. Specialized appendages on the anterior part of the worm form a mucous bag, which filters plankton from the moving water currents. When the bag is full, pumping ceases and the bag is detached and formed into a ball by means of a cup organ in front of the paddles; it is then passed gently forward to the mouth. Little parchment worm crabs, *Pinnixa chaetopterana,* are frequently cohabitants of parchment worm tubes. These tiny (only one-half inch long), oval-shaped crabs enter and exit the tubes at will, biting a larger hole in the end of the tube if necessary in order to leave. Parchment worm crabs feed on plankton filtered from the currents set up by the parchment worm.

The lugworm, *Arenicola cristata,* is another intertidal worm that leaves telltale signs on surface muds. Look for a hole surrounded by mud pellets and, a little distance away, a funnel-shaped depression. The lugworm lies deep below in an L-shaped tube with the descending shaft below the hole at the tail end of the worm and the head at the head of the tube below the funnel. Burrows may extend a

A bamboo worm community. Nassa mudsnails, *Nassarius vibex* (½ inch) and tiny spindle-shaped turret snails, *Mangelia plicosa* (⅛ inch), scavenge among the extended worm tubes while a bamboo-worm amphipod, *Listriella clymenellae,* secludes itself within a worm tube. An enlarged view of this ⅕-inch-long amphipod is shown.

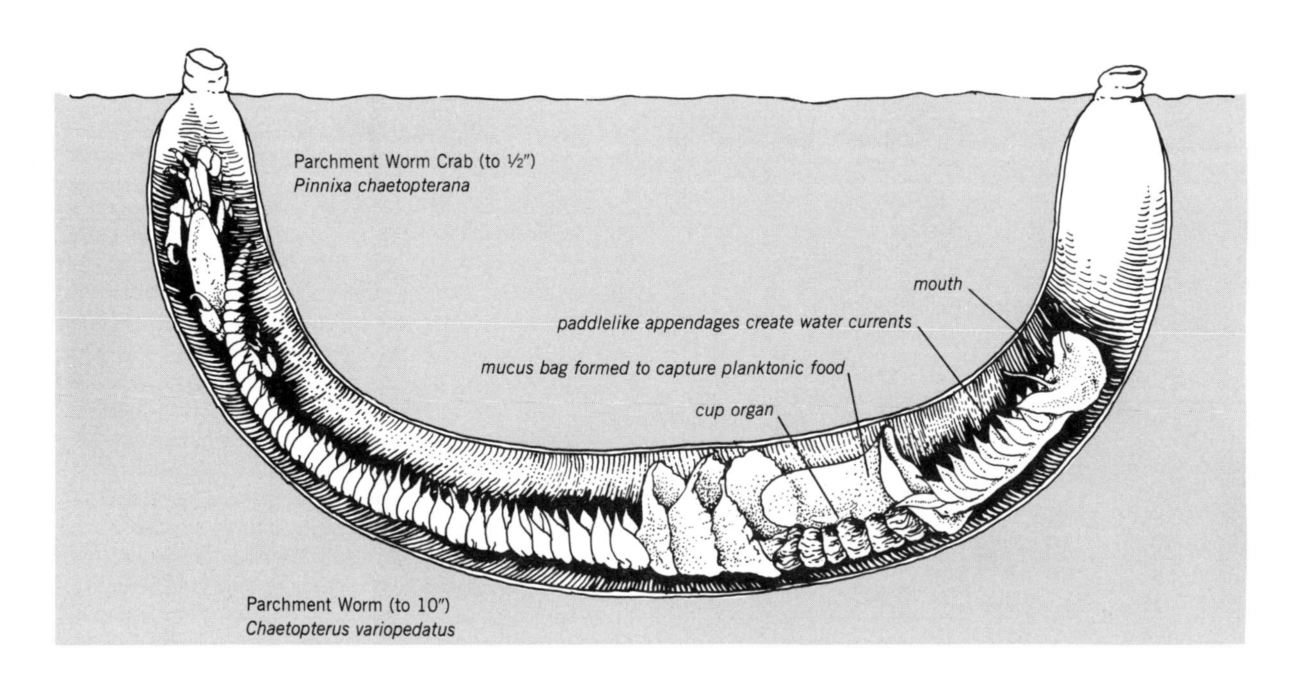

foot or more into the bottom. Oxygenated water is brought down the tail shaft and then forced up and out at the end of the mucous-lined horizontal burrow. The worm feeds by ingesting the muds at the buried end of the burrow, and, after organic matter has been digested, mud pellets are expelled and deposited at the surface around the shaft exit. If you manage to extricate a lugworm from its burrow you will discover a large (six inches or more), fat, blackish green worm with clumps of reddish gills along the mid-body—not the most beautiful worm in the Bay. Lugworms are distributed as far upstream as Annapolis.

One of the largest and most beautiful polychaete worms in the Chesapeake is the plumed worm, *Diopatra cuprea.* It, too, betrays its presence with long, debris-laden "chimneys" extending sometimes as far as two or three inches above the surface of intertidal flats near the low-water mark. Vacant tubes are often dislodged and strewn on the beach. They are easily recognized as soft but tough wrinkled skins with bits of shells, tiny stones, and other debris stuck to the chimney end. The tubes can extend as far as three feet. The worms themselves can be discovered where the tide is covering the flat. The bright red gill plumes surrounding the head are exposed as the worm sticks its head out of the top of the case in search of prey. Brightly iridescent plumed worms are scavengers and predators, more like the errant type of bristle worm than tube-building filter feeders. A plumed worm will occasionally leave its tube to capture food close by. Only the fast, adroit, and lucky beachcomber will be able to dig up a plumed worm. At the first sign of disturbance, the worm will quickly drop out the open bottom end of its tube.

Two other buried worms show brightly colored tentacles, which they wave in the currents of shallow tidal waters covering intertidal flats of the higher-salinity (Zone 3) regions. The fringed worm, *Cirratulus cirriformia,* and the ornate worm, *Amphitrite ornata,* are often found together, not only intertidally but also subtidally, in shallow waters and in soft mud under rocks and stones. The fringed worm has large, elaborate, orange-colored gill filaments extending in a flowing mass along most of its body. The filaments are thrust up through the muds to lie on the surface. Fringed worms are iridescent, yellowish to brown, and live in loose burrows in the bottom. The ornate worm, on the other hand, builds firm, sand-encrusted tubes. Its body structure is quite different from that of the fringed worm, as it has a bushy clump of bright red gills at its head and a

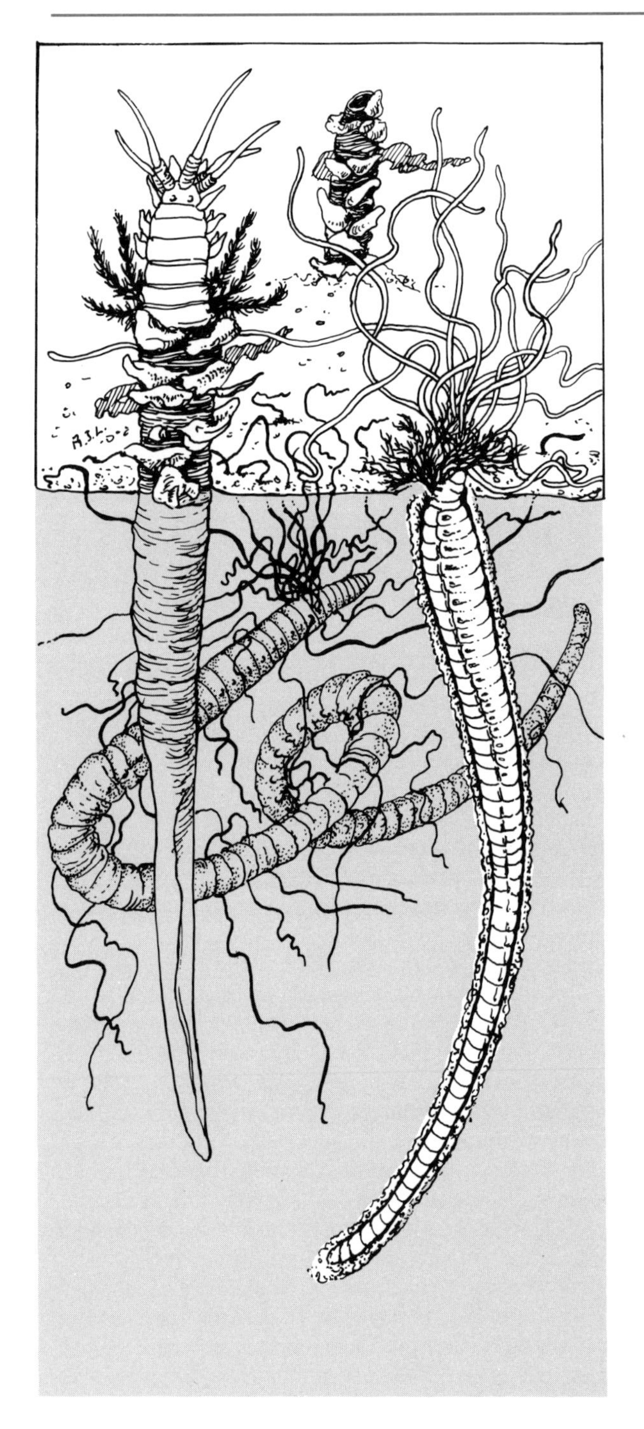

flowering spread of long, peach-colored tentacles just above the gills. The tentacles are constantly in motion and may spread out to a length as long as the worm itself.

Other Wormlike Creatures

A number of worms and wormlike creatures other than polychaete annelids may be encountered in intertidal flats. In the tidal freshwater and low-salinity waters where few species of bristle worms exist, their close annelid relatives, the oligochaetes, dominate. Earthworms are oligochaetes, and aquatic oligochaetes and their terrestrial cousins are look-alikes. Many species occur in the Chesapeake Bay system, but only the expert can identify them to species.

The acorn worm, *Saccoglossus kowalewskii,* is a very strange-looking worm. It is so fragile that it can rarely be retrieved in one piece. The acorn worm has no body annulations and is developmentally much more advanced than the annelids. Acorn worms grow to about six inches. They live in deep burrows in the fine sandy-mud flats of Zone 3. They are revealed by the stringy mud castings the worm places near its burrow openings. The acorn worm has a unique shape, with three distinct body regions: a pale pinkish proboscis; a short, bright orange collar; and a long, rumpled-looking orange to brownish body. Acorn worms are favored food to many fishes. When a freshly caught fish has a strong iodine flavor, chances are it has been feeding on acorn worms.

Burrowing sea anemones may be dug up at the low-water line on intertidal flats, but they will more likely be noticed when the tide is in and their daisylike flower heads suddenly appear and disappear on the shallow bottoms. When removed from the bottom they look more like "worms" than like the beautiful flowery sea anemones commonly attached to rocks and pilings. The longitudinal muscle ridges and tentacle-encircled mouth of burrowing anemones are completely unlike those of any worm. The small, burrowing anemone, *Edwardsia elegans,* is a typical estuarine species of the Bay, common in sandy-mud bot-

Some of the more spectacular Chesapeake Bay worms include the plumed worm, *Diopatra cuprea* (to 12 inches), within its debris-laden chimney; the fringed worm, *Cirratulus cirriformia* (to 6 inches), with its numerous gill filaments thrust through the muds; and the ornate worm, *Amphitrite ornata* (to 15 inches), within its sand-encrusted tube.

WORMLIKE BURROWERS OF THE CHESAPEAKE BAY

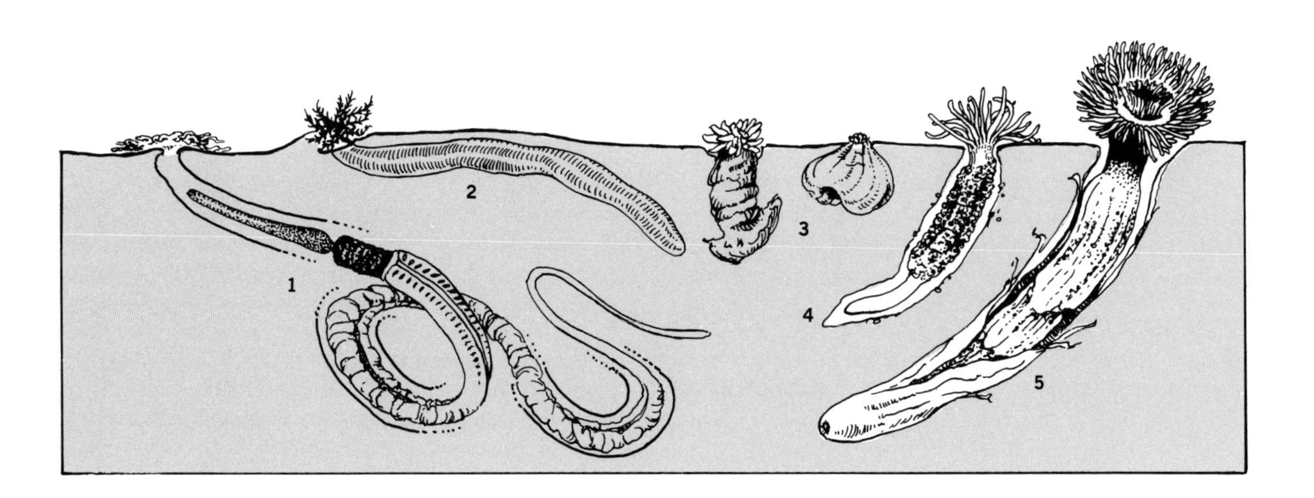

1. Acorn Worm, *Saccoglossus kowalewskii* (to 6″)
2. White Synapta, *Leptosynapta tenuis* (to 6″)
3. Sea Onions, *Paranthus rapiformis* (to 1″)
4. Burrowing Anemone, *Edwardsia elegans* (to 1¼″)
5. Sloppy Gut Anemone, *Ceriantheopsis americanus* (to 8″)

toms throughout Zones 2 and 3. It is small (little more than an inch long) and elongate, with a single row of about 16 tentacles; a smooth collar below a warty, deeply grooved, and muscular middle section; and a terminally constricted "foot," used for digging and anchoring in the bottom. In the higher-salinity regions of Zone 3, the sloppy gut anemone, *Ceriantheopsis americanus,* is occasionally found intertidally in silty areas within sloppy tube cases. However, it is typically a deepwater species. When freshly extracted from its case it is a beautiful deep plum color at the base of the tentacles, fading to pink and then to cream. It has many more tentacles than *Edwardsia*—up to 100 in two rows, the outer row being white to pink colored, the inner row deep peach—and is much larger, reaching a length of eight inches.

The sea onion, *Paranthus rapiformis,* is a burrowing anemone common only in the lowermost Bay regions; it lives mostly subtidally but can be found on the flat at very low tide. A small white anemone that anchors in the bottom muds by means of an expanded basal disc, it is generally attached to buried pebbles or shells. When its short tentacles are withdrawn, the sea onion actually looks more like a garlic clove than an onion.

Another bizarre inhabitant of lower Bay intertidal sandy-mud flats is the white synapta, *Leptosynapta tenuis.* It belongs to the sea cucumber family and is related to starfishes and sand dollars. The white, transparent body, six inches long, of the synapta lies mounded just under the surface, with a small, one-quarter-inch opening on the top of the mound out of which it thrusts its tentacles in search of bits of food. The white synapta looks much like the burrowing anemone but with 12 finely branched tentacles, which are different from the smooth tentacles of sea anemones, and with five white longitudinal muscle bands visible through the body wall. Synaptas can be quite abundant in some localities of lower Zone 2 and in all of Zone 3.

Clams

The most easily recognized animals hidden below the surface of intertidal flats are all the various types of clams that proliferate in the Chesapeake Bay. Literally hundreds of thousands live in closely packed communities, side by side with the dense populations of worms and other infaunal animals. Most people identify far more happily with the clams of the intertidal flats than with the worms, anticipating the wonderful taste of cherrystones, clam chowder, or steamed soft-shelled clams. A look at the annual harvest of soft-shelled clams and hard clams from the Bay (in good years running into the thousands of bushels) suggests the vast number of clams which must be present to support the fishery. Although soft-shelled and hard clams, along with oysters, may be the best-known bivalves of the Chesapeake, they are neither the most abundant nor the most widespread. Clams in the Bay come in many sizes and shapes—some minuscule, some large, some burrowing deep into bottom muds, others just under the surface. The young of larger clams look, at first glance, much like the adults of smaller species, but in general, burrowing bivalves of the intertidal flats will be among the easiest animals to identify using the shell guide in Chapter 2.

Burrowing clams generally have two characteristics in common: a fleshy extensible foot, used for burrowing, and a pair of siphons, which can be extended above the bottom surface muds. One siphon (the incurrent siphon) draws in water, with its dissolved oxygen supply, and microplanktonic or detrital food, and the other siphon (the excurrent siphon) expels waste products.

Soft-shelled clams, *Mya arenaria,* are widely distributed from low- to very high salinity regions of the Bay and tributaries, in all types of bottoms and both intertidally and subtidally to depths of 20 feet. Many people call them manninose, harking back to their old Indian name. New Englanders know them as "the clam," or "steamers." The densest populations of soft-shelled clams are in the mid-Bay region, where most commercial harvesting occurs.

Soft-shelled clams have long siphons encased in a thick, black membrane. These siphons are highly retractable, although they cannot be completely withdrawn into the shells as in most other clams. Set a few soft-shelled clams in warm water and in a few hours you will discover just how long their siphons can stretch—which gives a clue as to how deep they can be buried in the muds and still expose their siphons to the water. When the tide is out a peppering over the surface of variously sized round holes, some perhaps bubbling a bit, will tell you that there is a heavy population of clams below. Soft-shelled clams propagate by releasing eggs and sperm into the water column, where the eggs are fertilized. The eggs soon develop into free-floating larvae, each growing two tiny, transparent shells and a small foot. After a couple of weeks, the microscopic clam drops to the bottom and, by means of its foot, crawls over the surface until it finds some sand grains, whereupon a hairlike thread (the byssus) grows out from a gland on the foot and attaches firmly to the grains. Thus attached, the baby soft-shelled clam is not as liable to be tossed about by the waves. Most of the burrowing clams of the Bay develop in a similar manner.

While small, the soft-shelled clam can crawl and dig actively, but by the time it reaches adult size it can move only vertically and, if dislodged, cannot readily reburrow.

The hard clam, *Mercenaria mercenaria,* the other commercial food clam of the Bay, is not as widely distributed as the soft-shelled clam and is restricted mostly to areas where salinity is at least about two-thirds that of the ocean. Hard clams have thick, hard shells in contrast to the thin, easily broken shells of soft-shelled clams; thus their common name. People are often confused about the various colloquial names given to hard clams. Littlenecks, cherrystones, chowder clams, quahogs—all refer to the same species. "Littleneck" refers to the two barely visible siphons that can be seen in freshly shucked clams just below one of the cut muscles. These small siphons are separate from each other, and because they are short, the hard clam does not burrow deeply.

"Treading" for hard clams is a favorite pastime for many visitors to the beaches of the lower Bay. The clams are searched out with bare feet. When one is discovered the toes are curled over the clam and it is then drawn up against the other leg until it can be grasped and dropped into a floating basket. Cherrystone and littlenecks are the favored size for cocktail clams raw on the half shell. Cherrystones are named for Cherrystone Inlet on the Bay side of Virginia's lower Eastern Shore, a place where these succulent clams have traditionally been bountiful.

Some of the most plentiful clams throughout the Bay region are relatively small species that at first glance look either like small soft-shelled clams or small hard clams. Macoma clams, *Macoma* spp., are ubiquitous and typically estuarine species. Their chalky white shells are often found along the shore. Two species of macomas are shown in the shell guide. Macomas have long, thin, flexible siphons of unequal length which are fully retractable into the shell. The longer, inhalant siphon is poked out of the burrow to sweep across the soft bottom sediment and vac-

BURROWING CLAMS AND AMPHIPODS

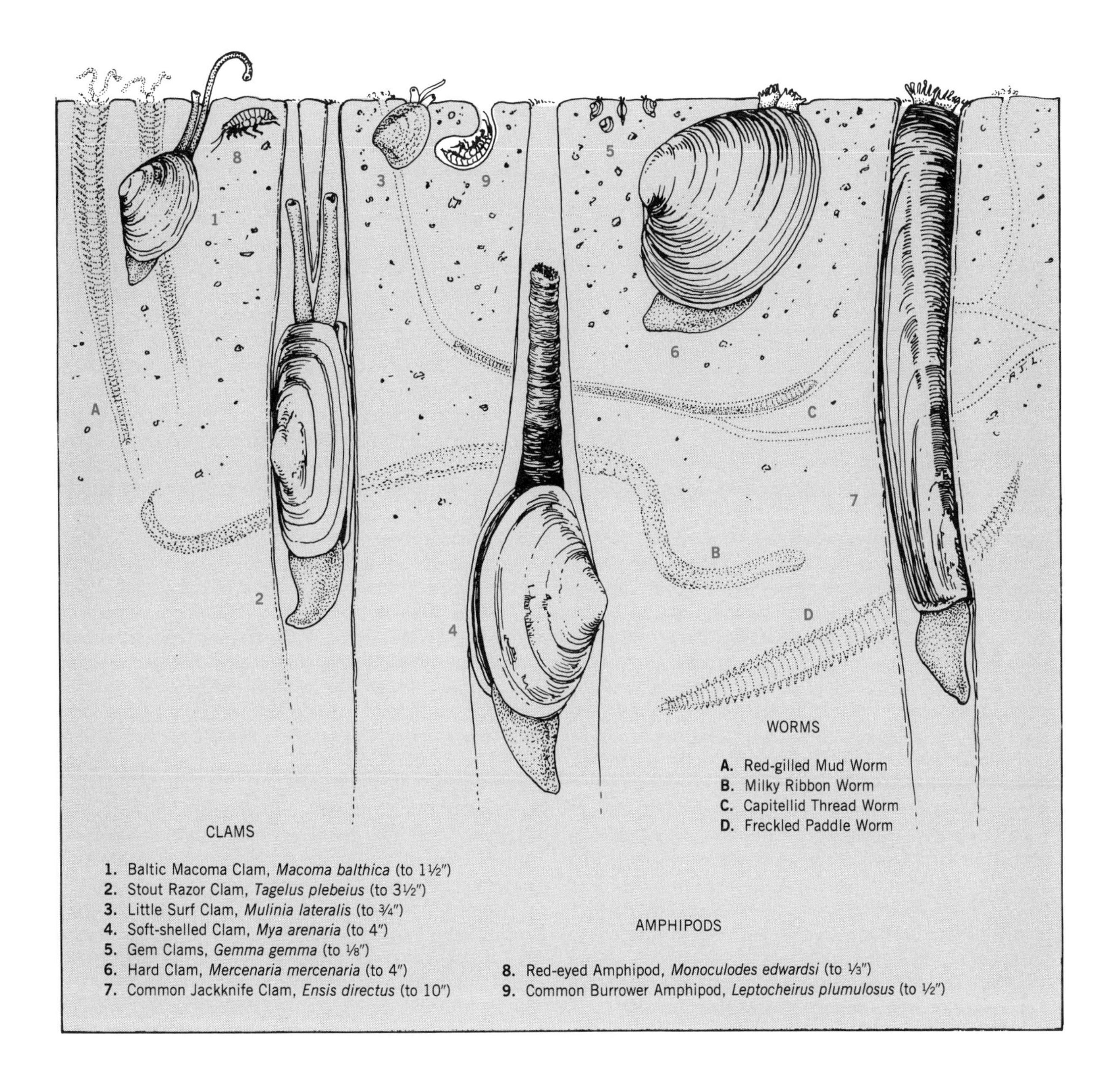

The Atlantic mud-piddock, *Barnea truncata* (to 2 inches), leaves distinctive borings in a hard peat clump. The shells of this clam gape widely at both ends.

uum up tiny bits of food. Macomas can also feed by siphoning in water containing microplankton. Macomas are rapid burrowers and frequently shift positions in the substrate. Some macomas lie horizontally rather than vertically in the bottom, burrowing sideways in search of food but moving rapidly down when disturbed.

The little surf clam, or coot clam, *Mulinia lateralis,* looks very much like a baby hard clam except for its smooth shell and the flat, sloping ridge on the back edge of the shell. Coot clams are found in dense beds, but they are short-lived and often disappear from an area as rapidly as they appeared. Coot clams as well as macomas are favored food for ducks and geese.

Gem clams, *Gemma gemma,* are so small that they are often overlooked, yet literally hundreds of thousands may populate a square yard of sandy bottom. No bigger than a lentil, they are one of the smallest known bivalves. Gem clams are rapid burrowers, but they stay very near the surface of the substrate. They brood their larvae in specialized internal pouches. When the larvae are released they do not swim but settle immediately to the bottom, where they are dispersed by currents and waves until they burrow.

The stout razor clam, *Tagelus plebeius,* is one of the most abundant larger clams in the Chesapeake Bay system. It looks like a squared-off soft-shelled clam. These deep-burying clams live in burrows that can extend as deep as two or three feet into the substrate. Stout razor clams live in permanent mucous-lined burrows, sheathed with compacted sand, in which the clam can ascend and descend by means of a robust but highly flexible foot. The siphons are separate, each lying within its own sheathed tunnel, and are capable of elongating up to six inches. Blunt tentacles at the open end of the siphons are able to draw in and protect the entrance of the burrow. It has been calculated that stout razor clams provide more biomass of living tissue in the estuaries of Virginia's portion of the Bay than any other species. They are also abundant in the mid-salinity waters of Maryland's Bay and tributaries, occurring both intertidally and subtidally in water up to 30 feet deep. The purplish razor clam, *Tagelus divisus* (see Shell Guide for illustration), is a smaller version of the stout razor clam, and can be found in the high-salinity waters of Zone 3.

One of the most unusual and easily recognized clams of the Chesapeake Bay is the common jacknife clam or razor clam, *Ensis directus.* Its very elongate, slightly curved shiny shell is a common beach shell along lower Maryland and Virginia shorelines. Occasionally, you may see a patch on an intertidal flat where half a shell is sticking upright out of the bottom. These clams seem to be trying to pop themselves right out of their burrows. This sometimes happens when they are attacked from below by a ribbon worm. In trying to escape they are forced out of their burrows. Normally, they lie completely below the surface in permanent compacted burrows extending as deep as those of the stout razor clams. While feeding, they lie very close to the top of the burrow with their short siphons extending out; if disturbed, however, they quickly contract their foot and move deep below. They have telltale keyhole-shaped siphon holes, but it is difficult to try to dig up or extract these clams, as they hold themselves so tightly that the shell will tear apart from the body first. A smaller version of the jackknife clam, the little green jackknife clam, *Solen viridis* (see Shell Guide for illustration), occurs in high-salinity regions near the mouth of the Bay.

Some of the most beautiful shells to be found along the shore are those of angel wing clams, which bore into hard muds, peats, and clay banks. If you come upon a hard chunk of clay or peat riddled with irregular cavities, you may be fortunate enough to find an angel wing captured within its solid home. When very small, the clam digs itself in by rocking the rough end of its shell back and forth. As it grows it becomes imprisoned, so that only its siphons can emerge from the small entrance hole. Three species occur in the mid- and higher-salinity regions of the Chesapeake Bay. The false angel wing, *Petricola pholadiformis,* and the Atlantic mud-piddock, *Barnea truncata,* are small, two-inch clams that bore into hard sediments. The largest, the angel wing, *Cyrtopleura costata,* grows to six inches, and its shell does indeed look like angel wings. It is found in softer, stickier sediments than the other boring clams inhabit, buried a foot or more under the surface.

Burrowing Crustaceans

Mantis Shrimp. One of the most incredible experiences a beachcomber or collector can have is to find a live mantis shrimp with its stalked, brilliant emerald green eyes and its long jackknife claws snapping back and forth like those of the better-known, garden variety praying mantis. Mantis shrimp, *Squilla empusa,* grow as long as 8 or 10 inches. They look like flattened lobsters with beautiful pale chartreuse bodies, each segment outlined in dark green with a bright yellow border. Few people have a chance to see these rather common creatures of the mid- to higher-salinity regions of the Bay, as they are mostly nocturnal and are well hidden below the surface in many-chambered burrows. Intertidal muds may be irregularly pockmarked by their large entrance holes, but it is difficult to capture a mantis shrimp, whose burrows have many exits. At night, mantis shrimp emerge to prey rapaciously on live shrimp, fishes, crabs, or even other mantis shrimp. Their claws are armed with recurved spines, which hold their prey fast—including a human finger! They live in deeper waters as well as at the intertidal edges of the low-water line; the molted shells of mantis shrimp are occasionally washed onto the beach. They are commonly taken by watermen dredging for crabs in the lower Bay.

Mud Shrimps and Snapping Shrimp. Burrowing mud and snapping shrimps live below the surface of many muddy areas of Zone 3. Unlike the predatory mantis shrimp, they seldom leave their burrows but feed passively on micro-

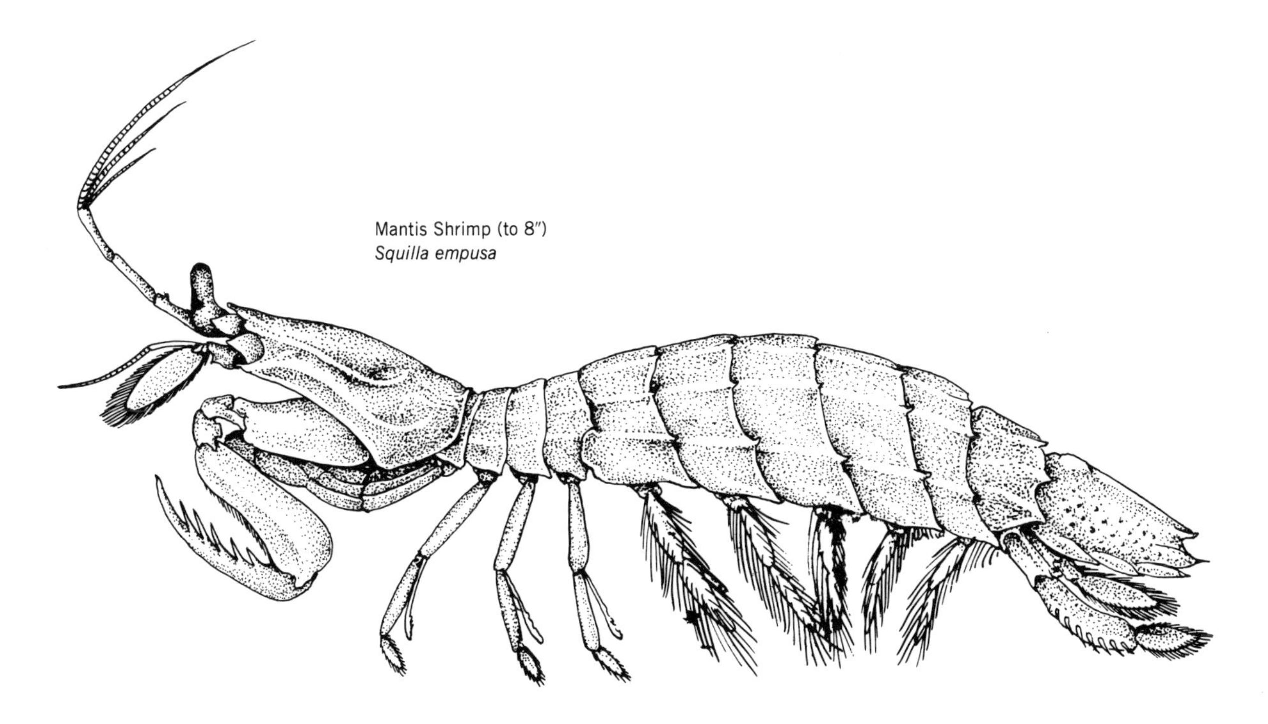

Mantis Shrimp (to 8″)
Squilla empusa

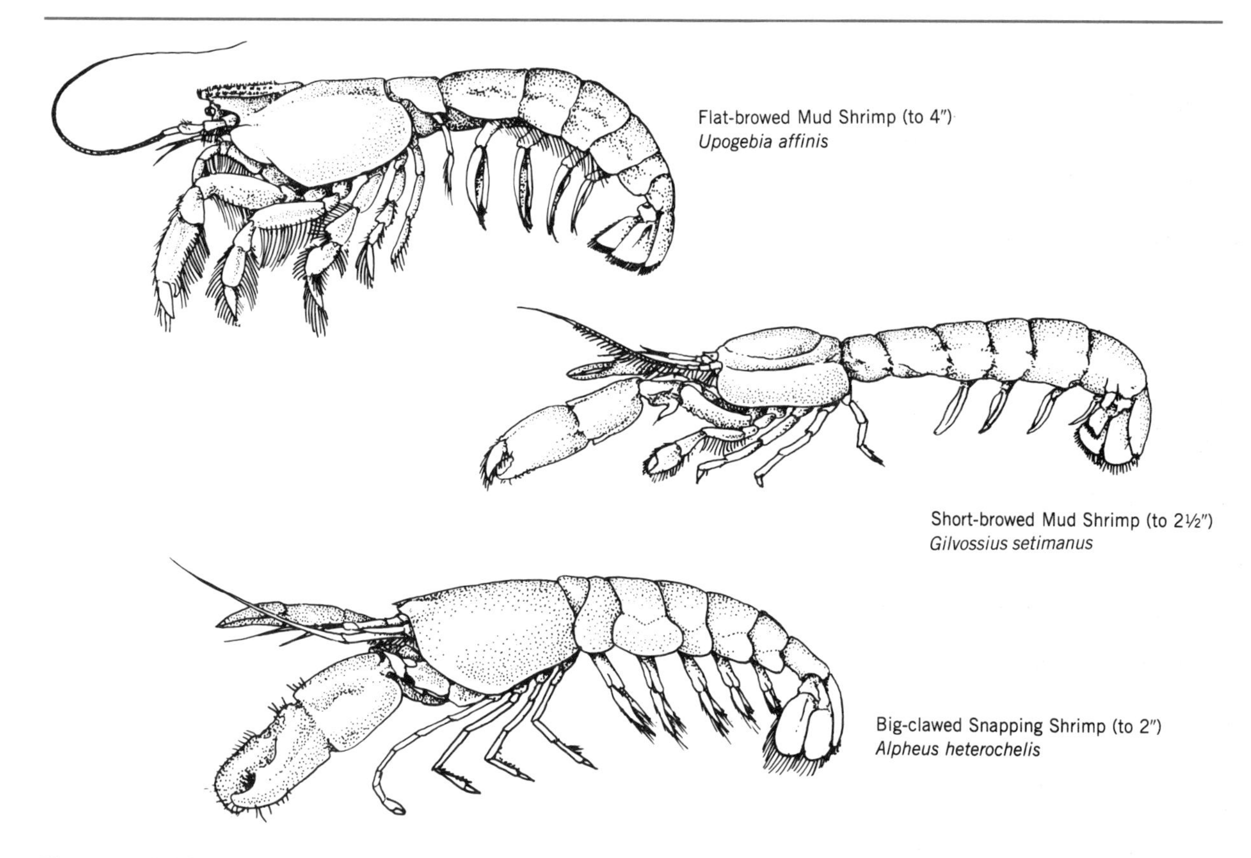

The two mud shrimps are easily identifiable by the top of the carapace: one is flat and hard and covered with clusters of short bristles; the other is smooth and rounded. Their claws and legs are also markedly different. The big-clawed snapping shrimp has a lobster-shaped body, a robust snapping claw, and spindly legs.

scopic bits of food brought to them by currents produced by their waving appendages. You must dig deeply with a shovel just above the low-water mark or in shallow water to see these strange little creatures. Mud shrimps often live communally with several others, each occupying one of the many-branched chambers, which may be two or more feet deep. The burrows have many openings to the surface, allowing water, with its vital food supply, to circulate freely. Mud shrimps are rather weak-looking, with a parchment-like shell of an indiscriminate, pale gray color. There are two species in Chesapeake Bay, the flat-browed mud shrimp, *Upogebia affinis,* and the short-browed mud shrimp, *Gilvossius setimanus.* The big-clawed snapping shrimp, *Alpheus heterochelis,* similar in size and appearance to the mud shrimp, hovers under shells and other sheltered places on mud flats. It gives its presence away by a distinct snapping sound made with its large claw.

Burrowing Amphipods. A number of amphipods burrow into intertidal and subtidal substrates. The red-eyed amphipod, *Monoculodes edwardsi,* and the common burrower amphipod, *Leptocheirus plumulosus,* are two such species. These are tiny bugs of the bottom often less than half an inch long. The single dorsal red eye of *Monoculodes* (which is actually two eyes fused into one) distinguishes it from other amphipods. You may find it by sifting fine bottom silts through your fingers in quiet waters near or just below the water line. The common burrower amphipod, on

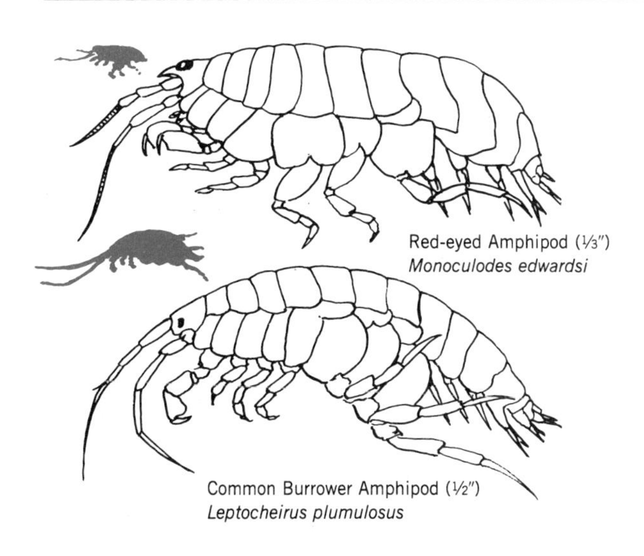

Silhouettes show actual size.

the other hand, usually lives in areas with a good current in a permanent tube constructed of sand grains and debris; it is one of the most abundant species in shallow- and deeper-water habitats. It lies head up filtering water and plankton through the open end of the tube.

WADERS OF THE FLATS

Wading birds common to the intertidal flats range from the dove-sized dunlins to the long-legged, glossy ibises, which are similar in size to some of the egrets commonly seen in the Bay area. Waders are wonderfully adapted for feeding on the profusion of animals harbored within and on the surface of tidal mud flats. Some waders arrive on the flats literally in thousands and begin probing frenetically for unseen quarry under the surface of sand and mud; other species appear in pairs or small flocks and stalk their intended meal in a most fastidious and discriminating manner. The rhythmic, mechanical pecking by so many short bills and long bills and the precise manner in which other species wield their chisel- and scythe-like beaks underscore the number and variety of animals inhabiting intertidal flats.

The dunlin, *Calidris alpina,* is a small shorebird only slightly larger than a sanderling. Dunlins appear to be neckless because of their hunched stance and are sometimes referred to as dumpy because of their shape. They have a black bill, slightly drooped at the tip, and black legs.

Dunlins, *Calidris alpina,* (8½ inches), and short-billed dowitchers, *Limnodromus griseus* (11 inches), probe for food along an intertidal flat.

In the summer they are gray with a reddish brown back and a black belly patch. In the winter, when they are commonly seen in the Chesapeake Bay, they are predominantly gray. They are common along the Eastern Shore from Tangier Sound south to the mouth of the Bay—particularly in the spring, when they are migrating to the Arctic to nest, and then in the fall, when they move south for the winter. Dunlins are usually found concentrated on mud flats and along estuarine shorelines rather than on sandy beaches, which sanderlings, their close relatives, prefer. Dunlins, like sanderlings, peck and probe in a rapid series of forays, using their short bills to feel for worms and mollusks under the surface of the mud.

Dowitchers, both the short-billed, *Limnodromus griseus,* and the long-billed, *Limnodromus scolopaceus,* are dark, chunky, medium-sized birds with long, straight bills and distinctive white or cream-colored eyebrows. Both dowitchers are similar in appearance with reddish brown coloring when in breeding plumage, which is generally not seen in the Chesapeake and mid-Atlantic coastal areas. More often, dowitchers seen in the Chesapeake region are dull gray above and white below. Both species show a wedge of white from their barred tails to their backs when in flight.

The two species of dowitchers are difficult to distinguish, and some experts listen for the "tu-tu-tu" call of the short-billed to separate it from the long-billed dowitcher's high, thin "keek" uttered either as a single note or as five or six rapidly accelerating notes. Short-billed dowitchers are common along the eastern and western shores and favor estuarine mud flats, while the long-billed dowitcher is more often found on freshwater mud flats. Both species have an almost comical manner of feeding, which has been aptly described as mechanically probing like a sewing machine. Their rapid up-and-down feeding is a distinctive dowitcher trait. The short-billed dowitcher feeds extensively on bamboo worms, amphipods, and clams, and the long-billed feeds on a variety of insects, spiders, mollusks, and plant seeds.

The glossy ibis, *Plegadis falcinellus,* is a colonial, or gregarious, species with long legs and a long, slender, downcurved bill. Its scientific name, which refers to its

Two American oystercatchers, *Haematopus palliatus* (18½ inches), feed on exposed oysters on a mud flat while a group of glossy ibises, *Plegadis falcinellus* (23 inches), search for worms and mollusks buried in the mud.

large, visibly curved bill, means "scythe" or "sickle." Glossy ibises are heron-sized birds, and when in breeding plumage in early spring to midsummer, are blackish brown with a glossy green and purple head and underparts. Their gray-green legs are marked with red joints. Nonbreeding adults have dark gray-brown heads, and their necks are flecked with narrow white streaks. Ibises fly with their necks extended, not crooked like the familiar herons and egrets. They often fly in groups, gliding and sailing in unison—even their wingbeats are timed together. The glossy ibis may be seen scattered throughout the Chesapeake Bay and its tributaries. Ibises are colonial nesters that roost in trees and shrubs, sometimes in mixed colonies with herons. They range from the Gulf of Mexico to New Jersey and regularly nest on Smith Island in Tangier Sound. Their nests are built of sticks and twigs in bushes and trees, usually over the water. The glossy ibis is typically found in marshes, swamps, and even flooded fields and wet pastures. It feeds on worms, mollusks, fiddler crabs, and snakes.

The American oystercatcher, *Haematopus palliatus,* with its large red-orange bill, black head, white wing and tail patches, and pink legs looks like no other bird in the Bay area. Its powerful beak is laterally flattened and triangular in cross section to prevent flexing. Adults are chicken-sized birds; their call is a shrill "kleep." Juvenile American oystercatchers differ somewhat in appearance from the adults—their back and wing feathers are edged in buff, which gives them a scaly or mottled look. In the spring, oystercatchers nest on the sandy shores of sections of the Chesapeake Bay and on rocky coasts as far north as New England. They range to the Caribbean in the winter. This species, like the glossy ibis and the brown pelican, is now commonly seen along the lower Eastern Shore and the mid-Atlantic coast during summer and fall. Generally they are in scattered pairs or small flocks feeding on mud flats, beaches, and exposed oyster bars. Oystercatchers are highly adapted for feeding on oysters and other mollusks. They are adept at stealthily stalking a gaping mollusk and plunging their bills between the open shells, severing the adductor muscle and preventing the mollusk from closing. The meat of the mollusk is then chiseled away from the shell and eaten. Sometimes, rather than stabbing its prey, the oystercatcher will resort to hammering the shell of the mollusk with a rapid series of woodpecker-like blows until the shell is shattered and the soft body of the mollusk is exposed. Oystercatchers can be seen probing in the mud and sand for small clams and mussels, and searching out fiddler crabs. They are prodigious feeders on almost any mollusk or crab.

● This litany of the animals of the intertidal flats might lead you correctly to conclude that every inch of bottom is teeming with life, particularly when you remember that only the commoner animals have been mentioned. It would seem that wherever you tread a creature will be met with. Intertidal animals are indeed present in monumental numbers, but many are very small and populations are sporadically distributed or patchy—dense in one area, absent from another. One species will overpopulate a spot in spring only to die off in summer, when another species takes over. But whatever the season, you can be sure that any bare-looking mud flat that has not been degraded is teeming with life, as evidenced by the voracious feeding of so many birds.

PIERS, ROCKS, AND JETTIES

Glance at almost any pier piling and you will notice that above the intertidal high-water and spray line the piling is devoid of plant and animal life, whereas, immediately below the water line, the piling is overgrown with a profusion of life. Similarly, the rocks of a jetty often show a darkened band of stubby growth and encrusting barnacles at the low-water line, sharply demarcated from the lighter-colored dry rock surface above.

Almost any hard surface along the shoreline presents a welcome face to the myriad forms of sessile aquatic life that must seek a firm base on which to grow. Attached life is found on just about any submerged object or structure—bulkheads, bottoms of boats, ropes tied to moorings, crab traps, shells, and pebbles exposed on the intertidal flat. Even the fronds of submerged aquatic plants furnish a base for firmly fastened sponges, hydroids, and other fauna. Often whole communities of attached organisms grow over other animals, which serve as substrates. The Chesapeake Bay oyster bar, with its thriving community of epifaunal life, is the ultimate example of a sessile community of organisms using one another as a place for attachment.

If a clean wooden board is set in Bay waters, within a matter of days a fine coat of "scum," composed of microscopic bacteria, algae, and protozoans, will invariably appear on its surface. This initial filmy coating seems to be necessary before other animals can settle. In a few more days, little calcareous knobs of young barnacles and scattered colonies of encrusting bryozoans and sponges appear. A bright green patch of sea lettuce or green-tufted seaweed develops, and the delicate stalks of hydroids intertwine with the petaled heads of sea anemones and the round clumps of sea squirts. Eventually, literally every inch of the board will be covered with some form of life. As this rich growth develops and spreads, tube-building worms and amphipods arrive and build fragile mud tubes, utilizing the fine sediments that have become entrapped among

4. PIERS, ROCKS, AND JETTIES

the animal and plant growths. Wandering clamworms and little crabs and snails graze over the surface, while flatworms attack the soft innards of the barnacles, leaving a trail of empty shells behind. Shrimps, small fishes, and blue crabs hover nearby, availing themselves of any available edible morsel.

These animals that are so quick to colonize and flourish belong to a complex of species known as sessile organisms. They are a diverse group from many phyla comprising hundreds of species that have certain characteristics in common. Most species produce planktonic larvae, which float about in the currents for a time. Some species are planktonic for only a few hours or days; others remain planktonic for weeks. However, in order for the larvae to survive they must find a suitable substrate and attach themselves within a short period after settling; otherwise they perish. The development of a planktonic larval stage allows species with nonmotile adults to colonize new fertile spots. Different animals will attach at different times and seasons depending on the various spawning phases of species in the Bay.

Another characteristic of sessile species is that most also reproduce asexually, insuring even greater chances for their survival. For example, a single bryozoan (the tiny individual animal of a bryozoan colony) larva, after attachment, can bud one or two new images of itself within a day or so, each of these in turn budding and growing, so that in a week or two there may be thousands of separate animals forming a single colony that had its start from a single larva. Still another characteristic of sessile forms is their rapid development to sexual maturity. Within just a few weeks' time after settling, a new generation of larvae is released and a new colony repeats both the sexual and asexual cycles at another site. It has been suggested that populations of sessile animals in the Chesapeake Bay have increased substantially with the arrival of man, each pier, bulkhead, and jetty adding new home grounds for these creatures.

Life can, however, be very stressful in these communities, with each individual vying for space and food with members of its own kind and also with other species. Sessile colonies are unable to move, so they are at the mercy of motile predators; they are also vulnerable to storms and waves, which wrench them from their moorings. Accordingly, there are different successions of animals through the seasons, with various species waxing and waning depending on life cycles, predation, or storms. Successions vary even from year to year; the same pier piling that was covered with barnacles one year may be covered with sea squirts the next.

Just as the sand beach and the shallow flat have various intertidal zones, a piling, breakwater, or boulder also has vertical intertidal zones, each zone typically occupied by different forms of life. Sea roaches and wharf crabs are found in the highest zones at the edge of the high-tide mark. Moving downward through the intertidal zone, barnacles are encountered first, then a gradually increasing variety and number of organisms—mussels, worms, tube-building amphipods, some bryozoans, clumps of sea squirts, sea anemones, and patches of green seaweeds with reclusive sea pill bugs nestled underneath. Subtidally, below the low-tide line, most of the intertidal organisms still flourish and are joined by a lush growth of innumerable hydroids, sponges, and brightly colored macroalgae, commonly called seaweeds.

Birds, although certainly not sessile animals, are seen in great numbers on piers, pilings, jetties, bridges, and navigational structures, which they utilize for nesting sites and sunning and resting areas. These structures also provide protection from predators and serve as vantage points for a bird's next meal. Gulls take advantage of the rich life beneath their feet, capturing small crabs scampering over rocks or pilings, or grazing on worms and algae attached to a bulkhead or other hard surface. Cormorants and pelicans can be seen resting on these various structures with their wings outstretched to dry their plumage and regulate their body temperatures. Barn swallows, generally not thought of as birds tied closely to the water, are ubiquitous and abundant throughout the Bay area. They sit in long rows on electric lines that stretch across rivers and dart effortlessly under bridges and piers everywhere.

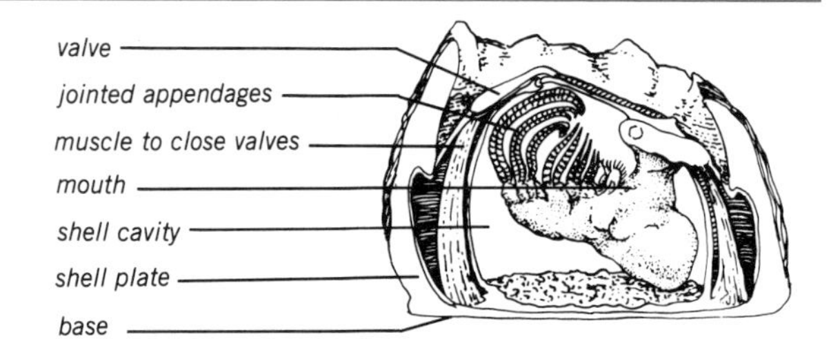

A cross section of the interior of a typical barnacle showing feathery appendages folded within.

BARNACLES

Anyone who has been to the shore is familiar with barnacles, but few people would guess that they are crus-

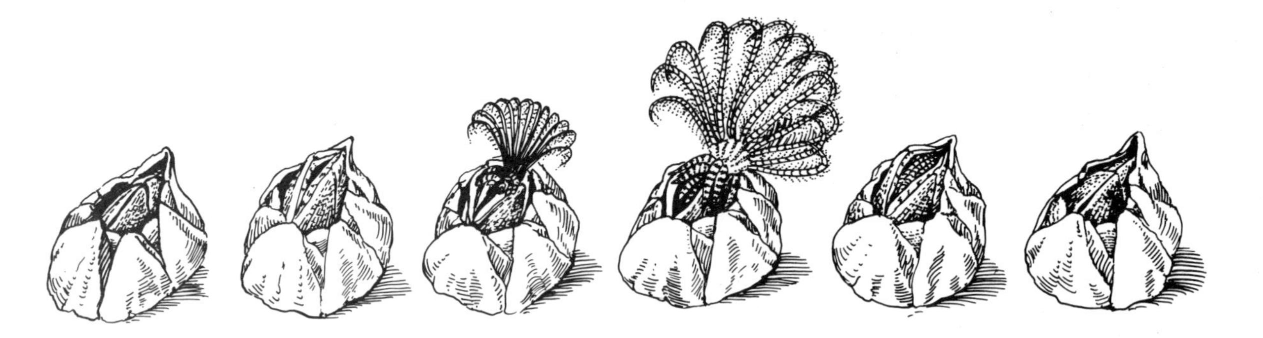

Sequential opening and closing of a barnacle.

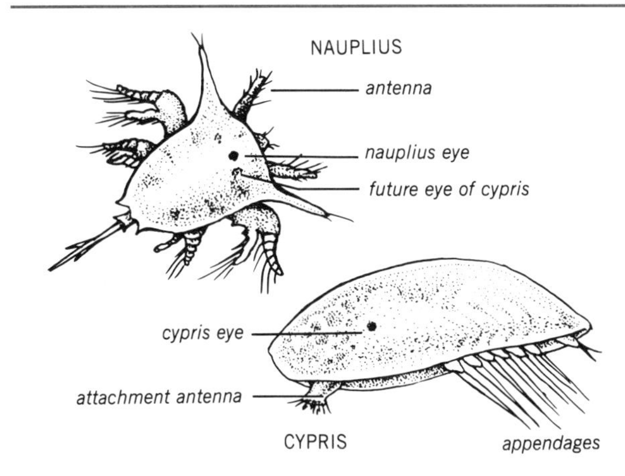

Two microscopic stages of barnacle larvae. The larger cypris is barely visible to the unaided eye. The nauplius has a single central eye, which is replaced in the cypris stage by two bilateral eyes.

taceans and are closely related to crabs. Most observers would assume them to be the empty shells of some type of mollusk. This conclusion is not entirely unwarranted, since exposed barnacles are tightly closed and show no sign of life. However, if you place a barnacle underwater, two "doors" will soon open and a fan of graceful feathery appendages will slowly emerge, unfold, and rhythmically undulate. Each pulsation sweeps tiny particles of food into the barnacle's interior. The feathery appendages are "jointed" and covered by an exoskeleton, thus indicating that a barnacle is not a mollusk but a crustacean. Louis Agassiz, the famous Harvard biologist, aptly described the barnacle as "nothing more than a little shrimp-like animal standing on its head in a limestone house and kicking food into its mouth." Its conical "house" is composed of six overlapping shell plates with an opening at the top covered with two valves (the doors) and a base, or flat bottom, securely cemented to a hard surface. Barnacles are often dislodged and only the circular flat bases remain as evidence of their former presence. A barnacle grows by adding calcium carbonate along the edges of each plate, which increases the size of the inside cavity. Meanwhile, the organism within must also grow by shedding its exoskeleton, like a blue crab or any other crustacean. These molted exoskeletons float on the surface of the water and can be so dense at times that a swimmer may think that he has discovered a new animal, never associating them with barnacles.

Barnacles are hermaphroditic, that is, each individual shares male and female organs; however, barnacle eggs must be cross-fertilized by separate individuals. Since barnacles cannot move, fertilization is accomplished by means of a slender sperm tube, which protrudes up and out of one barnacle and through the valves of its neighbor. Thus, it is advantageous for barnacles to congregate. The fertilized eggs are nurtured within the barnacle shell until they hatch into tiny larvae and are released into the water. During May and June in the Chesapeake Bay region, the plankton is thick with barnacle larvae. There will be literally thousands of larvae in each square foot of water in certain areas. These dense swarms of larval barnacles are favored food of young fishes and are consumed in such large numbers that relatively few actually settle and develop into recognizable barnacles.

Barnacle larvae pass through two stages. The first-stage larvae are called nauplii, strange triangular-shaped creatures that in a few days molt into the second-stage cypris larvae, which look like tiny transparent seeds. The cypris swim about for a number of days in search of a suitable place to attach. Cypris larvae have an affinity for settling in areas that have previously been occupied by other barnacles of the same species. Apparently, a chemical released by the older barnacles attracts the young to the same areas. The cypris attaches itself at its head region with a cement produced by special antennal glands and soon begins to produce the calcareous plates that will eventually surround and encase its body.

Although barnacles seem to be well armored and protected, populations are liable to substantial die-offs. Barnacles quickly succumb to any prolonged drying and to ex-

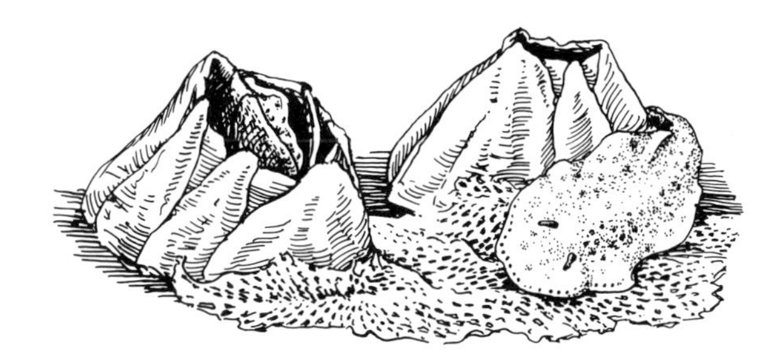

An oyster flatworm, *Stylochus ellipticus* (to 1 inch), creeps over the base of a barnacle; the edge of another oyster flatworm that is feeding can be seen within a gaping barnacle.

tremely cold weather and harsh winds, but perhaps their greatest enemies are predators, sponges, and bryozoans, which grow over them, smothering them in the process. The little oyster flatworm, *Stylochus ellipticus,* is a major predator of barnacles in the Chesapeake Bay. The flatworm approaches a feeding barnacle and quickly inserts its pharynx through the open valves. The barnacle, immediately disturbed, closes its valves, but the pharynx of the worm remains within and begins to feed. Eventually the barnacle weakens, gapes, and allows the whole worm to slide in and continue its feast. A careful search of empty barnacle shells in the low intertidal zone will often produce an oyster flatworm.

Three species of barnacles are abundant and common in the Bay and its rivers and range upstream into water that is almost fresh. The bay barnacle, *Balanus improvisus,* and the white barnacle, *Balanus subalbidus,* are the dominant species of the upper half of the Bay and its tributaries. The white barnacle tends to inhabit less saline waters than the bay barnacle. The ivory barnacle, *Balanus eburneus,* is more dominant in lower Bay regions. All three species thrive intertidally and subtidally on almost any firm substrate.

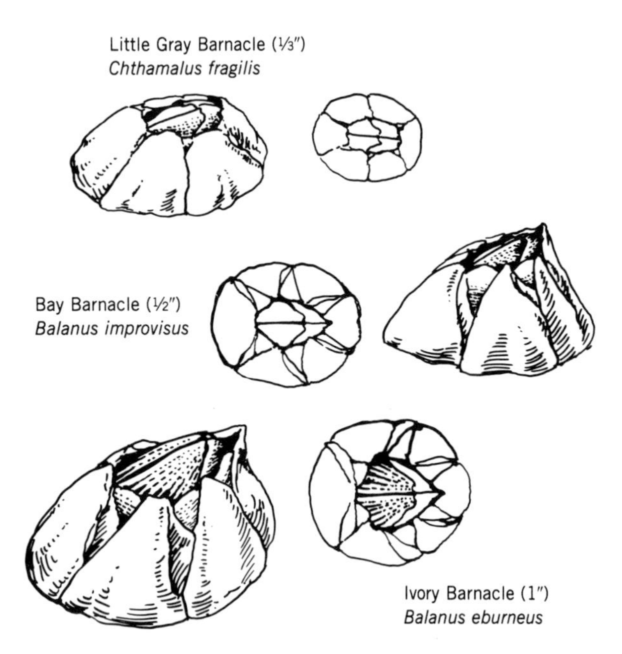

The three common barnacles of the Chesapeake Bay are identifiable by differences in the alignment of the shell plates, the type of base plate, and the sculpturing on the valves. The little gray barnacle has a brown membranous base plate and closely aligned shell plates. The ivory barnacle has fine longitudinal grooves on its valve, a marked triangular opening, and somewhat more widely spaced shell plates than the bay barnacle or white barnacle (not illustrated). However, these identifying characteristics are often obscured by overcrowding and fouling growths.

In the lower Bay, a fourth species, the little gray barnacle, *Chthamalus fragilis,* congregates in the high intertidal zone, where it is free from competition from other barnacle species and above the range of flatworms and other predators that generally do not move far above the water line. Colonies of little gray barnacles also attach to tall marsh grasses. This small barnacle can be distinguished by its surface, which is smoother than that of the other barnacle species, and its shell plates, which are closely aligned rather than separated. If a little gray barnacle is plucked off the substrate, its remaining base will be distinctly brownish and membranous in contrast to the white and calcareous base plates of the other species. It is somewhat difficult to distinguish among ivory, bay, and white barnacles, as their shapes are often distorted by crowding. However, a close examination will show marked differences: the ivory barnacle has wider spaces between the plates and tiny longitudinal grooves on the top surface of its valves. The bay and white barnacles have narrower spaces between the plates and valves without longitudinal ridges; however, it requires microscopic examination to see these subtle differences. The bay and white barnacles are small in diameter, generally less than half an inch across, whereas the ivory barnacle may grow to an inch in diameter; however, size is not a good criterion for identification, since colonies of ivory barnacles in the Chesapeake Bay are generally composed of smaller-sized individuals.

BUGS OF THE PILINGS

A swarm of sea roaches, *Ligia exotica,* scurrying over the dry surfaces of a piling, jetty, or sea wall is a common but rather unpleasant sight, for they look too much like obnoxious cockroaches. These little isopods (a little over an inch long) can be abundant on shaded areas of pilings or on rocks in the mid and lower Bay regions. They belong to

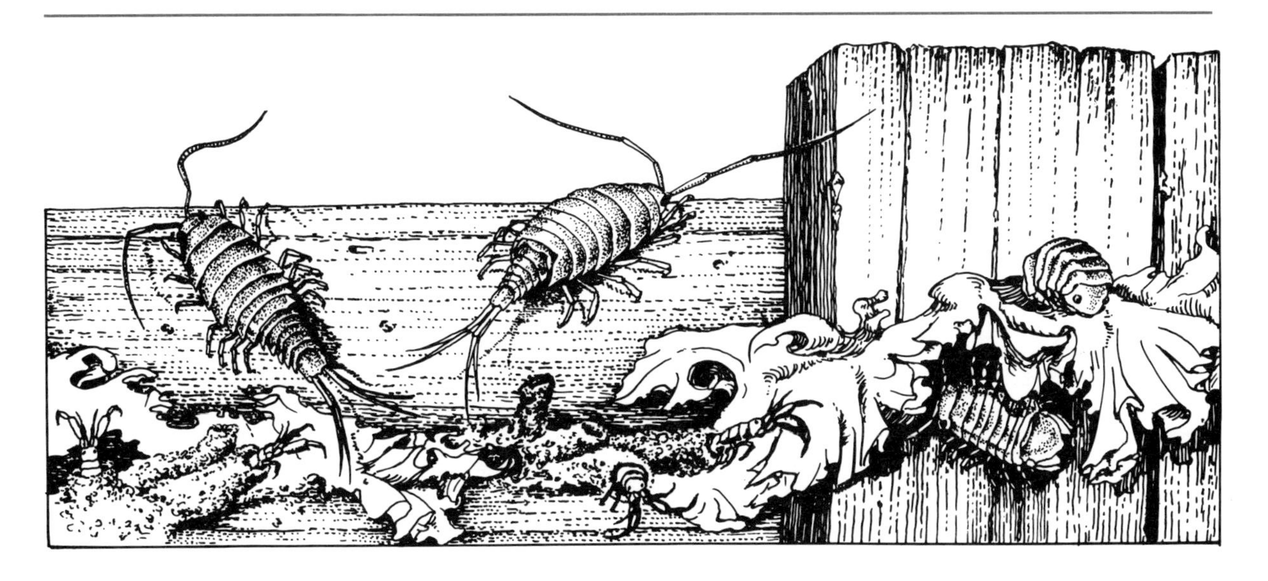

Sea roaches, *Ligia exotica* (to 1¼ inches), scattering over an exposed board; sea pill bugs, *Sphaeroma quadridentatum* (to ⅜ inch), grazing over a patch of sea lettuce; and slender tube-building amphipods, *Corophium lacustre* (to ⅛ inch), popping in and out of their soft mud tubes, form a busy community along the intertidal zone.

the terrestrial group of isopods and are not even mentioned in some books on marine isopods. They are, however, found only near the edge of the sea and move down the water's edge periodically to wet their gills. They occupy an ecological niche somewhat akin to that of ghost and fiddler crabs, but, unlike those crustaceans, they do not release their larvae in the water. The female sea roach broods its eggs in a pouch and releases its young on land as tiny but fully formed sea roaches, which race about as quickly as the adults.

Another isopod, *Sphaeroma quadridentatum,* the sea pill bug, is often encountered lower down on the pilings under intertidal algae or hidden among the barnacles and sometimes curled within the dead shells. It is a small, dark isopod, mottled with white and about half an inch long. The sea pill bug can readily be identified because of its habit of immediately rolling into a ball when disturbed (*sphaero* means "ball" in Greek). It is widely distributed throughout the Bay region into low-brackish waters (Zones 2 and 3).

Amphipods wander over the intertidal fouling organisms. Many are scuds, the same species found among the seaweed and debris of the intertidal flats. Tube-building amphipods proliferate where mud and silt collect in epi-

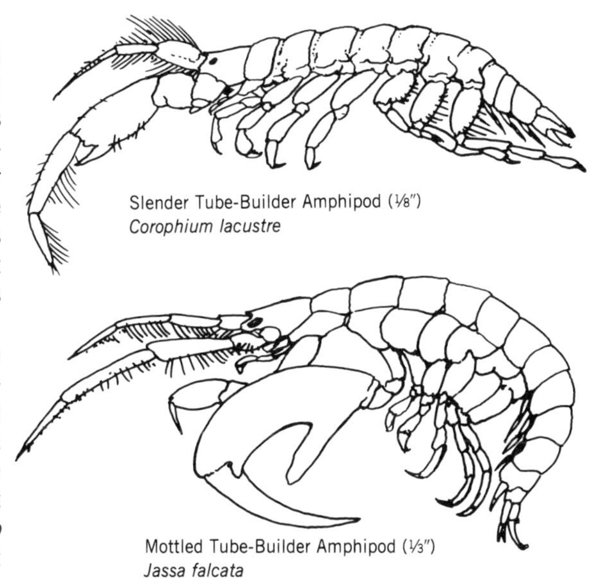

Two amphipods that live in tubes constructed of mud and sand grains commonly found on the hard surfaces of pilings, jetties, rocks, and shells in the Chesapeake Bay. The marked differences in their antennae and claws help in their identification.

faunal growth. Small holes of tube-building amphipods can be seen peppering the surface of any soft, muddy coating on almost any hard surface of the intertidal zone. These tube-builders are also found in the crevices of dead shells, strewn over flats, and at the base of marsh grasses or other aquatic plants. In the Chesapeake Bay, there are a number of tube-building amphipods, but most are too small to identify easily. The most abundant and ubiquitous is the slender tube builder, *Corophium lacustre.* Mottled gray and brandishing a pair of enlarged antennae, it has a body that is flattened from top to bottom, more like that of an isopod than the laterally flattened body typical of amphipods. Occasionally, the slender tube-builder leaves its mucous-lined tube to graze on bits of detritus, but it usually settles halfway back into its tube, moving water currents with its feet and straining bits of food out of the water with its large antennae. In turn, it is heavily preyed on by fishes, crabs, and ducks. Another amphipod, the mottled tube-builder amphipod, *Jassa falcata,* is larger and has huge pincer claws, making it somewhat easier to identify. It occurs only in the lower Bay. Look for it on pilings where there are strong currents and waves.

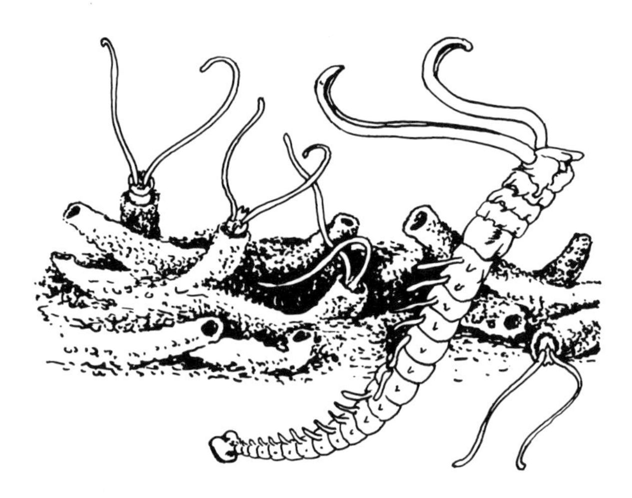

The probing, feeding palps of whip mud worms, *Polydora cornuta* (to 1 inch), extend from a matting of soft mud tubes. The worm in the foreground has been extracted from its tube. Note its wide fifth body segment, with a row of minute hooks, and its suctionlike tail, both of which are used to hold the worm within its tube. The fingerlike extensions are breathing appendages.

WORMS ON THE PILINGS

Many of the same species of wandering polychaete worms which live on the intertidal flats move over the epifauna of the pilings. Here they find rich colonies of favored prey both in the intertidal zone and underwater. Almost any probing of surface growths will reveal a clamworm or two. The soft, tortuous mud tubes of the whip mud worm, *Polydora cornuta,* may be in such profusion that they completely smother all other epifauna. This mud worm has been said to be the most abundant worm in the Chesapeake Bay, found on all types of bottom sediments and on the surface of any substrate. As many as 70 worms per square inch have been counted on a single oyster and may be the cause of substantial oyster mortalities. The two long, antennae-like palps of a whip mud worm may often be seen protruding from the open end of its tube, curling and unfurling as it probes for bits of food.

SEA SQUIRTS AND THE LIKE

Great clumps of sea squirts or sea grapes, *Molgula manhattensis,* protrude from pilings and sprawl over rocks; small clusters cover empty shells strewn over the flats; and oysters bought from the market are rarely without at least a few "grapes" attached to the shell. The densest populations on pilings occur at the base of the pilings rather than near the surface, although a sea squirt can certainly tolerate intertidal conditions. Children take great delight in squirting each other as soon as they discover that sea squirts, with very little encouragement, can eject a considerable jet of water. Sea squirts look very much like a bunch of yellowish green grapes. They are a widespread estuarine species, found from very low-salinity waters as far as the ocean. Sea squirts belong to an advanced class of invertebrates called tunicates, named after their outer-coating or "tunic," which is quite smooth and tough but is frequently covered with silt or other debris, or with encrusting bryozoans and tiny hydroids. The Chesapeake Bay species is found all along the Atlantic coast and can survive in such polluted areas as New York harbor, near the city after which it is named.

Sea squirts cannot tolerate much competition and often will not attach until other epifaunal animals have sloughed off, leaving a clear space in which they can thrive.

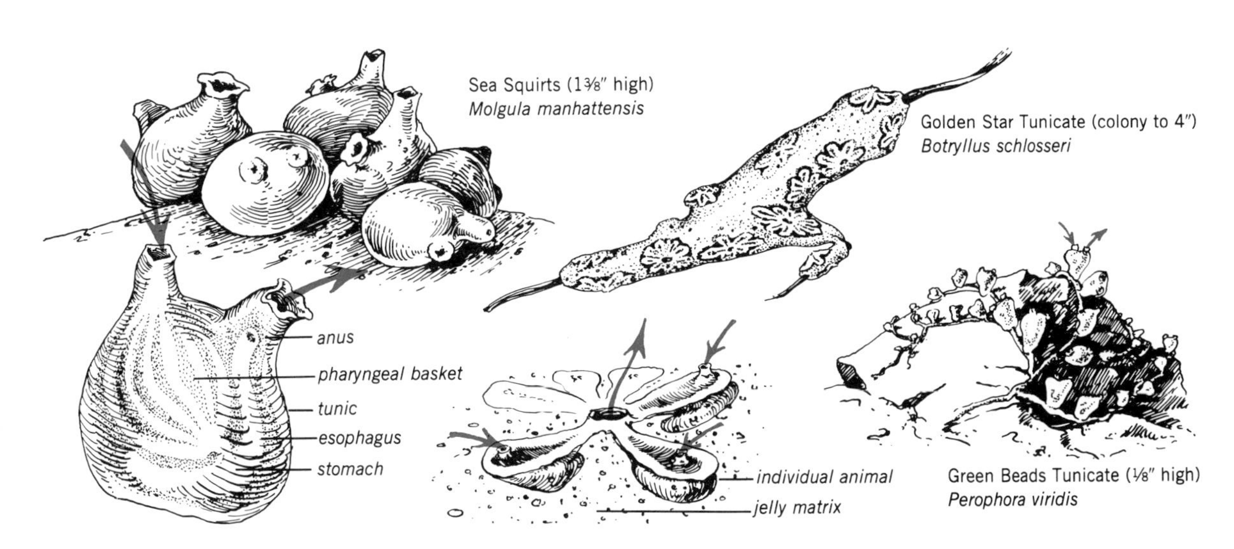

Three tunicates of the Chesapeake Bay. The internal organs of the sea squirt can often be detected through the semitransparent tunic. The pharyngeal basket, visible in this illustration, filters food particles from incoming currents. The arrows show the direction of water flow through a sea squirt, a green beads tunicate, and a colony within the matrix of a golden star tunicate.

Out of the water and with their two siphons (one incurrent and one excurrent) withdrawn, isolated specimens might be confused with sea anemones, as both resemble rounded gelatinous globs.

It was not apparent to scientists how high a position on the phylogenetic tree sea squirts occupied until their tadpole larvae were discovered. These minuscule transparent tadpoles have long tails with a primitive notocord (precursors of our backbone) and a slender nerve cord, which passes forward to an enlarged hollow brain. They even have a primitive eye. After a short free-swimming period, tadpole larvae settle out and attach to substrates by means of an adhesive organ on their head. After attachment, the tail, nerve cord, and notocord are absorbed until only a small nerve ganglion remains. The body distorts and bulges out and the larval mouth and pharyngeal opening are raised into two spouts—the siphons of the adult.

Sea squirts may grow in dense clusters, but each "grape" is an individual animal, whereas other tunicates are colonial. Two species of colonial tunicates may be found in high-salinity waters of the lower Bay and tributaries. The golden star tunicate, *Botryllus schlosseri,* forms smooth, rubbery masses over pilings or envelops strands of seaweeds and eelgrass. Clumps are beautifully and strikingly colored, with star-shaped patterns over the surface. Color is highly variable; sometimes the clumps are golden yellow with deep purple stars speckled with white or gold; at other times the background color is purple or brown with lighter colored stars. Each star is made up of colonies of 5 to 20 tunicates embedded in the jelly matrix, each individual having its own incurrent siphon at the top of the "petal" of a star and a common excurrent opening at the center of the star. Golden star tunicates are so unusual that there is little chance of not recognizing them. The colors quickly fade when they are removed from the water, but the unique star pattern remains discernible.

It is easy, however, to miss the tiny green beads tunicate, *Perophora viridis,* which grows like a vine over algae, pebbles, rocks, or shells such as mussels. Colonies are greenish yellow and are composed of small tunicates, only a tenth of an inch long, arising from a stolon network that creeps over the surface. The two small siphons of each grape can just be discerned with a hand lens.

MUSSELS

Many people are surprised to learn that the same delectable mussels steamed and served with garlic and wine sauce in a French restaurant can be collected in the Chesapeake Bay. However, this edible species, the blue mussel, *Mytilus edulis,* is generally abundant only at the mouth of the Bay. During the winter, blue mussel larvae are occasionally carried by currents higher into the Bay and set on pilings or rocks, but these in-Bay populations seldom survive the summer.

There is a mussel, however, the hooked or bent mussel, *Ischadium recurvum,* which attaches to almost anything hard—pier, rock, or shell—throughout the Chesapeake region. Hooked mussels are small, generally only one to two inches long, a rather dull-colored black or gray, but with a lovely shiny purple to rosy brown interior. The surface of the shell is distinctly ridged and curved. Mussels attach themselves firmly by means of strong thin threads called byssus threads, which can easily be seen under one side. Each thread ends in a little knob adhering closely to the substrate; these little round knobs are frequently the only telltale remnants of a mussel's previous attachment. Mussels can actually move along by pulling themselves forward on the threads, releasing some and

Blue mussels, *Mytilus edulis* (to 4 inches), cluster closely on a piling while hooked mussels, *Ischadium recurvum* (to 2 inches), nestle among a colony of sea squirts at the base; Atlantic ribbed mussels, *Geukensia demissa* (to 4 inches), lie half-buried in the bottom muds.

reattaching others. Needless to say, this is a very slow process and they do not move far.

In tidal fresh and low-brackish waters, clusters of the dark falsemussel, *Mytilopsis leucophaeta,* are common on almost any firm substrate (see Shell Guide). These are very small smooth-backed dark brown mussels, with a platform, or shelf, on the inside just below the hinge ends. Their classic "mussel" shape helps to identify them. The ribbed mussel, *Geukensia demissa,* common to many marshes of the Bay region, lives in a manner quite unlike that of the hooked mussel and is commonly found half buried in the bottom muck rather than attached to hard substrates. The blue mussel is the only species that is generally eaten. The other mussels, though edible, are usually not very abundant and often have an unpleasant taste.

SPONGES

Sponges may spread out over pilings and rocks in colorful patches of yellow, red, orange, purple, and green; some are flat encrustations, others grow up in thick, twisted fingers; some are fragile and flake to the touch, others are firm and dense. However, none of the sponges in the Chesapeake Bay are anything like the familiar household "bath sponge." Sponges are primitive colonial animals composed of microscopic individuals embedded in spongy tissue supported by tiny hard spicules. They are often difficult to distinguish from other rather unstructured growths, but if you break a piece off and rub it between your fingers, the "gritty" texture of the spicules will help you

SPONGES OF THE CHESAPEAKE BAY

Volcano Sponge (colony to 3″ wide)
Haliclona sp.

Sun Sponge (colony to 3″ high, 12″ wide)
Halichondria bowerbanki

Stinking Sponge (colony to 4″ wide)
Lissodendoryx carolinensis

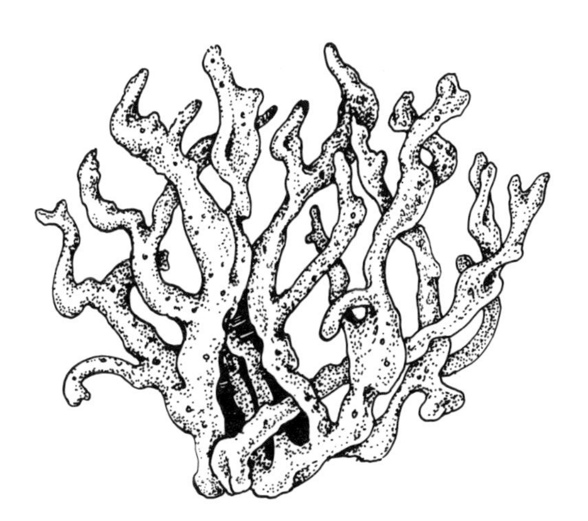

Redbeard Sponge (colony to 8″ high, 12″ wide)
Microciona prolifera

identify it as a sponge. It is often difficult to identify species, as the color and shape of any single species can vary considerably depending on environmental conditions. As one expert has commented, "the last shadow of doubt seldom disappears" when trying to tell one sponge from another without microscopic examination.

Openings of various sizes—or "oscula" and "ostia"—can often be seen on the surface of sponges. Water is pumped through the ostia into the center of the sponge, bringing food and oxygen to the individual cells within, while water with waste products is pumped out through the oscula. A fairly constant flow of water is necessary, so most sponges grow subtidally or at the very low intertidal zone, where they are exposed for only relatively short periods of time. Marine sponges are found well into the Bay and its tributaries up to mid-salinities, but they disappear in low-salinity waters. Freshwater sponges of the family Spongillidae may be found in the tidal freshwater and slightly saline stretches of Bay rivers and streams.

Sponges are home to a number of other animals, which seek refuge and protection within their folds and cavities. Break a large sponge open, and skeleton shrimp, various worms, mud crabs, and even small fishes such as gobies or blennies may often be discovered within.

One of the most widespread and abundant sponges in the Chesapeake Bay, and certainly the most brilliantly colored and easily identified species, is the redbeard sponge, *Microciona prolifera.* These sponges can grow into heavy masses, a foot or more in diameter, although smaller clumps are also common. They are the only sponge in the estuary that grows thick, fleshy, intertwining fingers. Their shape and bright orange or crimson color makes them readily identifiable. Young redbeard sponges begin as flat encrustations, but their brilliant coloration is usually enough to identify them.

The most conspicuous fouling sponge in summer in the Bay is the yellow sun sponge, *Halichondria bowerbanki,* often seen on pilings at low tide. Usually a bright yellow, it can also be buff- or cinnamon-colored. Its surface is often raised in irregular leaflike masses, and patches can spread a foot or more across the substrate and two to three inches upward. The sun sponge is not as tolerant of lower salinities as is the redbeard sponge.

The greenish yellow and very odiferous stinking sponge, *Lissodendoryx carolinensis,* is also common on intertidal pilings, spread around the base of sea squirts or hidden under bushy clumps of hydroids or bryozoans. It seems unable to clear a space for itself as well as some other species can. Its surface is raised in short tubercles with small oscula at their ends.

Fragile and delicate volcano sponges, *Haliclona* spp., may also be found on the intertidal pilings throughout most of the year. In the height of summer they tend to die off. Their surface is raised in chimneys capped with distinct open oscula. The chimneys may be slight mounds or be raised to a height of an inch or more, sometimes twisting and intertwining. Color is variable, from gold to tan with a delicate violet or lavender tinge, to dark lavender or pinkish purple. Volcano sponges are often found on eelgrass as well as on pilings and other hard substrates.

Green-striped Anemone (to 3⁄4")
Haliplanella luciae

SEA ANEMONES

Peer down into the clear waters at the surface edge of a piling and observe an open sea anemone and you will understand why it is called the flower of the sea. Unfortunately, these beautiful little creatures usually go unnoticed, although they are abundant on almost any piling. When they are exposed to the air their petal-like tentacles withdraw instantly into the stalk, and all that remains is an inconspicuous mound of jelly. When the tide returns to cover them again, they immediately unfold their constantly pulsating tentacles and begin reaching for food, capturing it, and directing it toward the mouth, which lies in the center of the circle of tentacles. Sea anemones are related to sea nettles, and like their cousins, they have tentacles armed with stinging cells, which quickly stun any prey that comes their way. Although always attached to some substrate, anemones are capable of surprising movement. Locomotion is accomplished by manipulating the pedal disk (the rounded flat base of the body stalk) somewhat in the manner of an inchworm. A sea anemone seen on the bottom of an aquarium in the morning can be halfway up the side by afternoon.

In the Chesapeake Bay two kinds of sea anemones are commonly found, on pilings, empty shells, oysters, rocks, and plants—almost anywhere. The white or ghost anemone, *Diadumene leucolena,* is perhaps the most abundant and widespread anemone in the Bay and is distributed into very low salinity waters. Stretched out, it is only an inch and a half long, so you must look carefully to spy it. The white anemone is pale and transparent, often with a pinkish cast. Some individuals take on a green or

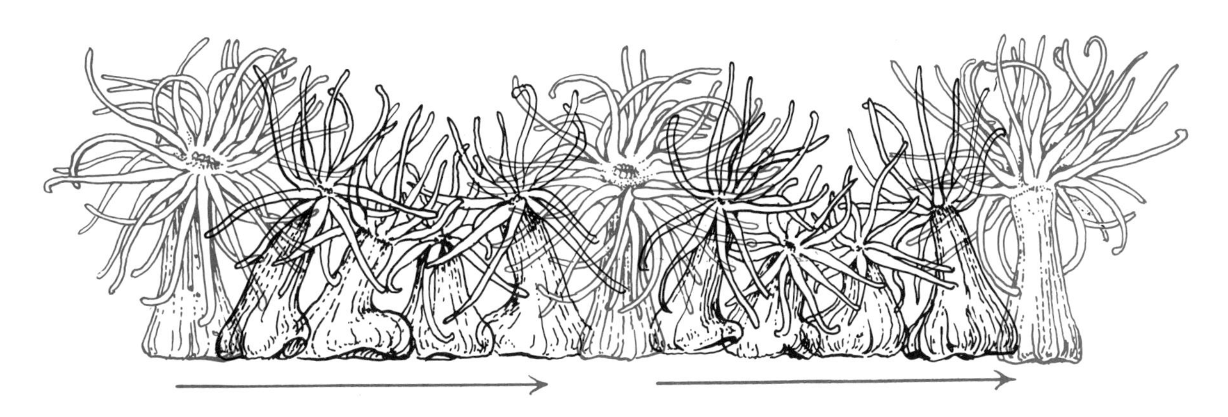

A white anemone, *Diadumene leucolena* (to 1½ inches), inches its way along by manipulating the pedal disk at the base of its stalk.

HYDROIDS OF THE CHESAPEAKE BAY

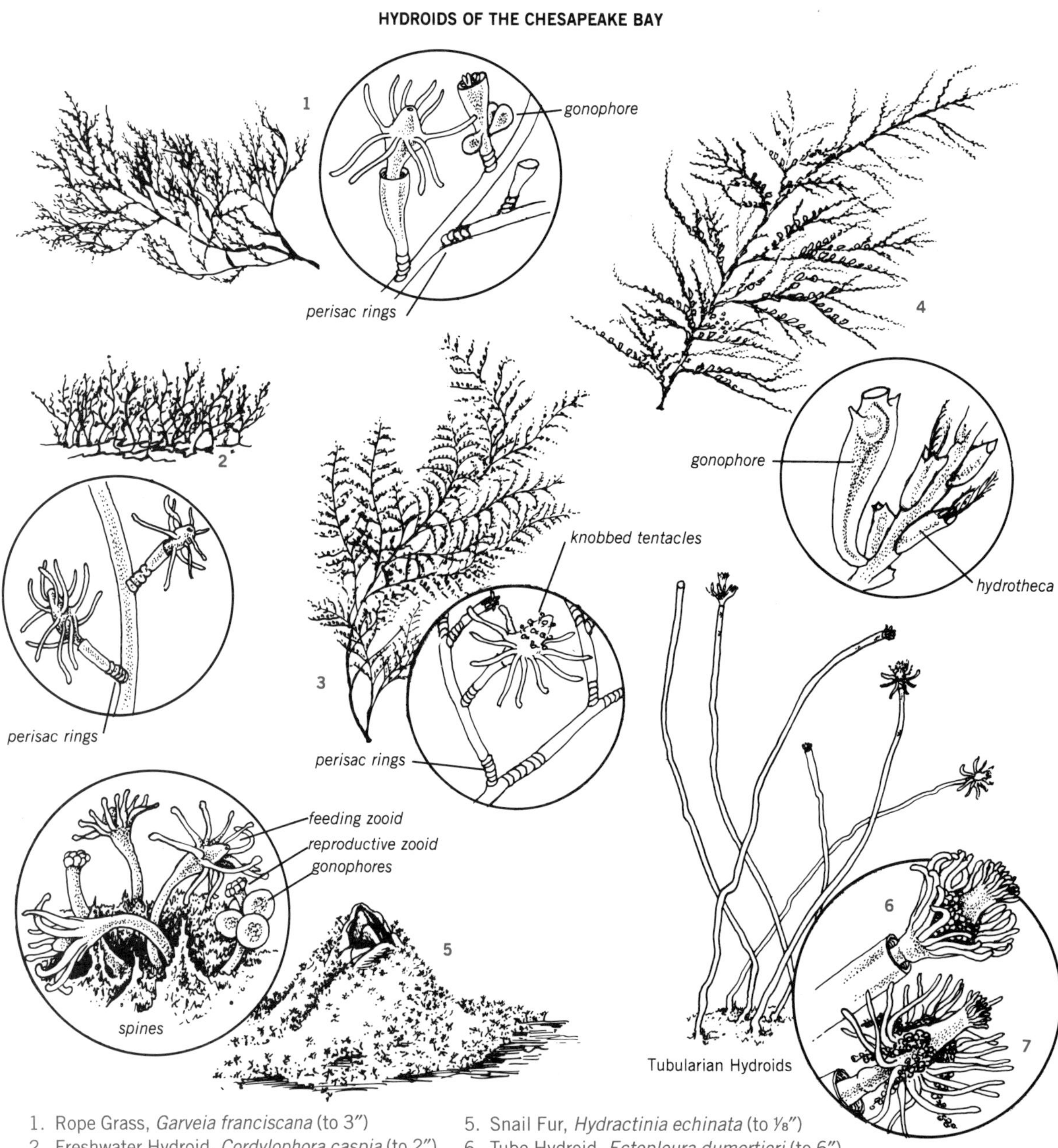

1. Rope Grass, *Garveia franciscana* (to 3″)
2. Freshwater Hydroid, *Cordylophora caspia* (to 2″)
3. Feather Hydroid, *Pennaria disticha* (to 6″)
4. Garland Hydroid, *Sertularia cupressina* (to 9″)
5. Snail Fur, *Hydractinia echinata* (to ⅛″)
6. Tube Hydroid, *Ectopleura dumortieri* (to 6″)
7. Pink-hearted Hydroid, *Ectopleura crocea* (to 6″)

brownish coloration from tiny symbiotic algae growing within the tentacles and body cavity.

The green-striped anemone, *Haliplanella luciae,* is often found along with the white anemone, but it is generally not as numerous, nor is it distributed as far upstream. The green-striped are more reclusive than the white, preferring small crevices and sheltered places. Dark green with bright yellow or red stripes, they are somewhat smaller than white anemones.

HYDROIDS—FEATHERY ANIMALS

What first appears to be seaweed is, more often than not, actually one of the many species of hydroids that flourish on almost any firm substrate in the Bay. Some hydroids look like long, delicately branching plumes; others resemble soft, bushy clumps or downy coatings over wood or stone. Often, the best place to catch a good glimpse of hydroids is on the shady side of a piling or large rock, just below the low-water mark, where the tentacle of each hydranth (the feeding end of each zooid) of the hydroid colony can be seen outlined against the darker background.

Hydroids are closely related to sea anemones. In a sense they are a collection of minuscule sea anemones, most of which are encased in a chitinous sheath (the periderm) with each individual connected to another. The hydranth of each zooid emerges to feed with tentacles armed with stinging cells, used to stun their prey. The food is then passed into the mouth and down to the interconnecting branches, where it is digested and passed throughout the colony. Some hydroids have elaborate branching stems, other simple stalks. Some have elaborate cuplike "homes" called hydrotheca which protect each hydranth. Most hydroids are attached to the substrate by a stolon network, from which arise the stems and branches of the periderm. Some hydroids get to be quite large, growing to heights of a foot or more, but most species range from just a few inches tall to only a fraction of an inch. These small species are almost impossible for the amateur to identify. Identification of larger species can also be difficult, but certain common types are quite recognizable when such characteristics as branching, general color, and shape are closely observed. Certain bushy bryozoans look much like hydroids, so study the drawings of bryozoans in this book carefully in order to learn to distinguish them from hydroids.

A hand lens will show up the intricate details of structure that are not obvious to the naked eye, verifying—if nothing else—that you have an animal at hand, rather than a seaweed. Like sea anemones, hydrants pop back into their protective sheaths as soon as they are taken out of water. If you place a hydroid-covered pebble or shell into sea water, you will soon be rewarded with the subtle beauty of hydroids, as each little hydranth emerges and unfolds its tentacles.

At certain times of the year many hydroids have noticeable globular sacs (gonophores) attached to the stems. Gonophores produce tiny miniature jellyfish called hydromedusae, which, when mature, are released and float free in the water. Translucent and usually less than one-quarter inch in diameter, hydromedusae are quite inconspicuous in the water. They produce planula larvae, which ultimately settle onto a hard surface, attach, and begin to grow stolons, stalks, and zooids and eventually develop into full hydroid colonies. Some hydroids do not produce hydromedusae (in fact, not all hydroids produce a polyp stage, but exist only as free-floating medusae) but do produce specialized actinula larvae, which, when released, float for only a very short time, soon dropping out of the water to crawl over the surface by means of long tentacular "feet." When a suitable spot is found they attach and begin to grow a new hydroid colony.

Snail fur, *Hydractinia echinata,* so often seen covering shark eye and mudsnail shells occupied by hermit crabs, also spreads over wood pilings and over barnacle

Identifying characteristics of some common hydroids of the Chesapeake Bay: (1) has a perisac ringed at base of branches and pedicals, a single whorl of 8 to 10 tentacles, and gonophores borne on pedicals; (2) has a perisac ringed at base of branches and pedicals and tentacles scattered over head of hydranth; (3) has a perisac ringed at base of and above the origin of each branch and pedical and 2 to 3 whorls of short knobbed tentacles above a ring of longer tentacles; (4) has hydranths within hydrothecae closely aligned to and alternating along stems and large bulbous gonophores arising from base of hydrothecae; (5) has feeding and reproductive zooids without a perisac; (6) has a top ring of 30 smaller tentacles, a bottom ring of 25 longer tentacles, and small gonophores closely attached to hydranth between; and (7) has two rings of 20 to 24 tentacles each with clusters of gonophores often hanging below tentacles.

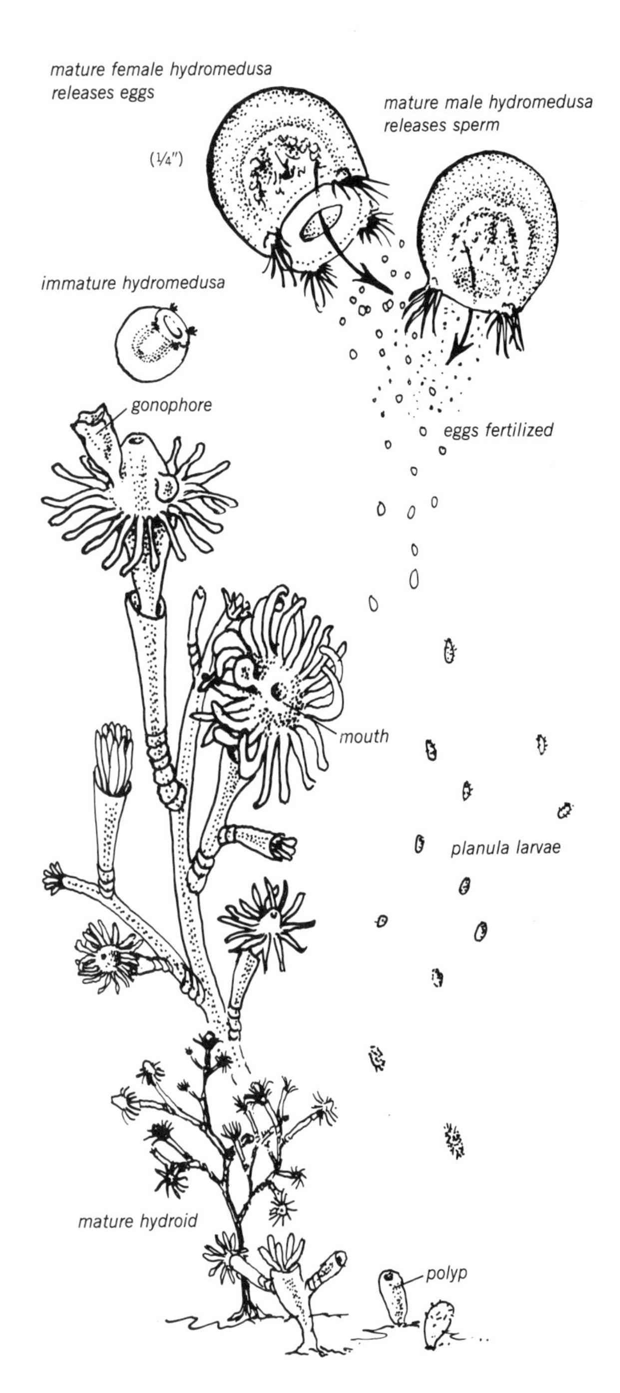

and oyster shells. You should be able to feel the roughness of the spines if you run your finger over the telltale dark brown crusts. The feeding hydranths, when viewed underwater, create an illusion of a soft pinkish cloud covering the surface. Snail fur is restricted to high-salinity regions of the lower Bay and lower Virginia tributaries.

Tubularian hydroids are among the more distinctive and easily recognizable hydroids. Their long whitish stalks are tipped with bright pink hydranths, which color the ends of the stalks even when withdrawn. The stalks are long rods, often growing four to six inches tall in tangled masses, but they also occur in simple colonies consisting of just a few stalks. Two species occur in the Chesapeake Bay. The pink-hearted hydroid, *Ectopleura crocea,* is apparently limited to areas near the mouth of the Bay, where it is the most conspicuous hydroid in summertime. The more widespread species, the tube hydroid, *Ectopleura dumortieri,* occurs well into the Bay to mid-level salinities (lower Zone 2). It flourishes in the spring and the autumn, often dying off during the summer. Although these species look almost exactly alike, the pink-hearted hydroid produces actinula larvae from grapelike clusters of gonophores, whereas the tube hydroid produces medusae from smaller gonophores.

Rope grass, *Garveia franciscana,* is a troublesome fouling organism that clogs watermen's nets and traps during most summers. This hydroid is an abundant species of the mid-Bay, where it forms thick, brownish, matted growths over almost anything in the water. The individual hydroid colonies are only a few inches high; however, colonies often twist themselves into ropes a foot or more long, hence their colloquial name. The freshwater hydroid, *Cordylophora caspia,* grows profusely, forming bushy whitish treelike growths an inch or two high. This is the only hydroid in the Chesapeake Bay which proliferates in lower-salinity waters and in tidal freshwaters (Zone 1 and upper Zone 2).

The life cycle of a typical hydroid with an asexual hydroid stage and a sexual medusa stage. Maturing gonophores budding from a parent hydroid eventually release free-floating hydromedusae, which in turn mature into males and females. The hydromedusae release eggs and sperm into the water, where fertilization occurs. The resulting planula larvae eventually settle, attach, and develop into tiny polyps, which divide and branch into full-grown hydroids.

The graceful feather hydroid, *Pennaria disticha,* is another distinctive hydroid abundant in the lower Bay and rivers (Zone 3) attached to pilings, ropes, and many other substrates. It is also common in eelgrass beds. It looks very much like a dark brown feathery seaweed about five to six inches long. A careful examination of the ends of each tiny branchlet will reveal a translucent tip with pink or vermilion hydranths at the open end.

Another distinctive hydroid of the lower Bay is a garland hydroid, *Sertularia cupressina,* called white hair by the watermen, who gather great unwanted bundles in their dredges and nets during winter and springtime. White hair differs from most other Bay hydroid species in that the hydranth capsules arise directly from the stems, giving them a distinctive saw-toothed appearance. Plumes of white hair grow to be quite large, extending to 10 inches or more. White hair dies out in warmer weather. It prefers deeper waters over sandy and shell-covered bottoms, where it attaches to almost anything—oysters, mussels, hard clams, spider crabs, and worm cases. Consequently, it will not be found by many shore visitors, even though it comprises perhaps the greatest biomass of invertebrate fauna in the lower Chesapeake Bay in winter.

WHIP CORAL

The beautiful deep purple or violet whip coral, or sea whip, *Leptogorgia virgulata,* is a spectacular resident of

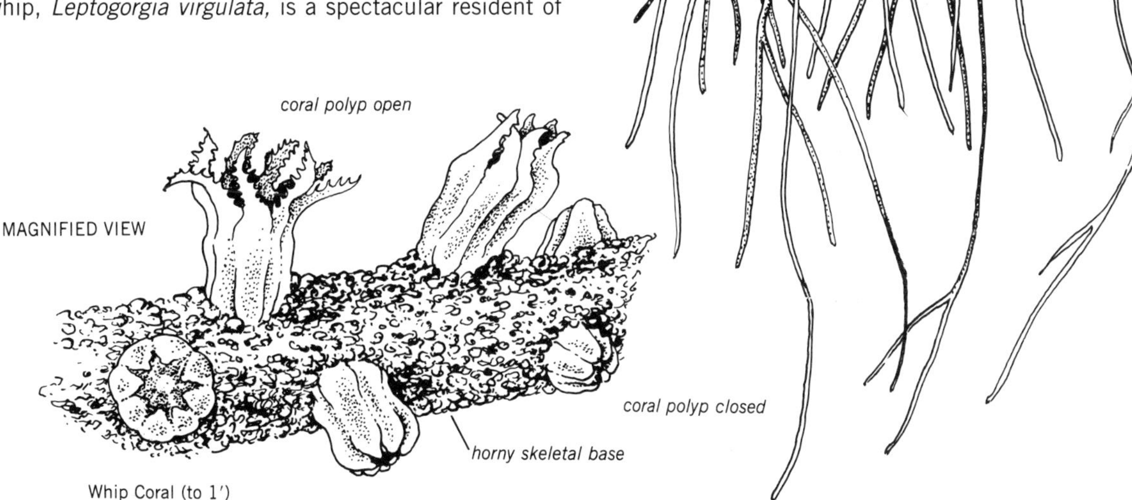

Whip Coral (to 1′)
Leptogorgia virgulata

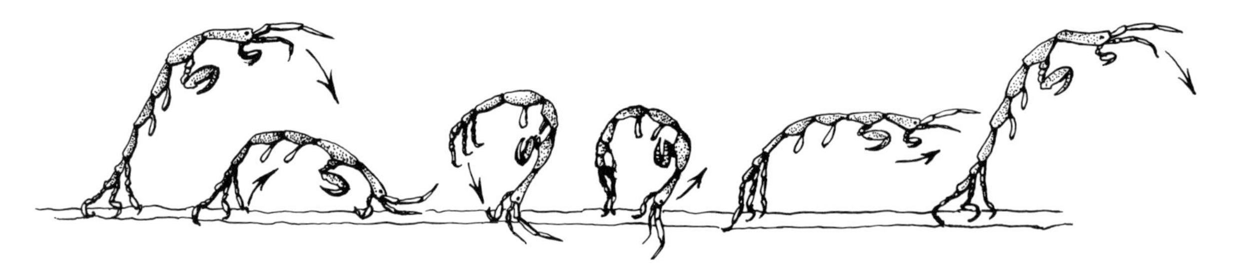

A skeleton shrimp sways up and down as it moves along like an inchworm.

higher-salinity regions of the Chesapeake Bay. It may be discovered hanging off a bulkhead or pier piling, but it also occurs in deeper waters. Its occurrence in the Bay is irregular; there will be long periods when the whip coral is quite abundant, followed by periods when it just about disappears. Closely related to sea anemones and hydroids, the whip coral is actually a colony of polyps connected by a horny skeleton. The eight transparent white tentacles of the individual polyps have somewhat saw-toothed edges, and when the polyps emerge they appear as white dots against the brightly colored skeleton rods. Whip corals within the Bay tributaries are usually purple, many-branched, and shorter—one to two feet—than those at the mouth of the Bay, which tend to be yellowish tan with single long, whiplike branches.

SKELETON SHRIMPS

Perhaps a hydroid you have collected appears to be writhing. The movement does not originate with the hydroid, but in a dense population of skinny, gangly amphipods, *Caprella* spp., called skeleton shrimps, or caprellids. Skeleton shrimps attach to the hydroid by means of hooked rear legs; their sticklike bodies constantly dip and sway side to side or back and forth. Their forward legs are free, folded as if in prayer, but they are actually poised to pounce on any passing copepod, algal cell, or bit of detritus that might come their way. Their peculiar shape camouflages them perfectly against hydroids, and they are also capable of changing color to match their background. It is truly a delight to watch a skeleton shrimp climb about the hydroid branches, pitching along like an inchworm. It bends down to clasp the branch with its front legs, releases its hind legs and moves them forward to meet its front legs, bending its body in a loop, and then releases its front legs and repeats the whole process. The females carry their eggs in large pouches, which droop like transparent raindrops from their midsection. The pouches are large enough

Two skeleton shrimps, *Caprella* sp. (to 1 inch), and a long-necked sea spider *Callipallene brevirostris* (body to 1/16 inch), are well camouflaged among the feathery branches of a hydroid.

to be seen with the unaided eye. Skeleton shrimps are also found on seaweeds, sponges, and other epifauna in higher-salinity regions of the Bay (Zone 2 and lower Zone 3).

Another bizarre creature found among hydroids is the long-necked sea spider, *Callipallene brevirostris.* It is an arthropod, but it is far removed evolutionarily from the terrestrial spider. Sea spiders have such tiny bodies that they look like no more than a collection of long, awkward legs attached together. The tip of each leg has curved hooks, which gives them a firm hold as they crawl over hydroids and other epifaunal growths. The long-necked sea spider is not as easy to find as the skeleton shrimps, being more secretive and better camouflaged against its surroundings. Once seen, however, it is unmistakable.

BRYOZOANS

Bryozoans are often referred to as moss animals. This is an appropriate name for certain species, which look exactly like brown marine mosses, but only a few bryozoans

BRYOZOANS OF THE CHESAPEAKE BAY

1. Cushion Moss Bryozoan, *Victorella pavida* (to ½″ high)
2. Hair, *Anguinella palmata* (to 3″ high)
3. Spiral Bryozoan, *Amathia vidovici* (to 2″ high)
4. Jointed-Tube Bryozoan, *Crisia eburnea* (to ½″ high)
5. Dead Man's Fingers, *Alcyonidium verrilli* (colonies to 1′ or more)
6. Freshwater Bryozoan, *Pectinatella* sp. (colonies to 1′ or more)

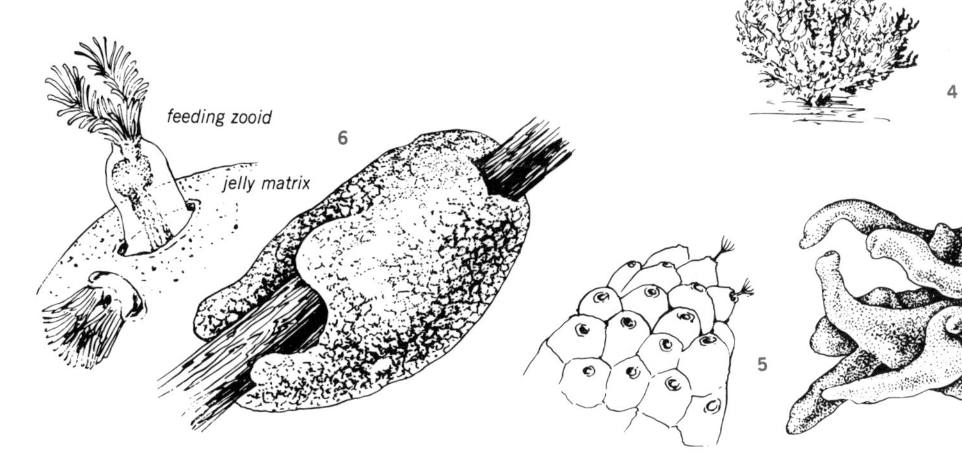

Magnified views show details of zooecia.

can be characterized as mosslike. This is a diverse group of colonial animals. Each colony is a collection of tiny individual zooecia, which are encased in variously shaped and structured "houses." They are as varied in form as the hydroids, making identification difficult for the uninitiated. The morphology of bryozoans is quite different from that of hydroids. For example, their tentacles are fringed with cilia and form a horseshoe shape over the mouth. As animals they are far more advanced than hydroids, but the differences are obscure without microscopic study. Knowledge of the general shape and characteristic growth patterns of the bryozoans that are generally found in the Chesapeake Bay will enable most careful observers to identify them. Most, although not all, bryozoans occur in one of two basic forms—either bushy colonies or calcareous encrustations. The bushy types take on various appearances, some looking like moss, seaweed, or hydroids. The encrustations form delicate, lacy patterns over almost any hard surface and grow in geometric shapes, like the cobblestone paving of a Mediterranean plaza.

The cushion moss bryozoan, *Victorella pavida,* is widespread in the Bay, but it is most abundant in the mid-Bay areas—a truly estuarine animal. It attaches to almost anything firm, forming soft, brownish, matted coatings over pilings, stakes, crab floats, shells, or barnacles and sometimes completely obscuring barnacles, except for their valves. The bottom of a boat in midsummer, in the mid-Bay region, can be completely cushioned within three or four weeks. Cushion moss is attached by a creeping basal stolon network. A hand lens will reveal the structure of the animal, which is quite different from that of any hydroid.

Another bushy bryozoan, called hair by watermen, is a common species that grows profusely on crab traps. Hair, *Anguinella palmata,* is often mistaken for a brown seaweed, even by crabbers who constantly handle it, because it is usually covered with soft brown silt, which obscures the long intertwining capsules of the zooecia. Hair does not survive in very low salinity waters. It usually grows to an inch or so, but sometimes reaches two to three inches.

The spiral bryozoan, *Amathia vidovici,* which inhabits the mid and lower Bay, looks like a soft, tufted, bushy growth with dark-colored zooecia winding around the

The lacy crust bryozoan, *Conopeum tenuissimum,* spreads and grows over other animals attached to the substrate, including barnacles, sea squirts, cushion moss bryozoan, whip mud worms, and slender tube-builder amphipods. Jelly spirals of sea slug eggs are laid on top while a clamworm tunnels its way through the overlying encrustations of the bryozoan colony.

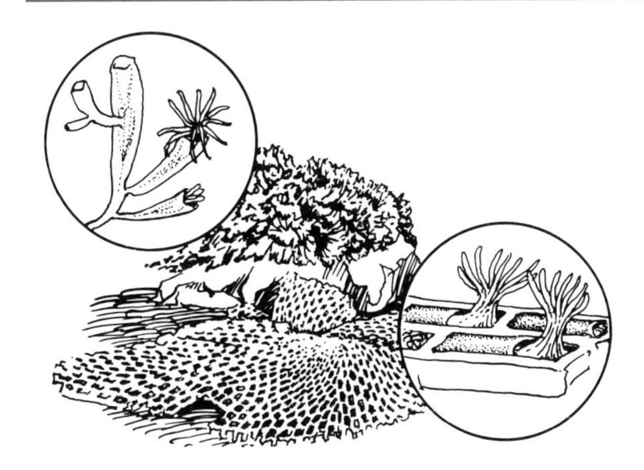

Individual zooids of a bushy bryozoan such as cushion moss and an encrusting type bryozoan differ in the construction of their homes (the zooicium), but each feeds similarly, capturing food with protruding tentacles.

stems at every branching node. Its structure is unique, and its pattern of growth can be discerned quite readily with a hand lens. Colonies grow to be two inches high, or higher, and attach to a variety of substrates, often among algae or hydroids.

In the saltier part of the Bay, small, twiggy-looking tufts of the jointed-tube bryozoan, *Crisia eburnea,* can occasionally be discovered attached to other bryozoans, seaweeds, or hydroids. They are less than an inch tall, but their white color and jointed appearance is eye-catching. Each "joint" separates a double row of cylindrical calcified tubes of the zooecia; the joints near the base are generally dark brown, the stems yellowish, and the tips transparent. Each tufted colony is attached by a single stem.

The yellowish, fleshy dead man's fingers, *Alcyonidium verrilli,* may at first appear to be the yellowish phase of a redbeard sponge because of its growth pattern, size, and color. However, this unusual bryozoan is very smooth and rubbery to the touch, without any of the grainy, sandpapery texture of sponges. When examined out of water, its surface is finely puckered, each indentation marking the site of an inverted zooecia. When observed closely underwater, the tiny tentacled heads can be seen emerging to feed.

Another rubbery bryozoan is often found in the tidal freshwater areas of rivers and streams of the Bay system. The freshwater bryozoan, *Pectinatella* sp., grows in huge fleshy masses, two feet or more in diameter. It attaches to sticks or wooden stakes and to plants along the shore. It looks like a yellowish mass of jelly, the surface mosaicked with dark brown radial patterns. Each brown streak is the site of a line of zooecia embedded in the jellylike matrix. Unlike the marine bryozoans, in which each zooid is a completely separate animal, the body cavity of all the zooids of this freshwater bryozoan are continuous in the matrix. The freshwater bryozoan dies off in the fall, leaving behind small encapsulated balls of tissue, called statoblasts, which tolerate freezing and other harsh winter conditions and emerge in the spring to start a new colony. These statoblasts are often transported on the muddy feet of birds and small mammals to other ponds and streams.

As confusing as the bushy group of bryozoans can be for the expert as well as the amateur, encrusting bryozoans are unmistakable, for nothing similar grows in the Bay. However, to identify a species you must carefully study the overall shape of the individual calcified boxes within which each zooid lives. The lacy crust bryozoan, *Conopeum tenuissimum,* is by far the most abundant and typically estuarine species found throughout Zone 2. Its delicate colonies cover just about any suitable substrate—pilings, rocks, shells, and aquatic plants. It can completely cover oyster shells and prevent oyster spat from setting. The coffin box bryozoan, *Membranipora tenuis,* is another common Chesapeake Bay species. Colonies expand so rapidly that they often overgrow and smother lacy crust colonies even though lacy crusts are many times more abundant than coffin boxes. Both species grow luxuriously, and when they run out of open, flat surface they develop uplifted frills and curls. Barnacles and sea squirts are frequently covered with coffin box bryozoans, but not with lacy crust bryozoans.

To appreciate the meticulous beauty of encrusting bryozoans, turn a colonized shell or stone to the light until shadows etch the design. Each colony originates from a single planktonic larva, the aptly named ancestrula, which, after settling, immediately begins to form two boxes, one at each end and each containing zooecia. Each new box develops new buds on the sides and ends and grows in a radial pattern until a circular colony results. As adjacent colonies expand and meet one another they may abruptly change their direction of growth. Zooids mature within a few weeks, at which time they produce and release larvae, which repeat the cycle at another site.

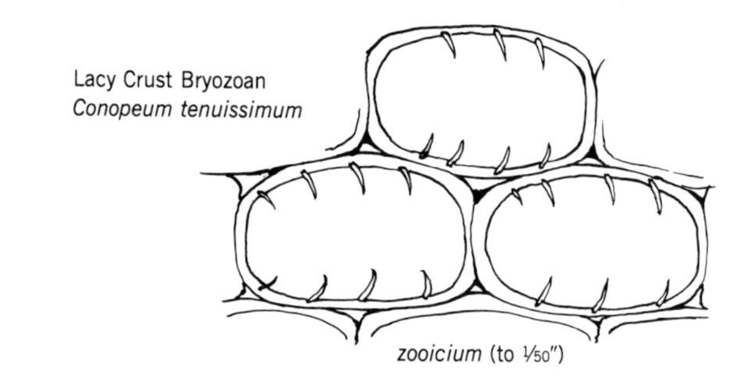

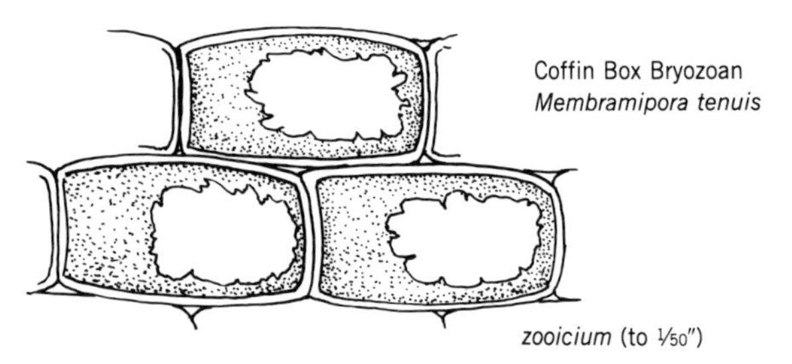

Detailed view of two common encrusting bryozoans. The tiny spines on the zooicium of the lacy crust bryozoan vary in number and are at times absent. The size of the top opening on the zooicium of the coffin box bryozoan is highly variable, sometimes small and at other times so large as to leave the top surface of the zooicium almost entirely uncovered.

TERMITES OF THE SEA

At times, a perfectly sound-looking piling will suddenly collapse, its interior completely riddled with and undermined by shipworms. The exterior of the wood may have shown no evidence of the thriving community of shipworms within. Shipworms flourish in submerged wooden structures in oceans and seas throughout the world. They have been known for centuries as termites of the sea and have always been a bane to mankind. Records show that as early as 412 B.C., bottoms of ships were covered with a mixture of arsenic, sulphur, and oil to prevent shipworm infestations. Creosoted pilings, antifouling paints, and the widespread use of fiberglass have greatly reduced shipworm destruction; nevertheless, there are still enough untreated or poorly treated wooden structures and enough wooden flotsam and jetsam available for shipworms to survive quite well.

There are hundreds of species of shipworms throughout the world, but only one is common in the Chesapeake Bay—Gould's shipworm, *Bankia gouldi,* which was first described from specimens retrieved from Norfolk Harbor. Shipworms are found throughout the Bay into brackish waters, but they do not survive in low-salinity or tidal freshwater.

Shipworms are mollusks, not worms, and are closely related to angel wings and other boring clams that inhabit hard mud bottoms. A shipworm is a bivalve with two small shells, each about one-third inch long, located at one end of their long, soft, wormlike bodies, which are protected and supported by the surrounding wood. The front edge of each shell has a series of sharply filed toothed ridges. Boring is accomplished by rotating the shells against the head of the tunnel and rasping off tiny bits of wood. If you break open a piece of driftwood, you can often see white porcelain-lined tunnels, about the size of a pencil, following the grain of the wood. Some wood is packed solid with tunnels, one right upon the other. The tunnels never intersect. If the wood has only recently washed ashore, you may be able to

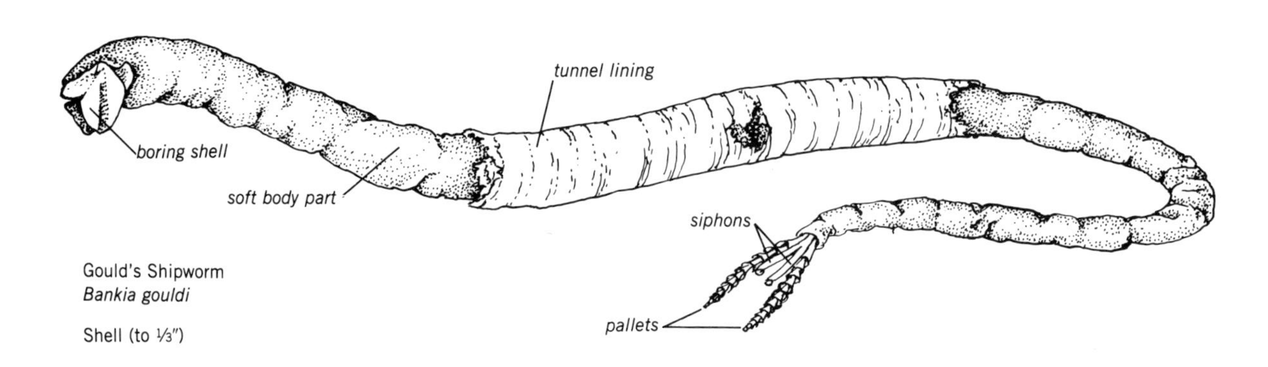

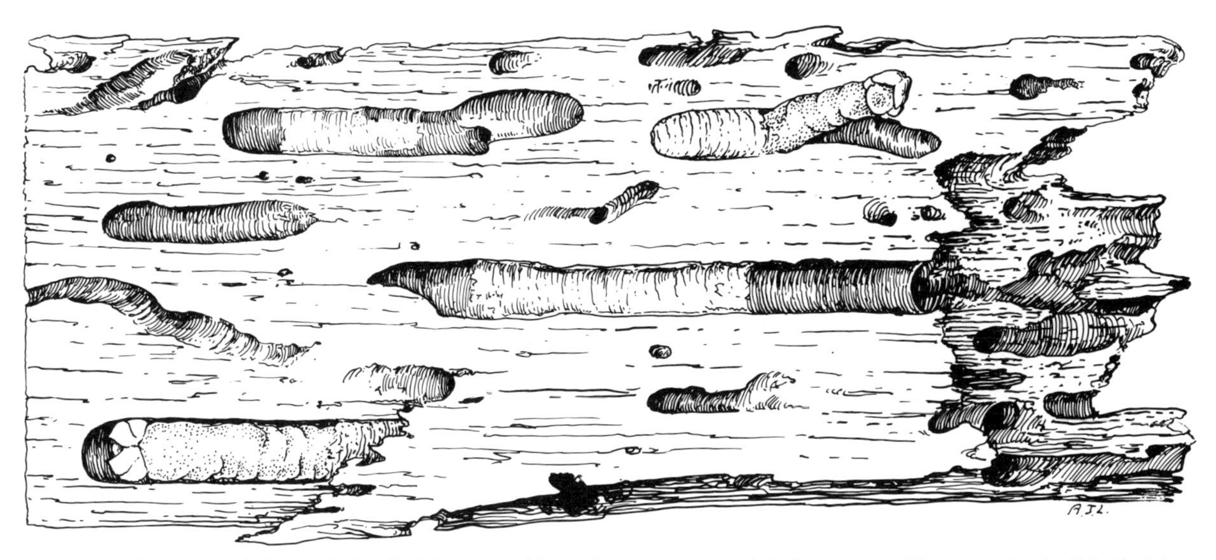

A fragment of shipworm-infested timber freshly retrieved from the water may contain live worms still encapsulated within their burrows. Generally, pieces of infested wood found on the beach will show only the characteristic network of burrows.

pry out two small shells. At the head of each tunnel two strange-looking calcified "shields" may be discovered close to where a tunnel breaks to the surface of the wood.

Like other clams, shipworms have incurrent and excurrent siphons. In shipworms they are positioned at the end without shells. The tunnel at this end of the shipworm is opened to the waters through a very small hole. The two siphons extend out into the water through this hole, but when disturbed, the siphons are retracted and the hole is plugged by the shield-shaped pallets. The pallets of Gould's shipworm are intricately constructed of layers of chitinous cones, much like a stack of sugar cones at an ice-cream parlor. The exterior openings are so minute they would ordinarily go unnoticed, yet 10 or more can be located within a square inch of heavily infested wood. It would be quite a sight to observe a mass of shipworm siphons underwater, as they writhe and twist like a bunch of snakes. Every so often you would see narrow rodlike pellets forcefully ejected from the siphons. These pellets are the remains of the tiny wood filings removed by the rasping shells. The wood was eaten, some parts apparently digested, and then passed through the shipworm and out the excurrent siphon. Shipworms, however, do not depend solely on wood, but also feed on planktonic food brought in through the incurrent siphon.

Like most mollusks, shipworms have planktonic veliger larvae, which remain in the water column for about two or three weeks before settling. If they are lucky (and many are not) they land on untreated wood. A larva walks around by means of a small foot, exploring nooks and crannies of the surface and eventually secreting a byssus thread for attachment. The larval shells then metamorphose into small rasping tools, and the baby shipworm tunnels into the wood and develops into an adult. The original tiny entrance hole remains as its only connection with the outside world.

AERIAL GLEANERS

Birds, especially large sea- and shorebirds, are usually the most evident signs of animal life along the shoreline. They sometimes gather in large, mixed flocks; they probe and soar, searching for food. They sun themselves on piers and navigational structures, frequently preening their plumage to rearrange and oil their feathers, which are so

Gulls alert on a jetty. A great black-backed gull, *Larus marinus* (30 inches), overlooks all at left; a juvenile laughing gull, *Larus atricilla* (16½ inches), and an adult in breeding plumage stand in the middle; a laughing gull in its winter plumage crouches behind a rock with a ring-billed gull, *Larus delawarensis* (17½ inches), in breeding plumage in the background.

vital for flight and maintenance of body temperature. Most of their energy, however, is devoted to hunting for food for their nestlings and for themselves. Some species, like gulls, will eat practically anything: insects, garbage, small mammals, bird eggs and hatchlings, berries, grain, fish, and various available invertebrates. Barn swallows eat prodigious numbers of insects while on the wing. Other birds, such as cormorants, brown pelicans, and ospreys, generally eat only fish. Their calls range from the familiar raucous call of a gull to the high-pitched chirping of an osprey or the twittering of swallows. Brown pelicans and cormorants are usually silent.

Gulls (some refer to them as "seagulls," which is not considered proper by ornithologists), are typically found around sea walls, bulkheads, and pilings; in fact, any structure which gives them a resting spot and a place for observation is a place where gulls will be found. Don't overlook bridges—gulls will often hover, seemingly without effort, over these structures. They utilize the rising thermals from the heated pavement of the roadway to maintain their position, which is a clever way of conserving energy while searching for one more scrap of food. Gulls are generally colonial nesters and are often crowded closely together on the nesting grounds. They maintain their gregariousness beyond the nesting area and are often found in large numbers, apparently displaying little or no inclination to territoriality.

The laughing gull, *Larus atricilla,* aptly named after its call, which sounds like a series of loud "ha-haahs," is the most common and abundant gull in the Chesapeake Bay during summer. This is the gull that follows behind the farmer as he plows the ground in the spring. There may be hundreds of laughing gulls hovering and probing for insect grubs in the newly plowed soil. Laughing gulls take three years to attain their breeding plumage: a black hood, white underparts, and slate or pale gray wings and back. The bill is deep red and the legs and feet are black. They lose the black hood in the winter, retaining only some gray around the eyes and the back of the head. Bonaparte's gull is similar to the laughing gull in that the adult is also hooded only in the summer. This small, ternlike gull is sometimes seen in small, discrete groups in the Chesapeake Bay during winter. Laughing gulls nest in the mid-Atlantic area from late May to early August; they dig shallow scrapes, sometimes lined with grass or small twigs, on the sand between dunes or in salt marshes. They lay an average of three olive-buff eggs, which hatch in about 20 days. Like most gulls, they are colonial nesters; sometimes they even nest with terns and black skimmers. They winter as far south as Peru and the Amazon Basin.

The laughing gull is the smallest of the four most common species of gulls in the Chesapeake Bay area. The next largest is the ring-billed gull, followed by the herring gull and the largest gull of all, the great black-backed gull, which has a maximum length of some 30 inches.

The ring-billed gull, *Larus delawarensis,* is also a three-year gull and is the most abundant winter gull in the Bay. In early spring, it may be seen intermingled with laughing gulls in newly plowed fields. The ring-billed gull has a characteristic black ring near the tip of its yellow bill. The three-year-old adult in breeding plumage has the typical gull coloration of a pale gray back or mantle (some call it silvery gray) and a white head and underparts. Its legs are yellowish. Ring-billed gulls are easily confused with herring gulls because of their color and their similar loud, mewing call; but the black ring on the bill and their smaller size will easily distinguish them, at least in the breeding phase. In the summer, ring-billed gulls are more abundant around freshwater areas of rivers and lakes. In the winter, they are quite common throughout the Chesapeake Bay and coastal regions. There are always a few around the Bay, however, during the summer. Ring-billed gulls, typical of most gulls, are not finicky eaters; they feed on fish, insects, worms, bird eggs, small rodents, and garbage.

The great black-backed gull, *Larus marinus,* with its massive yellow bill and huge wingspan, which sometimes reaches five feet, is an impressive bird. Great black-backed gulls, like herring gulls, require four years to reach sexual maturity. Their breeding plumage at that time cannot be mistaken for that of any other gull in the area. The upper part of their wings is charcoal black, their underparts are white, and they have pink or flesh-colored legs. The call of a great black-back is similar to the herring gull's, except that its "keeow" is somewhat deeper. The great black-backed gull also has a red spot on its yellow bill which chicks peck at to elicit food from the adult bird.

Great black-backed gulls, like herring gulls, also drop clams, oysters, and other mollusks, while on the wing, onto bridges, roadways, and piers to crack open their hard shells. Broken shells often litter the waterfront and are certain evidence that herring gulls or black-backed gulls have been feeding on the mud flats and oyster bars nearby. Great black-backed gulls are voracious feeders, even feasting upon significant numbers of eggs and chicks of other bird species in addition to their usual food. Great black-backed gulls often nest with herring gulls in large colonies north of the Chesapeake Bay. They are common throughout the Bay, particularly in the mid- and lower Bay, and their range appears to be expanding.

The barn swallow, *Hirundo rustica,* may be the most familiar of all the swallows. These sparrow-sized birds have deeply forked tails. Their upper parts are steely blue; they have creamy buff underparts and a cinnamon or rust-colored throat and forehead. The females and immature birds are paler, but are similarly marked. They have large, gaping mouths, sometimes referred to as frog-shaped, which are well adapted for catching insects in flight. These wonderfully agile flyers course through the air in a continual quest for insects. Suddenly they veer and head for the water, flying horizontally just a few inches from the surface; just as suddenly they are wheeling overhead, twittering and squeaking in a bubbly sort of way.

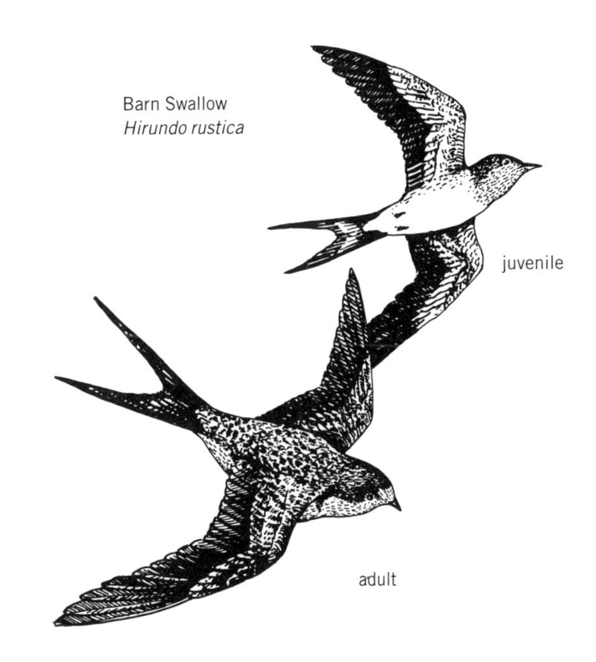

Barn Swallow
Hirundo rustica

Barn swallows are common throughout the Bay and its tributaries beginning in early April, when they arrive from their wintering grounds in the Caribbean and South America, until they depart in late summer or early fall. They construct nests of mud pellets reinforced with plant materials and lined with feathers. Swallows have tiny, weak feet and are unable to carry mud in their claws. However, it is interesting to watch the males land on soft mud, often at the water's edge, scoop up mud in their capacious beaks, and fly to the nest, where the female works the mud and plant matter into pellets. Ultimately a nest is formed and securely cemented to the side of a building, or to the underside of a pier or bridge. The female lays an average of four brown spotted white eggs and incubates them for two to two and one-half weeks; three weeks later, the young are ready to begin flying. Barn swallows are creatures of open areas, such as over water, fresh and saltwater marshes, grain fields, and pastures, and are generally not seen in forested areas. Like gulls, they often follow behind a farmer's tractor and feed on the clouds of stirred-up insects.

The double-crested cormorant, *Phalacrocorax auritus,* is a goose-sized bird with a long neck and a hooked bill. The adults are black and have a greenish sheen in the right lighting. Young birds are brown above and buff-colored below. The crests for which these birds are named are so inconspicuous that they are generally not seen. The throat pouch, or gular membrane, is orange the year around; when the bird flutters this membrane, it causes air to be forced through the lungs and a network of sacs, which helps regulate the cormorant's body temperature.

Double-crested cormorants are the most common and

A brown pelican, *Pelecanus occidentalis* (48 inches), with the light colored head and neck of a nonbreeding adult rests on a dock while a breeding adult with darker neck glides gracefully over the water. Double-crested cormorants, *Phalacrocorax auritus* (32 inches), sun themselves with heads upright; one spreads its wings to dry.

abundant cormorants along the Atlantic coast. They often fly in V-shaped formations like Canada geese but with their necks characteristically kinked in flight. Look for them standing upright or with their wings outstretched on navigational aids, and on pilings and piers. Their plumage is not water-repellent, so they often bask in the sun, holding their wings open to aid in drying their feathers and in regulating their body temperature. Cormorants swim low in the water and look very much like loons; cormorants, however, swim with their bills tipped up. This is an easy way to separate a cormorant from a loon, especially from a distance when the bird in question is silhouetted. Cormorants are gregarious birds and will hunt in small groups, flying low over the water in loose skeins searching for fish, their primary prey. When they locate a school of fish, they land on the water and plunge head first below the surface of the water, where they swim by using their wings and short, webbed feet. Cormorants' eyes are adapted for both aerial and underwater vision. They are almost entirely silent. Double-crested cormorants are abundant and common throughout the Chesapeake Bay and along the Atlantic coast and also inland on lakes and rivers.

Another fish eater, the brown pelican, *Pelecanus occidentalus,* looks like no other bird in the Bay region. Its massive bill, great, elastic throat pouch, heavy, large body, and expressive eyes allow this bird to be immediately identified without question. The breeding adult is chestnut brown on the back of the neck, with yellow on its forehead and at the base of the neck. Juveniles are grayish brown. The brown pelican characteristically rests its bill on its breast when sunning on a piling or pier. These short-legged, rather ponderous birds, with wingspans exceeding six feet, are aerial acrobats. They often fly in V formation or in a long line, synchronizing their wingbeats with one another and then, alternately, gliding. When brown pelicans fly against the wind they tend to fly close to the surface of the water, their wing tips almost touching. Pelicans soar gracefully high above the water with their heads drawn back toward their bodies—then, suddenly, one or more will veer and plunge dive into the water with wings partly folded and head and neck retracted. They quickly bob to the surface, as they are quite buoyant, and point their bills downward to allow as much as 10 quarts of water to stream quickly from spaces between their bills. They then point their bills upward to swallow their catch. They feed greedily on schooling anchovies, silversides, and finger mullet. It has been estimated that a brown pelican can eat well over one hundred pounds of fish in its first nine weeks of life.

Brown pelicans range all along the southeast Atlantic and Gulf of Mexico coasts. Until recent years they were considered a rarity in the Chesapeake Bay and along the coastal waters of Maryland and Delaware. Pelicans are commonly seen in the mid- and lower Chesapeake Bay during the summer. Look for them sunning on the rock jetty on the bayside of Smith Island or soaring gracefully over Smith Point at the mouth of the Potomac River.

Pelicans, like ospreys and bald eagles, went into serious decline in the mid-1950s and early 1960s due to the wide use of DDT and other forms of pesticides. The pesticides interrupted the physiology of the birds, which resulted in thin-shelled eggs which were easily broken by the adult birds. Rachel Carson's book *Silent Spring,* published in 1962, alerted the world to the dangers of pesticides in the food web when she cited the research of her coworkers demonstrating that DDT was concentrated in the organs and flesh of fish, and described the consequences of that insidious and pervasive poison. Society heeded the warnings of Carson and others, and now DDT and similar pesticides no longer seriously contaminate fish and birds. In fact, brown pelicans, ospreys, and bald eagles have significantly increased in abundance since the 1960s. The brown pelican regularly nests near Ocean City, Maryland, and along the Delaware coast.

The osprey, *Pandion haliaetus,* is a fish hawk and is widely distributed throughout the world on lakes, rivers, and estuaries, and along seacoasts. There are five major breeding areas in the United States: the Atlantic Coast, Florida and the Gulf Coast, the Pacific Northwest, the western interior, and the Great Lakes; the osprey populations all in all total around 8,000 nesting pairs. In the Chesapeake Bay region there are approximately 2,000 nesting pairs of ospreys, about 25 percent of all the ospreys in the contiguous United States. They are common throughout the Bay and its tributaries. Ospreys generally mate for life after the age of three years. They are migratory in this area and return to the Chesapeake around mid-March. The male and female usually return to the same nesting site and either rebuild the existing nest or construct a new one fashioned from branches snapped from trees. Cornstalks, shoreline debris, and even broken fishing rods are often woven into their nests, which can get to be as large as four to five feet in diameter or more. Ospreys build their nests

An osprey, *Pandion haliaetus* (25 inches), feeds its young. The chicks are settled well down in a large twig nest constructed within the pipe rim of a navigational marker.

on tree snags, artificial nesting platforms erected by nearby landowners, the tops of electric poles, and navigational aids throughout the Bay. Soon after the nest is readied, the female lays from two to four whitish-colored eggs marked with brown or olive. The eggs hatch in approximately 38–40 days, and some 50 days later the young are ready to try their wings. Along about late August the ospreys start to head south to the Caribbean and South America, where they spend the winter. By the time the first primal honks of returning Canada geese are heard in early September, the ospreys have all but vanished.

The osprey belongs to a group of birds collectively referred to as raptors, which means "to seize"; the group includes eagles, hawks, and vultures. Some ornithologists include owls in this group as well. Ospreys are large, hawklike birds, chocolate brown above, white below, and with a distinct dark eye stripe across their white heads. The dark wrist patches at the angle of the underwings are a good identification mark as they fly overhead. The breast feathers of the female are often tipped in brown, which gives the appearance of a ruffled necklace. The young look very much like the adults except that their feathers are edged in buff, making them look speckled, and their eyes are reddish-orange rather than the yellow of the adults. The piercing, chirping call of adult birds is frequently heard, particularly when they are wheeling high in the air searching for their next meal.

Ospreys feed almost exclusively on medium-sized fish about 6 to 10 inches long. Biologists estimate that it requires a minimum of three pounds, and by some accounts as much as six pounds, of fish per day to feed the nesting female and three young. Ospreys hunt by soaring high over the water and searching for schooling fish near the surface of the water. When an osprey sights its prey, it will often hover aloft by rapidly beating its wings to maintain position over the fish—then suddenly it will dive with folded wings and plunge talon first into the water. Just as suddenly it seems to explode out of the water. More than likely, it will have made a successful kill; and with slow, labored, deep wingbeats, it heads for the nest, carrying the fish head first to streamline its flight. An osprey will also occasionally fly low over the water on a horizontal path, skim the water's surface, seize a fish in its talons, and head upwind for a controlled landing at the nest. If ospreys fish in the upper Bay waters and tributaries where the water is only slightly salty or even fresh, they will likely feed on catfish and other freshwater species; in higher salinity waters, they feed on menhaden and other herrings, mullet, and several other species of fish.

● The roster of animals found on piers, rocks, and jetties includes some of the most widely dispersed forms of life in the Chesapeake Bay, forms that occur in all habitats covered in this book except sand beaches. Even here their remains may be seen—a piece of driftwood covered with dead barnacle shells and traceries of lacy crust bryzoans, or a clump of faded but still recognizable redbeard sponge. It is important to remember that this "habitat" is not always a discrete location, such as a wharf site or a wall of riprap, but is often merely a stone, a shell, a tin can, or a buoy, covered with two, three, or more types of fauna. Each object is a mini-habitat, a small replica of the larger substrate sites.

SHALLOW WATERS

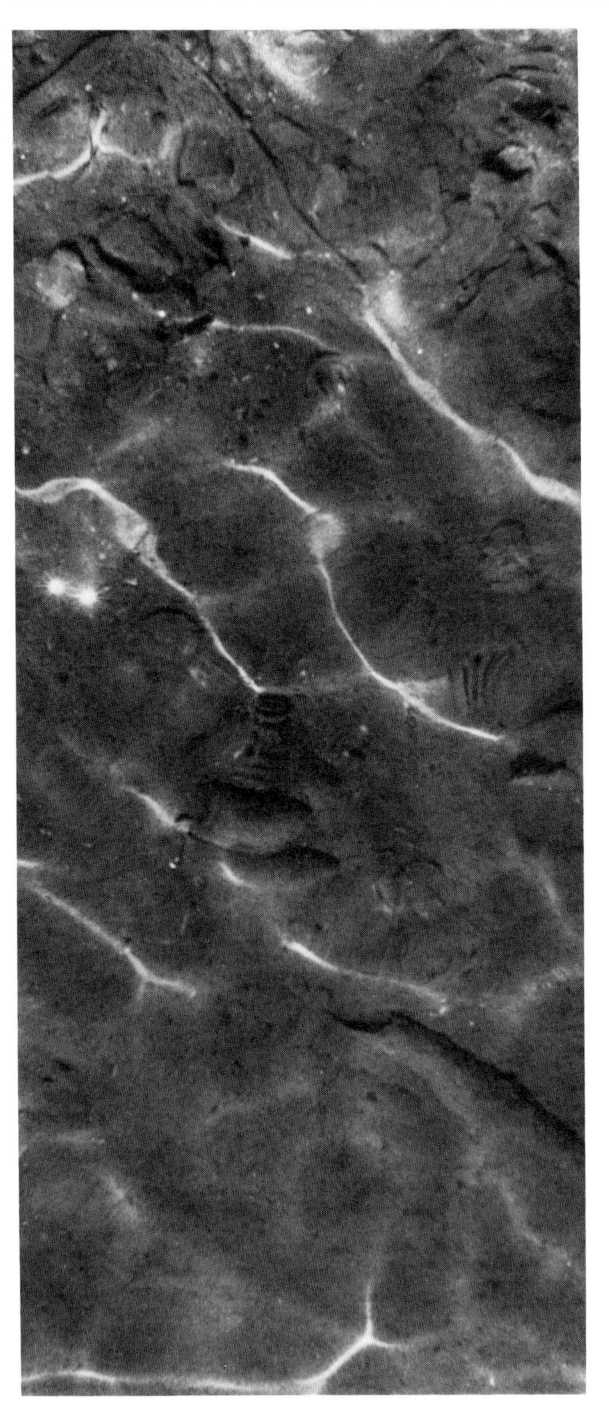

The swimmer or wader moves through a very special part of the estuary. Shallow waters are a different habitat from those we have talked about up to now—truly part of the sea, where water is always present, whatever the tides. It is a place rich in life, but much of its life is too small to be seen, buried from sight or vanishing when disturbed. Thousands of miles of shallows border the Chesapeake Bay and all its streams and rivers. At high tides, the shallow habitat might encompass entire creeks, embayments, or tidal ponds. It is difficult to set a rigid border between a shallow- and a deepwater habitat because they form a continuum. Generally, at depths of about 6 to 10 feet the characteristic physical properties of the shallow-water zone change to the typical estuarine patterns of flow and movement of the deeper areas.

The shallows are a fairly rigorous place to live, perhaps not as harsh as the intertidal zone, but still exposed to extreme environmental changes of both short and long duration. In summer, the sun can superheat the water to the bottom, with little moderation from incoming tides. In winter, the shoreward portion of the shallow zone is frequently covered with ice, so it becomes much colder than the deeper channel areas, where the surface air temperature exerts less influence. Winds and storms impact severely on the shore zone, churning the water to the bottom and suspending great clouds of sediments. Torrential rains may wash tons of soil off the land, turning the waters into thick brown soup. In spring, runoff from adjacent land and from upstream watersheds can be so great that almost the entire upper Bay is muddy. These periods when deeper waters are loaded with heavy suspended sediment are normally seasonal, occurring mostly in spring, when spring rains bring peak flows into the estuary. The heavy influx of sediments and the toxic materials combined with those sediments have profound effects. The shallow-water zone, however, is subjected to high sediment levels far more con-

5. SHALLOW WATERS

sistently throughout the year, with each rain storm adding more material from the uplands. High sediment levels in the water are not beneficial to marine life, as they reduce light needed by algae and plants, place stress upon filter-feeding animals, and smother benthic fauna. Sediments in the shallows are continuously shifted by currents and tides from the shore zone toward deeper waters, where the water volume is greater, thus diluting the sediment concentration.

Rain storms can abruptly lower salinities and flood the shallows with freshwater. Most estuarine species can adapt to these sudden short-term changes, but marine species that have migrated into the Bay and moved into the shallows may suddenly find themselves in the midst of the equivalent of a freshwater stream in salinities they cannot physiologically tolerate.

Shallow habitats are contiguous with other habitats. The entire intertidal flat changes to a shallow-water habitat when the tides come in, accompanied by a multitude of floating and swimming estuarine animals. The whole world of the piling and jetty community is a shallow-water habitat at high tide, as is the flooded marsh, low-lying shorelines, and forested wetlands beyond. Wanderers from the shallows explore the land edges at high tide searching for food, and overhead many species of birds scan the shallows and the shoreline looking for their next meal. Here too

are snakes and turtles and highly adapted insects, all of which live part of their lives on land as well as in the water. Some shallows are thick with stands of submerged aquatic plants, such as eelgrass, widgeon grass, and pondweeds, but this is a special habitat, discussed in Chapter 6.

Almost all the worms, clams, snails, and other burrowers and wanderers of the intertidal flat also live in abundance in the subtidal shallows. Most intertidal animals restricted to the lowest part of the flats are much more abundant in the shallows. Of the intertidal fauna, only the semiterrestrial fiddler crabs, wharf crabs, sea roaches, and the like are seldom seen in the water. A diverse epifaunal community of barnacles, hydroids, sea anemones, mussels, and bryozoans, as well as seaweeds, are attached to bits of shell, sticks, or stones in subtidal shallow waters. Added to this potpourri of benthic life are certain species, such as snails, crabs, and bivalves, that are essentially subtidal and rarely occur in the intertidal zone. However, the benthos is only one segment of a shallow-water community. This chapter introduces another vital portion of Bay life—all the life forms floating or swimming within the waters of the estuary. Here is a world of fishes, shrimps, jellyfishes, and many other creatures.

FISHES

Little Fishes along the Shore

Few observers who have spent any time at the water's edge can have failed to notice a flash of silver and catch a glimpse of a school of small fish as they dart away. From the surface it is almost impossible to identify the fish; most fishes look pretty much the same from above. One of the easiest and most rewarding ways to learn about the fishes of the shallows is to pull a small minnow net or seine through the shallows on a clear summer day. You will be amazed at the variety of small fishes caught in the net. Some of them will be the young of larger fish; others will be the full-grown adults of smaller species.

Come spring, many small fishes move shoreward to congregate in the shallows, where they remain throughout the summer months, feeding on the smorgasbord of worms and amphipods, seaweeds, shrimps and clams, and all the other creatures so abundant there. The fishes in shallower waters are somewhat protected from attack by striped bass, bluefish, and other larger predatory fishes that roam the deeper waters. Shore fishes are abundant in unvegetated shallows, but they often migrate into adjacent weed

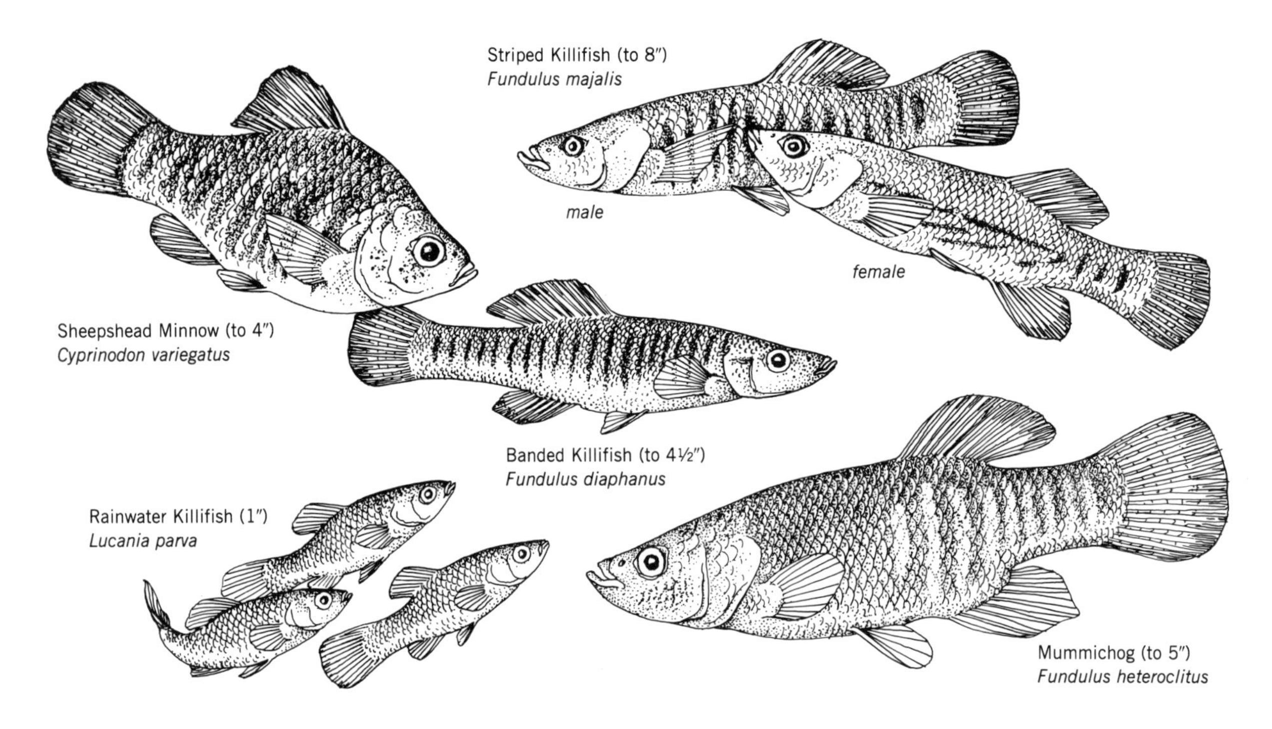

beds and marshes, where there is greater protection and an even richer supply of food. As the weather turns cold in the autumn, fish leave the shallows and move toward the deeper channel waters, where water temperatures are warmer during the winter months. Some species migrate completely out of the Chesapeake Bay into the Atlantic Ocean; others move upstream to bury themselves in bottom muds.

Killifishes—The Bull Minnows. These minnows-of-the-estuary obtain their name from the Dutch word for river, *kill;* and as most rivers in Holland are estuaries, the name is appropriate. Chances are that any minnowlike fish found along the shores of the mid and lower Chesapeake Bay is a killifish. However, in the upper Bay and upper tributaries one cannot be so certain. Many species of freshwater minnows migrate downstream into tidal fresh- and low-brackish waters and are found along with killifishes. Killifishes are chunky, with rounded or squared-off tailfins and a lower lip that juts beyond the upper one. The typical freshwater minnow is more streamlined and has a forked tail. Killifishes are generally greenish brown, often with a brassy gleam and with dark vertical or horizontal stripes of various shapes and sizes according to species, age, and sex.

Schools of killifish swimming over muddy shallows, particularly at the edges of marshes, are most likely to be mummichogs, *Fundulus heteroclitus. Mummichog* is an Indian word meaning "going in crowds." Various age groups, from newly hatched larvae to adults, school close to shore. They grow to five or six inches long but are usually smaller. Mummichogs seem to be real homebodies and in summer may range along a hundred feet or so of the shoreline and only a few yards offshore. At high tide, mummichogs move into adjoining marshes to feed. Striped killifish, *Fundulus majalis,* are similar in size to mummichogs but generally have darker bandings (the body stripes of the mummichog are more silvery). Female striped killifishes have horizontal bars; males have vertical bars. The young of both sexes have vertical bars, which can be very confusing. Striped killifishes prefer sandy bottoms just off beaches where they swim right up to the very edge of the water line. If stranded by the receding tide, they quickly flip back into the water.

The banded killifish, *Fundulus diaphanus,* is common in the tidal fresh- and lower-salinity brackish waters of the Bay regions and are often found along with mummichogs. Banded killifishes have many brilliantly colored silver-blue bands and more streamlined bodies than other killifishes. The tiny rainwater killifish, *Lucania parva,* is only about an inch long at full size and is often taken to be the young of other species. Unlike the other killifishes, it has no stripings or bar markings.

The sheepshead minnow, *Cyprinodon variegatus,* or variegated minnow, as it is also called, is a brightly colored fish with a stubby deep body. Males are a beautiful, iridescent peacock blue during the spring and early summer spawning season.

Silver Streakers—Anchovies and Silversides. Yes, anchovies live in the Chesapeake Bay, and they are very similar to the salted and canned variety on the grocery shelf. In fact, they are one of the most plentiful fishes in the Bay, along with silversides, another group of small schooling fishes. Both anchovies and silversides are favored prey for larger predatory fishes, such as striped bass and bluefish. They are small fish and are distinctly marked with a bright silver band along the body. In contrast to the stocky killifishes, they are slender and streamlined. An anchovy is easy to distinguish from a silverside. Open its mouth with your fingertip. In an anchovy, the gape will be huge and the fish will look like its head is bisected. Silversides, conversely, have very small mouths. This is a foolproof method, although there are other differences—anchovies are soft and almost transparent and have a single dorsal (back) fin. Silversides are firm and opaque, and they have two dorsal fins. These species move in to the shallows in warmer seasons. Silversides are typically an inshore species and often mingle with killifishes. Anchovies are more widespread; juvenile stages school close to shore, whereas the adults roam in deeper waters. The bay anchovy, *Anchoa mitchilli,* is the most ubiquitous and abundant of the two anchovy species in the Chesapeake. Three species of silversides—the Atlantic silverside, *Menidia menidia,* the inland silverside, *Menidia beryllina,* and the rough silverside, *Membras martinica*—occur in the Bay. It is difficult for the amateur to separate the three species—a task better left to the experts.

Needlefishes. Frequently a long, narrow fish, the Atlantic needlefish, *Strongylura marina,* can be spotted patiently stalking killifishes or silversides along the shore. Its extended scissors jaws are lined with tiny teeth and present formidable and effective weapons for capturing small fishes and shrimps, their favored prey. The Atlantic needle-

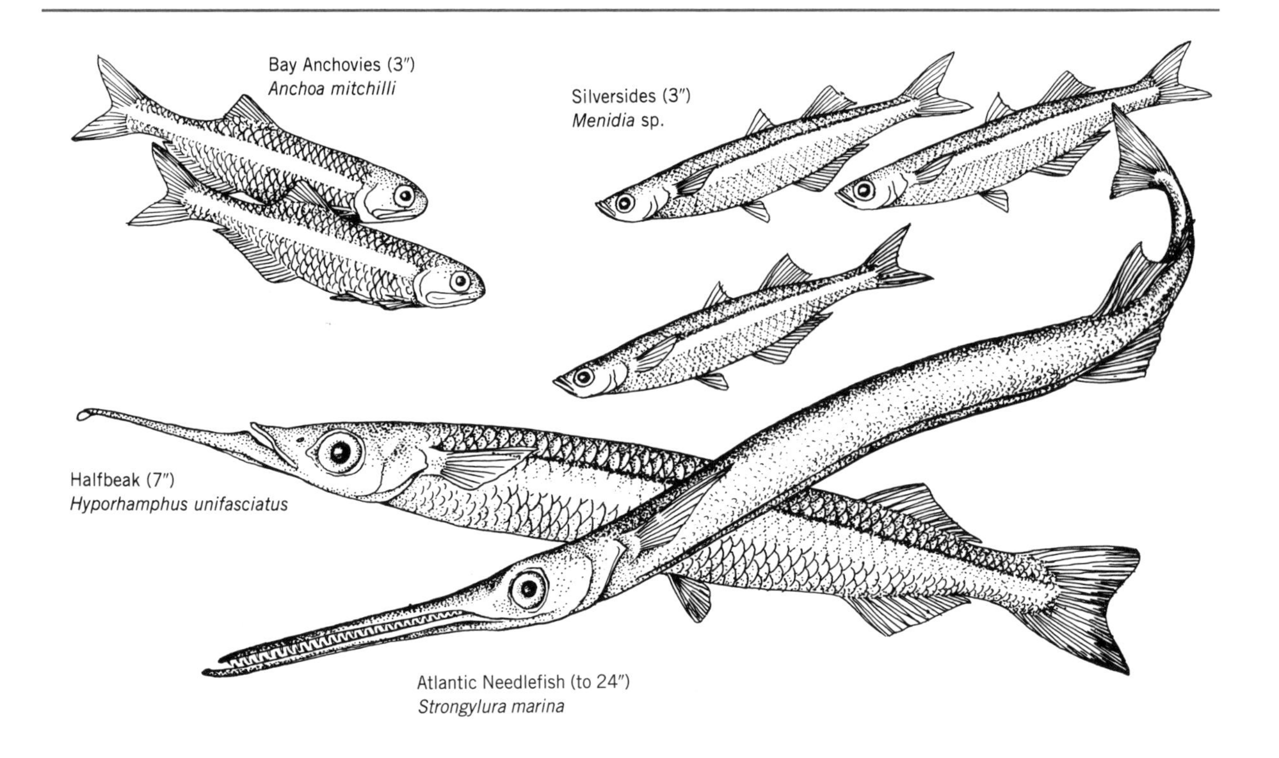

fish, a marine fish that enters the Bay in large numbers from the ocean in early spring, spends the warm season in the estuary moving upstream even into tidal freshwater. These fish move into the shallows and are found singly or schooling at the surface around docks, into marshes, and along beaches. Atlantic needlefish of all ages are found in the Bay, from early young, which do not yet have elongated jaws, to large adults two feet long. Atlantic needlefish have a broad silver stripe along each side, a squared tail, and a long upper jaw, which is not quite as long as the lower jaw.

Occasionally a really strange looking "needlefish," the halfbeak, *Hyporhamphus unifasciatus,* may be spotted cruising the surface and may even jump completely out of the water if startled. These fish are sudden and quick, and difficult to capture. The halfbeak looks like a needlefish whose upper jaw has been broken off. Its lower elongated jaw sports a crimson, black-bordered fleshy flap, presumably used as a lure. Halfbeaks are very slender and are smaller than needlefish, their maximum length being only 10 inches or so, whereas needlefish grow to over two feet. They move well into the Bay and have been found as far inland as Baltimore Harbor.

Fishes That Come from the Sea to Spawn—Herrings and Shads

Every spring, the banks of the upper rivers and streams of the Chesapeake Bay are lined with anglers fishing on schools of river herrings and shads, which make their annual spawning runs at this time. Herrings and shads are anadromous fishes; that is, they are ocean-living species that must migrate into freshwater to spawn. It is believed that, in some mysterious way, each individual follows a path from the ocean that returns it to the stream where it has spawned. Millions enter the Bay each year. Many never reach the spawning grounds because the rivers are blocked by dams or they are caught by the fishermen who await them with pound nets and gill nets in deeper waters or with dip nets and hook and line in shallow streams. Shad and herrings have declined in the Bay due to a number of causes, including changes in water quality and habitat, as well as overfishing. This is an all too frequent commentary on the world's fisheries. Some anadromous species spawn in the open areas of large rivers; others migrate beyond tidewater, fighting their way through riffles and rapids up the smallest streams. The adults move downstream after spawning, and by summer most of them have left the Bay for the ocean. Meanwhile, the young hatch and grow rapidly through the spring and summer in the tidal fresh- and brackish waters. As they develop, they gradually migrate downstream. Schools of small river herrings and shads are abundant in shallow waters until fall, when most migrate out into the ocean. They range the Atlantic coast in huge schools for three to five years, until they mature. The shadbush blooming along the shores of the Chesapeake signals the arrival of spring and the return of the spawning shads and herrings.

Shads and herrings look so much alike that only seasoned watermen who see large groups together are likely to name them accurately. They are thin, silvery fish, with a lovely opalescent glow on their sides. These fishes have large, smooth scales, which slough off easily, and a saw-toothed, thin-edged belly. There are six species of shads and herrings in the Bay, each with a number of descriptive names, some of which refer to more than one species. The American shad or white shad, *Alosa sapidissima,* is the largest, best known, and most delectable of the group. Shad roe, the eggs of the female shad, is a Chesapeake delicacy. Hickory shad, *Alosa mediocris,* also called hickory jacks or tailor shad, are claimed by old-time watermen to be sports, that is, a cross between white shad and herring, but they are definitely a true and separate species. The hickory shad grow as large as white shad and are similar in appearance except for their jutting lower jaw. Historically, the number of hickory shad has been small in the Bay. There are two species of river herrings in the Bay: the alewife, big-eye, or branch herring, *Alosa pseudoharengus;* and the blueback herring, glut herring, or again, alewife, *Alosa aestivalis.* River herrings are smaller than the shads, and not so deep-bodied, and they generally have a single spot on their shoulder, rather than a series of spots like the shads. Schools of river herrings run up into tiny rippling streams.

Not all herring and shad of the Bay are anadromous. The gizzard shad or mud shad, *Dorosoma cepedianum,* is often abundant in tidal fresh- and brackish water. It is generally not eaten and is considered a nuisance trash fish, which clogs the nets of commercial fishermen. It does,

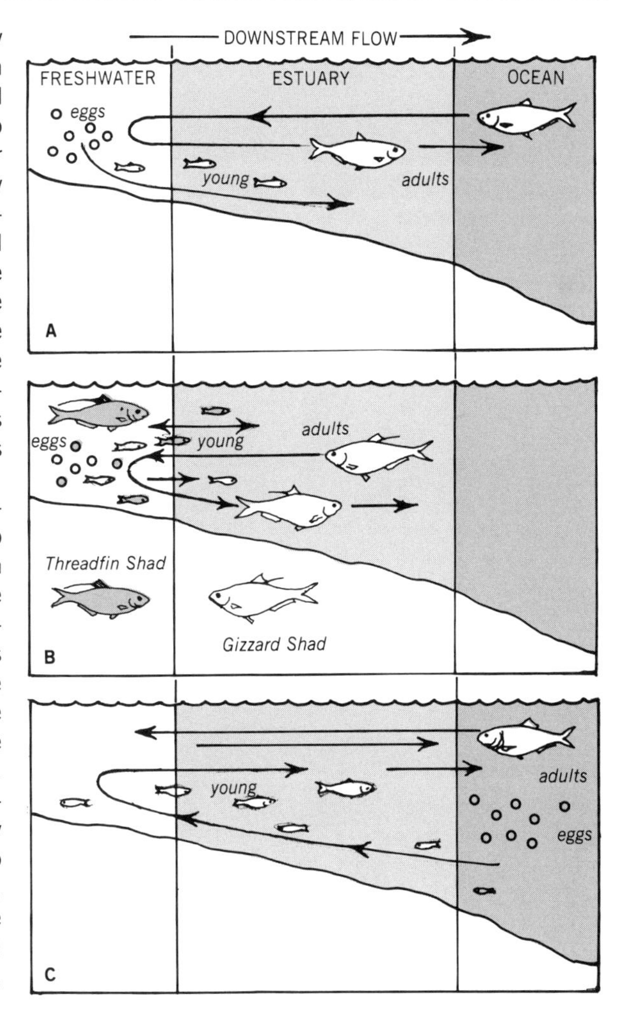

Diagram A. The anadromous species of the herring and shad family of fishes—alewives, blueback herrings, and American shad—move up from the ocean in spring to spawn in freshwater; their young move downstream into salty water as they grow and eventually move into the ocean to join the adults.
Diagram B. Adults of the freshwater species, threadfin shad, rarely move into estuarine waters; their young, however, are common visitors in low-salinity waters. Gizzard shad spend most of the year in the estuary but migrate upstream in spring to spawn in freshwater.
Diagram C. Menhaden spawn in the ocean, but both adults and young migrate into the estuary, moving well upstream even into tidal freshwater.

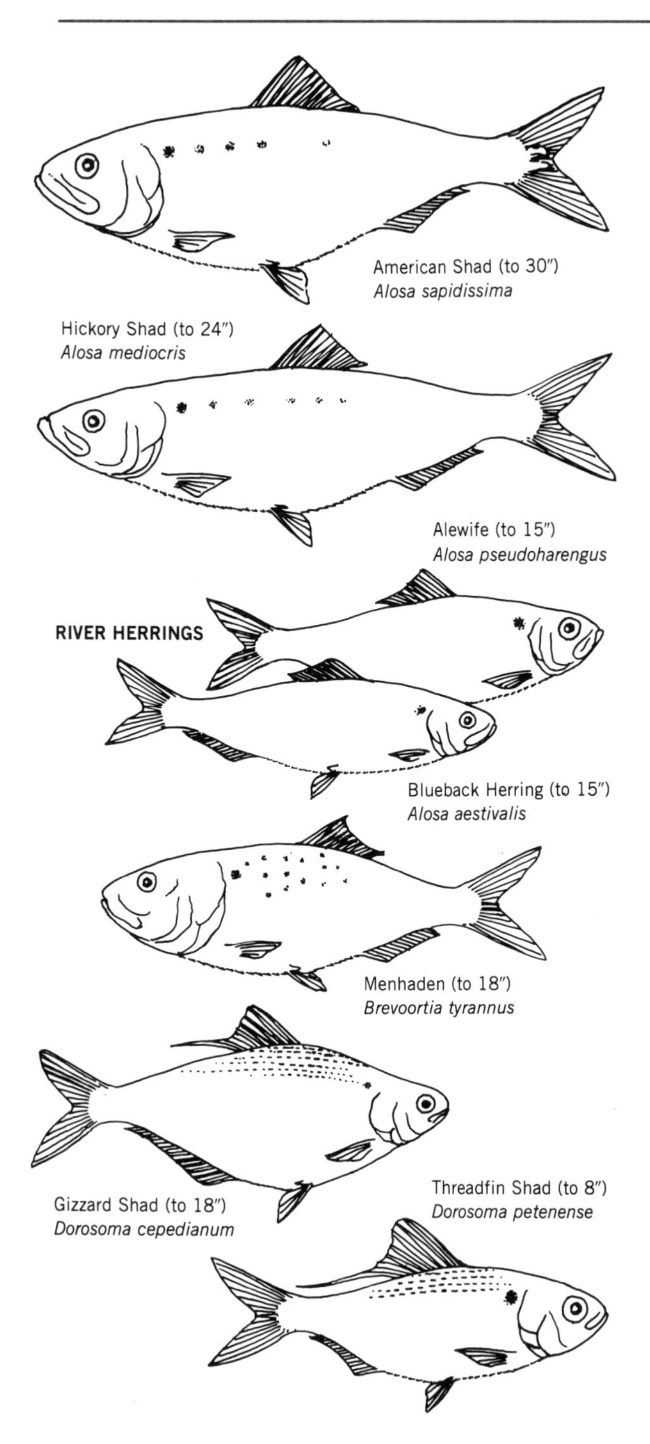

American Shad (to 30") *Alosa sapidissima*

Hickory Shad (to 24") *Alosa mediocris*

Alewife (to 15") *Alosa pseudoharengus*

Blueback Herring (to 15") *Alosa aestivalis*

Menhaden (to 18") *Brevoortia tyrannus*

Gizzard Shad (to 18") *Dorosoma cepedianum*

Threadfin Shad (to 8") *Dorosoma petenense*

however, serve as food for many of the more desirable fishes. A gizzard shad is a somewhat aberrant looking herring with a small, scaleless head, a rounded nose overhanging its mouth, and a thin thread extending from the end of its dorsal fin, which is actually the elongated last fin ray. Gizzard shad are essentially freshwater fish, often found in lakes and ponds, but in the Chesapeake Bay they behave more like anadromous species. They spend most of the year downstream in moderately salty water and migrate upstream to tidal freshwaters to spawn. The young of another freshwater fish, the threadfin shad, *Dorosoma petenense,* moves downstream into the brackish waters of the Bay region. They look much like river herrings, but have an extended last dorsal fin ray, longer than that of the gizzard shad, reaching almost to the base of the caudal fin as they mature.

Menhaden—A Most Abundant Fish

During the summer months, throughout the Chesapeake Bay, dense schools of herring-type fish called Atlantic menhaden, *Brevoortia tyrannus,* are seen more often than any other fish. It is mesmerizing to watch a school moving rapidly just below the surface, fins and tails just ruffling the surface as each individual swims in close unison, following a single lead fish. After a period of time, the leader turns and melds into the center of the school and a

One must look closely to tell shads and herrings apart. American shad and hickory shad are the largest of the family; both usually have a dark shoulder spot followed by a series of paler spots. The lower lip of the hickory shad juts beyond the upper lip. American shad have toothless lower jaws; hickory shad have lower jaws with numerous small teeth. River herrings have a single shoulder spot, sometimes none, and are smaller than the shads. The eye of the alewife is proportionately larger than that of the blueback herring (its diameter greater than the length of the snout), and the lining of its abdominal cavity is black rather than pale. Menhaden have large heads and a dark shoulder spot plus scattered lighter spots over the back. Gizzard shad have a small head with a rounded snout overhanging the mouth. The last ray of the dorsal fin extends approximately halfway to the base of the tail. Threadfin shad are the smallest of the group, with the last ray of the dorsal fin long and filamentous, extending to the base of the tail, and with a distinct shoulder spot as well as a series of dark stripes over the back. The young of most of these species congregating in shallow waters have not yet developed the distinguishing characteristics of the adults and are correspondingly more difficult to identify.

new lead fish takes over. Menhaden are one of the least-known fish to the public, yet they support one of the oldest and largest fisheries along the Atlantic coast. Menhaden are so oily they are almost totally inedible, but this very quality makes them a rich source of chicken food and fertilizer, of oil for paints and cosmetics, of tempering products for steel. Menhaden are the "breadbasket" of the oceans and bays, prodigious in number, and favored prey of bluefish and striped bass, tunas and sharks, seatrouts and mackerels. Menhaden are probably the fish Indians urged the pilgrims to plant with their corn seeds. They called this fish "Munnawhatteaug," meaning "that which manures," a name corrupted over the years to menhaden. They are also known as bunkers, skipjacks, mossbunkers, pogies, fatbacks, and quite often and confusingly, alewives or elwives.

Menhaden are not anadromous fishes. The heavy runs of menhaden that enter the Bay in droves do not come to spawn like their close cousins, the herrings and shads, but to feed upon the rich plankton soup of the estuary. They return to the ocean to spawn. Menhaden swim rapidly with their open mouths straining hundreds of gallons of water and filtering out tiny planktonic organisms with their feathery gills. Soon after hatching, the larvae are caught by the currents and are carried into the Bay, where they move upstream into low-brackish waters to feed and grow in the same nursery areas that nurture small herrings and shads. Young menhaden tend to stay close to shore in the shallows; older ones cruise the deeper, open waters. Fishermen keep a sharp lookout for menhaden schools and for the sudden churning of water as a school of snapping bluefish attacks them. When they sight the frenzied feeding and diving gulls they race to the spot, knowing they will return home with coolers filled with "blues."

Striped Bass and White Perch

Striped bass, or rockfish, *Morone saxatilis,* and white perch, *Morone americana,* must also seek freshwater to spawn. They are "halfway" anadromous, or semianadromous, as some biologists have labeled them. White perch and most striped bass do not migrate all the way from the

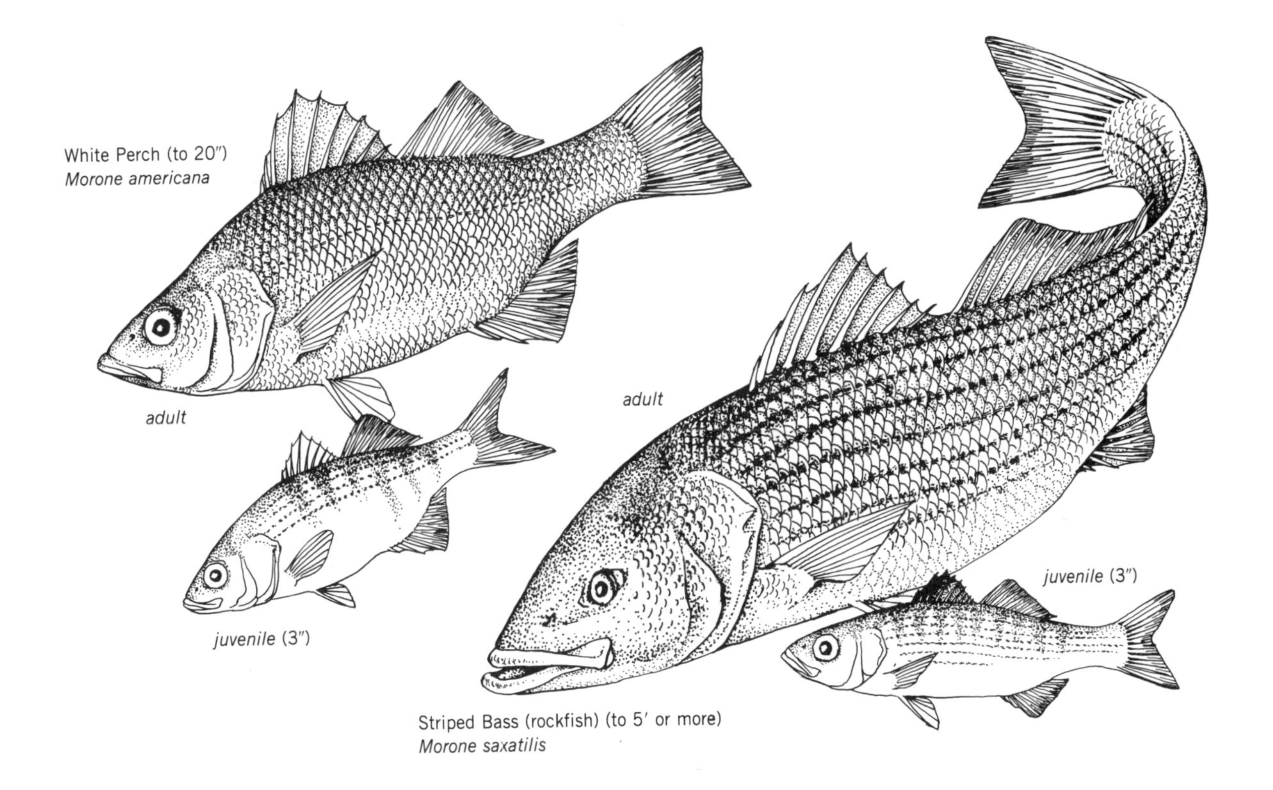

Striped Bass (rockfish) (to 5′ or more)
Morone saxatilis

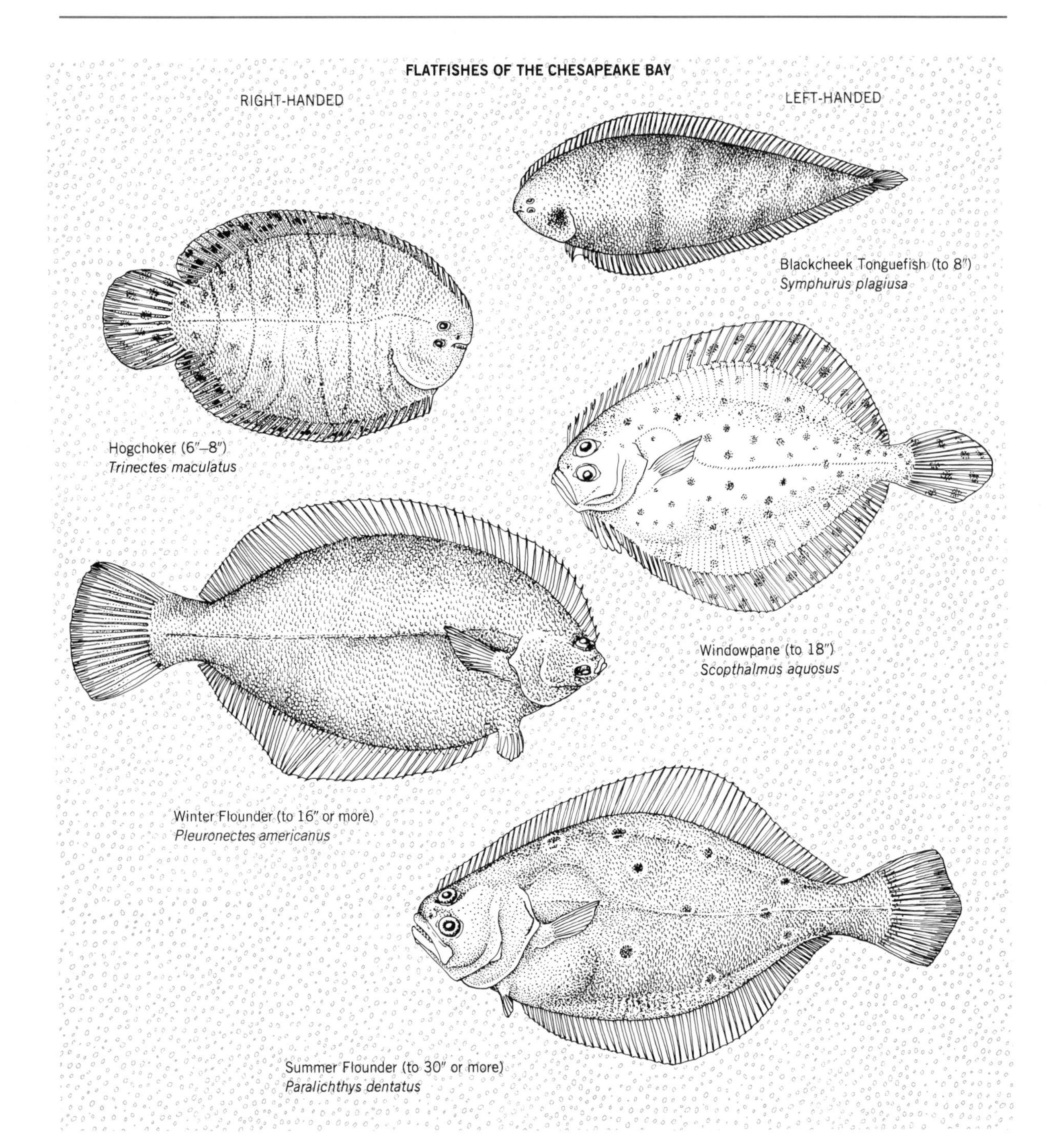
FLATFISHES OF THE CHESAPEAKE BAY
RIGHT-HANDED
LEFT-HANDED
Blackcheek Tonguefish (to 8″)
Symphurus plagiusa
Hogchoker (6″–8″)
Trinectes maculatus
Windowpane (to 18″)
Scopthalmus aquosus
Winter Flounder (to 16″ or more)
Pleuronectes americanus
Summer Flounder (to 30″ or more)
Paralichthys dentatus

ocean, like the truly anadromous species, but from the estuarine waters lower down the Bay and rivers. However, a small percentage of the spawning runs of striped bass in the Chesapeake Bay come from populations entering from the Atlantic Ocean. White perch and striped bass are closely related and are the only two naturally occurring members of their family in the Bay. In spring, after spawning, adults migrate back downstream toward the shoal and surface waters to spend the summer feeding.

Meanwhile, in the upstream spawning areas, the eggs hatch and the larvae mingle with herring and shad larvae that were spawned in the same regions, and, like them, the young white perch and striped bass, as they grow, migrate downstream to brackish waters, where they congregate in the shallows throughout the summer. The appearance of small white perch and striped bass a few inches long shows their close family relationship. At this stage, both species have longitudinal stripes overlaying a series of vertical bars along their sides.

White perch are truly estuarine species, never found in the ocean. Each major river system of the Bay apparently maintains its own separate population. In contrast, striped bass are more adventurous; some remain within a single tributary, but others move throughout the main stem and other tributaries of the Bay. In winter, both white perch and striped bass move to deep, channel waters to await the warming of spring and a renewal of the upstream spawning migrations. Some striped bass leave the Bay entirely in late winter and spring and move into Atlantic coastal waters, migrating north, some as far as Nova Scotia. Many a striper caught in New England began life in the Chesapeake Bay. In the early 1970s, striped bass went into a sharp decline. Stringent management regulations were applied in the Chesapeake Bay and along the Atlantic Coast, and now the striped bass is again flourishing.

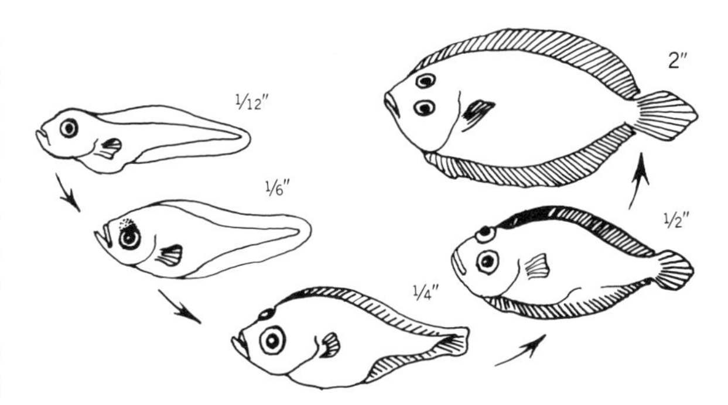

The larval development of a summer flounder showing the migration of its eye from one side, over the top of its head, to the other side.

Flatfishes—Bottom Dwellers

Flatfishes have bodies well adapted to living close to the bottom. Their pancake-flat bodies bear both eyes on one side. Thus, they can lie half buried in the bottom sands or silts, dark side up, always alert for passing prey. Most flatfishes can change the color pattern of their dark side to match the color and pattern of the substrate. It is almost impossible to detect a flatfish in shallow waters, but they are there by the thousand.

In the Chesapeake Bay, the hogchoker, *Trinectes maculatus,* a bottom flatfish, is one of the most abundant and ubiquitous fish of the estuary and is distributed from tidal freshwaters to the mouth of the Bay. It is a small, right-handed sole no more than six inches long. Hogchokers are said to be "right-handed" because their mouth and eyes are on the right when viewed from their dorsal, or top, surface. During the larval stage the normally placed left eye migrates over the top of the hogchoker's head to a position next to the right eye. In left-handed flatfish the opposite movement occurs. Hogchokers are rather useless as human food because they are so small and bony; however, their flesh is quite tasty. Their unusual name comes from earlier times, when farmers used to feed them regularly to hogs, which had a difficult time swallowing them because of their rasplike scales and rigid, bony bodies.

Young of the winter flounder, *Pleuronectes americanus,* also a right-handed flatfish, are common in shallow waters in the summer. They are the young of adults that entered the Chesapeake Bay during the previous winter to feed in deeper waters and spawn in shallow waters well upstream into the mouth of the Potomac River, and in the upper Bay at least as far as Annapolis. Summer flounders, or fluke, *Paralichthys dentatus,* a left-handed flatfish, also enter the Bay and, as their name suggests, remain during the warmer months after the adult winter flounders have left for the ocean. Summer flounders spawn in the ocean, but their young move into the estuary and can be caught in shallow shoreline waters throughout the warmer months. Larger winter and summer flounders are caught in numbers in deeper, open water by commercial fishermen. Occasionally, the lucky angler also captures them. Since most

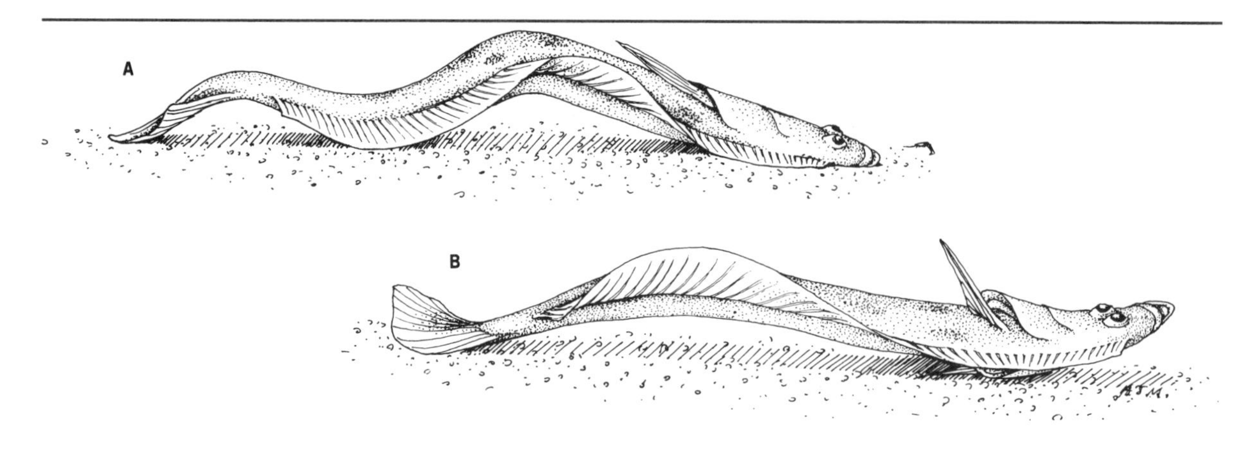

A summer flounder feeding on the bottom first sights its prey (A) and then engulfs it (B).

anglers are warm-weather fishermen, summer flounders are the usual recreational catch in the Chesapeake Bay.

It is easy to tell whether a particular flatfish is left- or right-handed. Hold it flat, with the brown side up and the belly side toward you. (The belly side is the edge where the gills open.) If the head faces to your left, the fish is left-handed; if it faces to your right, the fish is right-handed. The summer flounder has a large mouth, which gapes beyond the eyes, and its back is spattered with rounded spots. The hogchoker lacks pectoral (side) fins and has narrow stripes across a dark back and a spotted or blotched blind side. The winter flounder has a distinct pectoral fin just behind the gill opening and a separate, small pelvic fin on the belly.

The blackcheek tonguefish, *Symphurus plagiusa,* and the windowpane, *Scophthalmus aquosus,* are left-handed flatfishes. The blackcheek tonguefish is a small—less than eight inches—little-known fish, which is rarely seen by the fisherman, although it is not uncommon. The body is elongate and terminates in a sharp taper. These fish are colored an inconspicuous dark brown with dark crossbars. The windowpane is an ovate fish with a large mouth, dotted with small brown spots on an olive to slaty brown background. It is found over sand and mud bottoms and occasionally occurs up the Bay as far as Kent Island, Maryland.

A Nursery for Ocean Fishes

Almost three hundred species of fishes have been recorded within tidal waters of the Chesapeake Bay and its tributaries, of which approximately half are ocean fishes that enter the Bay, mostly in spring and summer, to feed and return to the oceans in autumn. Some of these ocean fishes are mere stragglers that wander into the Bay more or less accidentally. Others, such as bluefish and seatrout, are predictable and regular visitors. Most marine fishes common to the Bay spawn in the ocean, but the hatched larvae and juveniles enter the Bay at an early age. Here they grow rapidly on the dense populations of invertebrates and small forage fishes, such as anchovies and silversides. Adults of ocean fishes tend to remain in deeper waters, but the young are prone to move inshore. There is always the chance that a fisherman casting his line from the beach will land some completely unfamiliar fish.

Drums—Spot, Croaker, and Silver Perch. The most abundant ocean species in the shallows is the little spot, *Leiostomus xanthurus,* a favored and feisty sport fish found throughout the mid and lower Bay. Schools of young spot can often be seen hovering around pilings and jetties a foot or two below the surface, occasionally flipping their bodies and flashing their distinct shoulder spot. Menhaden have a similar shoulder spot, but their surface swimming behavior sets them apart. Spot have sloping heads with angled oblique stripes across their back and a slightly forked tail. They have a distinct lateral line running from behind the gills along the body and onto the tail fin. Menhaden, in contrast, have deeply forked tails and no stripes or lateral line. Spot are members of a large family of fishes called drums, a name referring to the characteristic loud drumming or croaking sounds made by most species of the family. The sound is produced by the rapid contracting of a special muscle that runs along the sides of an air-filled sac, the air bladder, which acts as a resonator. Atlantic croakers, *Micropogonias undulatus,* or hardheads, as they are also called, make the loudest noise of any of the drums. Small croakers called pin heads, about eight inches long, congregate close to shore in the lower half of the Bay.

Croakers are not as widely distributed or as abundant these days as are spot. Thirty years ago they were far more numerous, and Bay natives still remember when a day of fishing was always rewarded by a panful of croakers. What has happened to them? Scientists have various theories. Some believe that very cold weather kills the juveniles that enter the Bay in fall and winter. Others point to predation by increasing numbers of striped bass and bluefish, or simply to a natural cyclic fluctuation. The causes are uncertain, but it would not be surprising if they became plentiful again. In fact, as this revision went to press, croakers had once again returned to the Bay in large numbers and were being caught all over the Bay. A croaker is highly luminescent and has a pinkish glow when fresh from the water. It has a slightly pointed tail, vague oblique stripes across its back, and a fringe of tiny chin barbels.

The silver perch or sand perch, *Bairdiella chrysoura,* is a small drum similar in appearance to a white perch and often mistaken for one. Silver perch frequent the shallows of the lower part of the Bay. The larvae and very small juveniles migrate farther upstream into low-salinity river and stream waters, but as they grow they move back downstream, congregating in higher salinities. Silver perch can be distinguished from white perch by a slightly pointed tail fin, rather than a forked one, by a lateral line that extends into the tail fin, and by yellowish ventral fins.

Young of other drums, such as weakfish, spotted seatrout, red drum, and black drum, will move into the shallows, as well as young of black sea bass, bluefish, and many other species of the deeper, open-water habitats.

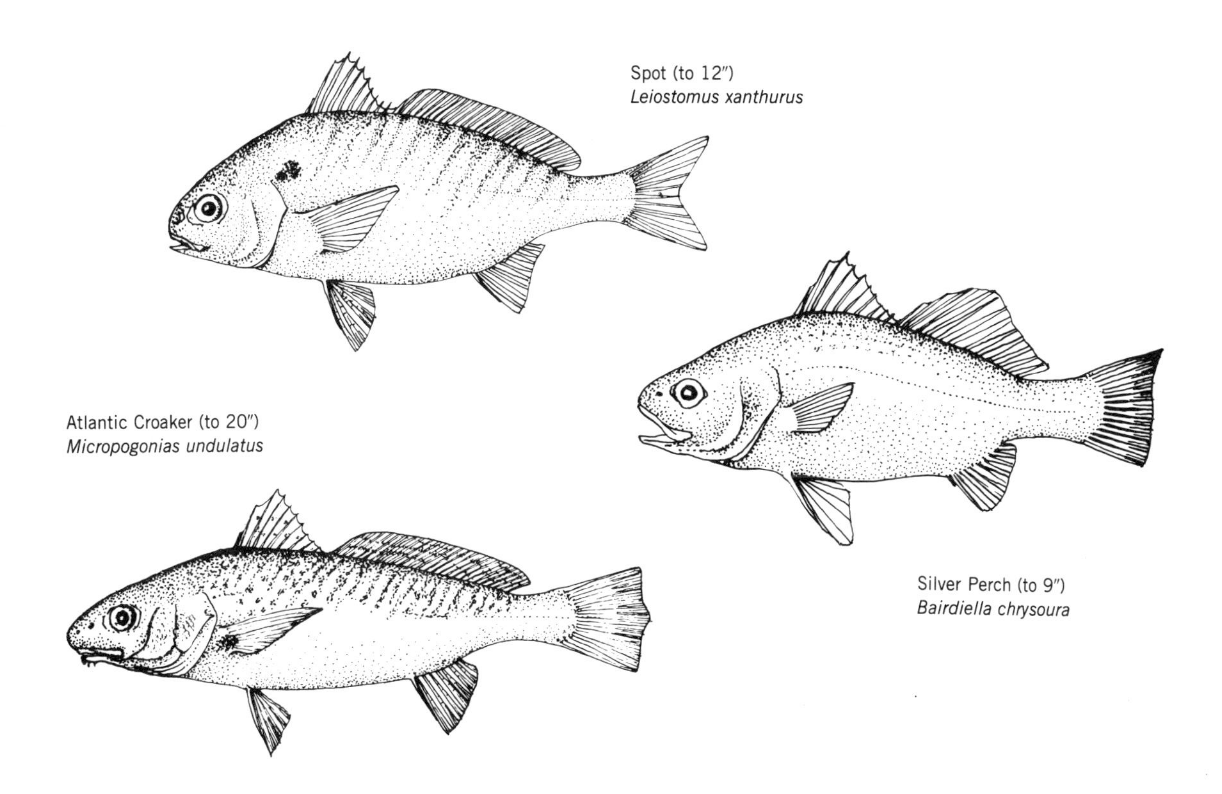

Spot (to 12″)
Leiostomus xanthurus

Atlantic Croaker (to 20″)
Micropogonias undulatus

Silver Perch (to 9″)
Bairdiella chrysoura

Silver Jumpers—Mullets. Sometimes, during summer, schools of fish will be seen jumping out of the water as they follow the shoreline. This behavior is certain to mark a school of mullet that has moved into the Bay from the ocean. Mullets are common bay fishes in some southern states, but in the Chesapeake they appear only sporadically. Schools of small adults 12 inches or longer and smaller juveniles are found well into the Bay, sometimes to low-brackish-water areas. Mullets are bottom feeders on mud and algae and tend, therefore, to move toward shore and into tidal creeks and flats when they enter the Bay. They are a good eating fish but not favored sport fishes, as they do not respond well to bait; they are better captured by nets or seines. Two species occur in the Chesapeake Bay: white mullet, *Mugil curema;* and striped mullet, *Mugil cephalus.* Both are long, round-bodied, blunt-nosed fish, with small mouths and a pectoral fin placed high on the back. The principal difference in appearance between them, as one would suspect from their common names, is that one is striped, the other not. Their general body color is a dark olive-green above, and silvery below.

Eels and Elvers

On a warm spring night in April or May, in many a small stream, thousands of elvers may be spotted by flashing alight on the surface of the water. They swarm in the shallows close to the shore, intent on making their way upstream, directed always toward freshwater. By day, they will have disappeared, having burrowed into mud or hidden under rocks awaiting darkness before renewing their journey. Their persistence is amazing. They will slither over rocks and even slide over wet grass bordering the stream to overcome a particularly difficult obstruction.

Many elvers reach the freshwater streams and lakes beyond the tidewaters of the Chesapeake Bay; others remain in the estuary. These tiny, snakelike fish just a few inches long are the young of the American eel, *Anguilla rostrata.* They have traveled thousands of miles from their spawning grounds in the Sargasso Sea, an area of the Atlantic Ocean east of the Bahamas and north of the West Indies. In contrast to anadromous fishes, eels are catadromous; that is, they must leave freshwater and return to the sea to spawn. Elvers develop into eels and remain in the Bay and in the freshwater rivers and streams before they mature and migrate back to the sea to spawn. Meanwhile, they have spent at least 5 years and even as many as 15 to 20 years in the estuary. Apparently females migrate farther upstream than males. Many landlocked eels never leave, some attaining 35 to 40 years of age. An eel in an aquarium in Sweden lived to be 88 years old. When ready to return to the sea, eels participate in one of the most dramatic fish migrations known. Sexually mature eels, some three feet long, emigrate from all coastal rivers and estuaries along the Atlantic coast, ultimately converging over the Sargasso Sea spawning grounds to spawn and die. Newly hatched eels are bizarre transparent, leaf-shaped creatures called leptocephali. Prevailing ocean currents sweep the leptocephali toward the coast. It takes them about a year, as they drift, to metamorphose into a more eel-like shape. By the time they enter the coastal estuaries they are pigmented brown and have been transformed into elvers.

Eels are literally everywhere in the Chesapeake Bay system, in shallow shoreline waters, in deep and swiftly moving channels, in tiny creeks, and in large tidal ponds. They feed at night on almost anything, nosing out buried clams, stalking unprotected soft crabs, or attacking small fishes. Macoma clams are a favored food of eels. They are attracted by the clams' siphons sticking out of the bottom as they probe for food. The siphons apparently look much like a squirming succulent worm to the eel. Eels are less active during the day, often burying themselves in bottom muds. Many people mistake eels for snakes, but they are true fishes, with long uninterrupted fins continuous along the back, tail, and belly. Their flesh is delicious and is considered a delicacy to the Japanese and to Europeans. Elvers are also eaten; they are the *anguilles* on many a Spanish menu. Unfortunately, to the average American they are slithering, unappetizing creatures, better used for crab bait than eaten.

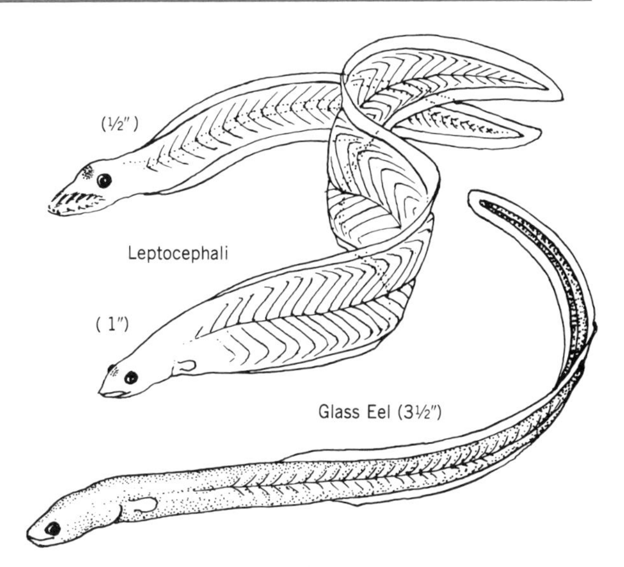

Larval stages of the American eel, *Anguilla rostrata.*

Freshwater Intruders

Certain freshwater fishes of the rivers and streams flowing into the Chesapeake Bay tolerate slightly salty waters and often migrate downstream. They tend to congregate in shallow streams and protected coves of the larger rivers of the estuary. A few species move into deeper channel waters as well. In spring, freshwater fishes move upstream above tidewater to spawn, and afterward some return downstream. The most extensive downstream migration generally occurs during winter months, but at this season the fish congregate in deeper waters and will not ordinarily be encountered in shallow-zone habitats.

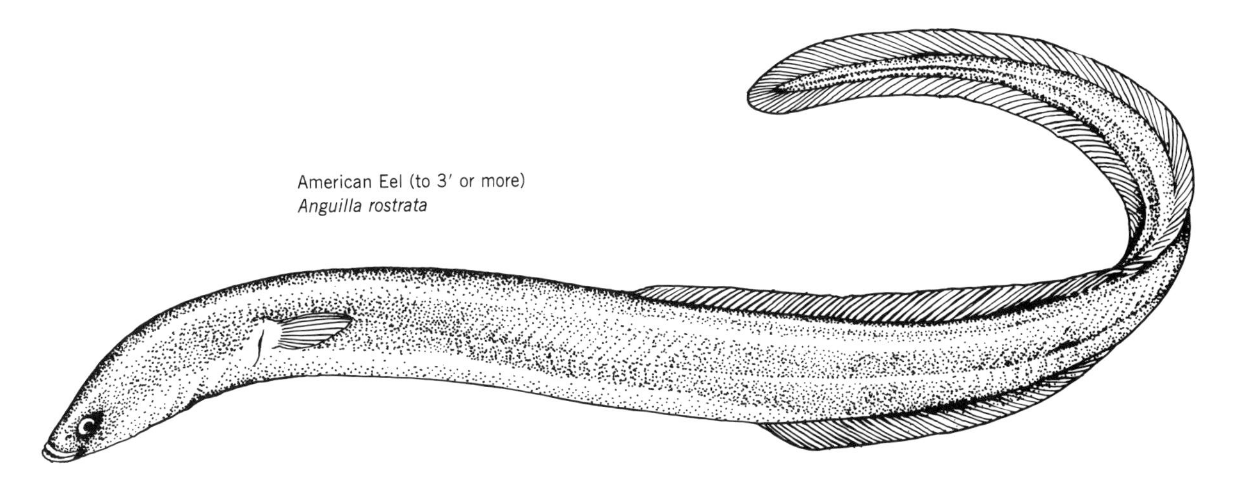

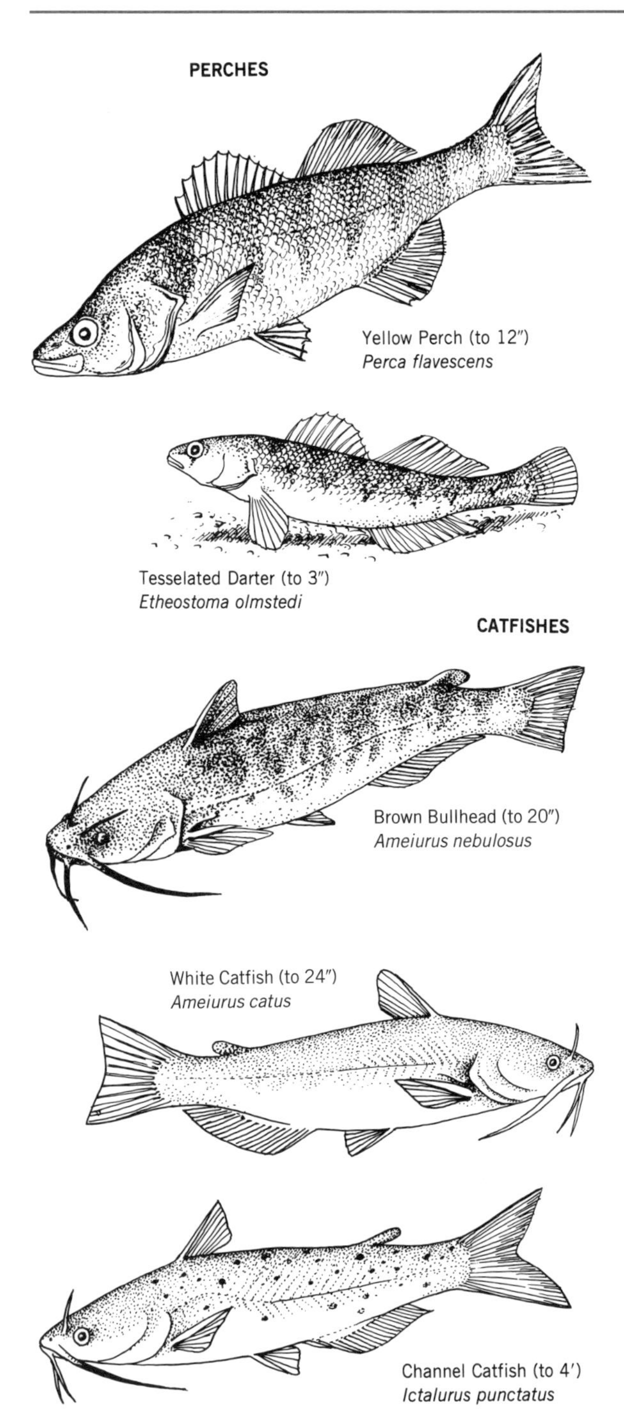

Perches. Probably the best-known freshwater fish of the estuary is the yellow perch, *Perca flavescens.* Yellow perch are freshwater fish so acclimated to brackish waters that in the Chesapeake Bay region they behave more like semi-anadromous white perch and gizzard shad. They spend most of the year in the upper part of the estuary and return to freshwater in late February and March to spawn in small shallow streams. Each spring fishermen formerly thronged to known spawning streams, where yellow perch were so densely congregated that it was difficult not to catch a few fish. However, water quality conditions have deteriorated in the spawning streams, the water has apparently become more acidic, and yellow perch have declined. Yellow perch are about the same size as white perch—a perfect pan-sized fish. Their spawning areas mark the upstream limit of yellow perch distribution in the Chesapeake. After spawning, adults migrate downstream, leaving behind long, amber-colored, gelatinous, accordionlike strands of eggs, which adhere to fallen branches and other debris along the banks of shallow streams. Yellow perch are colorful fish, bright yellow to gold with six to eight dark vertical bands. Spawning males are even more brilliant, with intense orange-red fins.

The tessellated darter, *Etheostoma olmstedi,* is a small relative of the yellow perch. It ventures downstream into the same brackish areas sought out by the yellow perch. Tessellated darters are found in fast-flowing waters, in riffles, where they may be dislodged from under rocks and pebbles. The tessellated darter is a small, sprightly fish, two to three inches long. It jerks quickly along the bottom from one spot to the next in search of food. This rapid movement gives darters their obvious name. The eyes of tessellated darters are set close together on top of their heads and their sides are irregularly marked with V-shaped blotches.

Catfishes and Bullheads. Catfishes and bullheads, lacking scales, are smooth-skinned and have long, catlike "whiskers." Few people fail to recognize this group. Catfishes have forked tails, and bullheads are catfishes with squared tails. All should be handled with care, as the pectoral fins have heavy, sharp spines. Three species—channel catfish, white catfish, and brown bullheads—are common downstream into the brackish waters of upper Zone 2. Catfishes rely on their long whiskers or barbels to find food by touch and taste in muddy waters. In spring and early summer, they move upstream to spawn. The males guard and herd the offspring until they are about half an inch long. Occasionally, dense schools of dark-colored young catfishes move into tidewater, where they can be seen as swirling black balls at the surface.

Channel catfish, *Ictalurus punctatus,* is a streamlined, grayish fish immediately identifiable by its brownish black spots. It has a sharply forked tail and dark chin barbels. Channel cats are the most active of the catfishes and prefer deeper channel waters. The brown bullhead, *Ameiurus nebulosus,* and the white catfish, *Ameiurus catus,* are found in the shallows as well as in open clear areas, and in deeper waters. Both are bottom feeders and are more lumbering in habit than channel catfish. The brown bullhead is a mottled brown or black with dark chin barbels and has a square tail and sharp barbs along the spine of its pectoral fin. The white catfish is smoothly grayish brown with a white belly, slightly forked tail, and pale chin barbels.

Minnows and Shiners. Minnows, shiners, dace, chubs, and carp all belong to the largest family of fishes in the world, the Cyprinidae. There are over 1,500 species of cyprinid fishes, and almost 200 species occur in the United States. Most are small, silvery fishes less than six inches long, so similar in appearance that few people bother to identify them, lumping them all together as minnows or gudgeons. The carp, *Cyprinus carpio,* the exception to

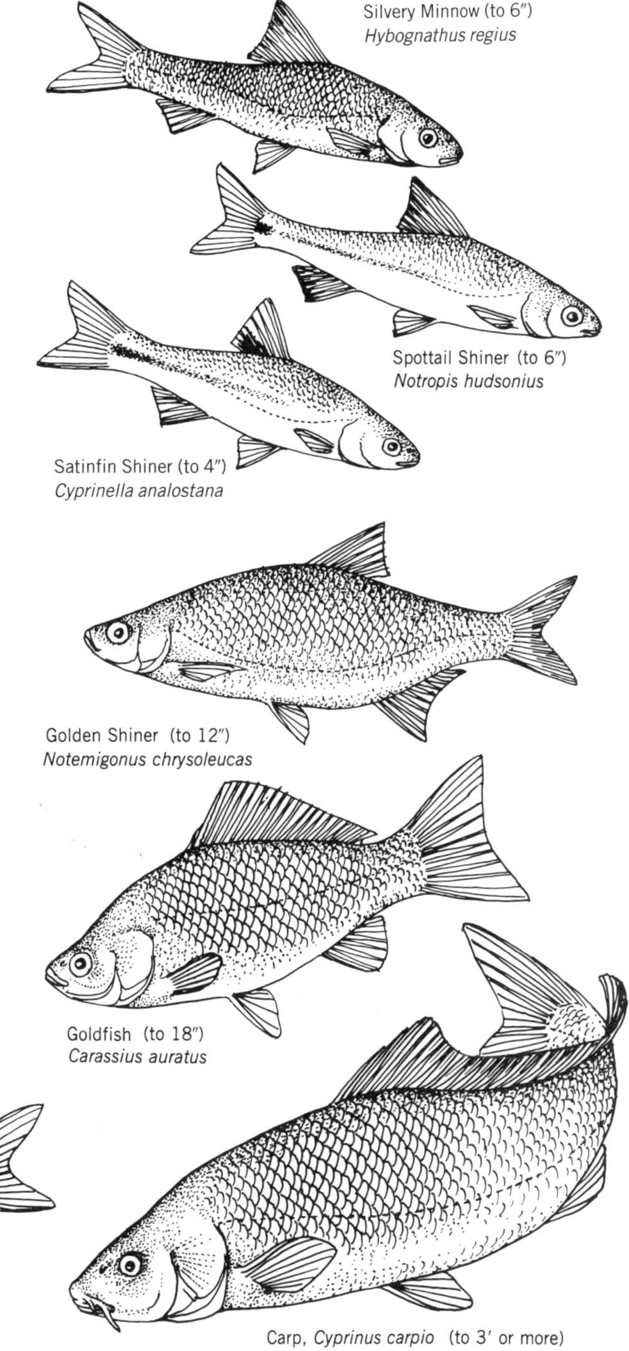

Silvery Minnow (to 6″)
Hybognathus regius

Spottail Shiner (to 6″)
Notropis hudsonius

Satinfin Shiner (to 4″)
Cyprinella analostana

Golden Shiner (to 12″)
Notemigonus chrysoleucas

Goldfish (to 18″)
Carassius auratus

Creek Chubsucker (to 10″)
Erimyzon oblongus

White Sucker, *Catostomus commersoni* (to 20″)

Carp, *Cyprinus carpio* (to 3′ or more)

this large group of small fishes, reaches three feet in length and may exceed 50 pounds. Carp are effectively fished for with dough balls, worms, or practically any other bait. They are not a popular food fish in this country, but are highly prized by the Chinese, the Japanese, and by many Europeans. Carp were introduced into the United States many years ago and have adapted extremely well. They have large, distinct scales and much longer dorsal fins than minnows have.

People are often surprised to learn that goldfish, *Carassius auratus,* the same aquarium species sold in pet stores, have been collected from a number of locations in the Chesapeake Bay. From time to time they are apparently dumped into streams from home aquariums and fare quite well in the wild.

Of the dozens of smaller minnow species of Maryland and Virginia, only four are common to brackish waters—the golden shiner, silvery minnow, spottail shiner, and satinfin shiner. The golden shiner, *Notemigonus crysoleucas,* with its rich green-bronze color, is the largest and most attractive of the group. In spring, males are golden and emblazoned with brilliant orange-red fins. Golden shiners grow to over six or seven inches and are sometimes caught by anglers in turbid waters. Young golden shiners have dusty bands along their sides, as do other types of minnows.

The silvery minnow, *Hybognathus regius,* and the spottail shiner, *Notropis hudsonius,* are small silver minnows that are often assumed to be the same species; both are referred to as gudgeons. The silvery minnow is greenish above with silvery sides and may have a bright dusky band along the sides toward the tail. The spottail minnow is similar in shape and general coloration except for a dark band along the midline of its back, visible when the fish is seen from above. When the spottail is fresh from the water, a narrow silvery band can be detected along its side, the silver color sometimes overlying a darker band. A dark spot of varying size and intensity is located at the base of the tail fin. The satinfin shiner, *Cyprinella analostana,* is somewhat easier to identify; the dorsal fin is located back on its body behind the pelvic fin and, when depressed against the back, overlaps the anal fin. It also has a dark side-band toward the tail and a distinguishing dark pigmented spot on the back edge of the dorsal fin. Minnows often live in the same areas as mummichogs and other killifishes, but they are quite different in general appearance, minnows being slenderer, more streamlined fish with forked tails.

Suckers. Suckers are closely related to minnows, but they have soft, protrusible mouths and fleshy, sucking lips on the underside of their head. The mouth is especially adapted for feeding off the bottom. Half a dozen species are found in Chesapeake streams above tidewater, but only two species frequently move downstream into brackish regions. The white sucker or common sucker, *Catostomus commersoni,* can be quite numerous in many larger streams and in smaller streams in springtime, when it spawns. White suckers are elongated round fish of a rather indiscriminate, smooth coloring without any distinctive markings. The young, 2 to 3 inches long, have splotches that fade as they grow. The little creek chubsucker, *Erimyzon oblongus,* is a small sucker, of 10 inches or less, that frequents more sluggish streams and swamps that feed into brackish waters. It has the distinctive sucker mouth. Creek chubsuckers are coppery or brownish green, often with vague spots along the side. They have no lateral line, in contrast to white suckers, which have a distinct lateral line.

Sunfishes. The bright little saucer-shaped sunfishes, speckled and brilliantly colored, are a group of beautiful freshwater fishes. The family includes not only sunfishes, but also crappies and the voracious large freshwater basses, some of the most sought-after freshwater sport fishes. A dozen or more species of sunfishes are distributed in Maryland and Virginia, but, as with other freshwater families, only a few are common to tidal waters. The pumpkinseed, *Lepomis gibbosus,* and the bluegill, *Lepomis macrochirus,* are highly adventuresome and move well down into estuarine streams. Both congregate in shallow protected coves of the larger tributaries in summer; in winter, they tend to school in deeper channels. They are sparkling, multicolored fishes—blue and green, yellow and orange. They are much alike in appearance, but the pumpkinseed may be identified by the bright orange crescent border on its ear flap and a series of blue and orange stripes across the lower part of its head. The bluegill has a black earflap and a dark blotch at the base of its dorsal fin. These two species are able to interbreed, so confusing specimens occur.

Crappies are also saucer-shaped but are not as brilliantly speckled as sunfish. They are limited to upper tidal regions, more typically in weedy areas than in open shallows. The black crappie, *Pomoxis nigromaculatus,* also called calico bass, or speckled bass, is widespread in most streams of the Chesapeake Bay. The white crappie, *Pomoxis annularis,* is far less common. Largemouth bass, *Micropterus salmoides,* are at times abundant in sluggish streams, particularly in the shallows and weed beds. It, too, wanders downstream into moderate salinities but not generally into the mainstream of larger tributaries as sunfishes do. Smallmouth bass, *Micropterus dolomieu,* are not nearly as common, nor do they move as far into tidewater. They are absent from Eastern Shore streams. They may be differentiated by the position of their eye and mouth: in smallmouth bass, the lower jaw does not extend beyond the eye; in largemouth bass, it extends well behind the eye.

Pickerels and Others. Pikes or pickerels are long, slender fishes with broad, spatulate snouts. The lurk in shallow weeds then suddenly dart out and capture an unwary prey. They are voracious feeders, eating fishes up to almost half their own size, as well as frogs, crayfishes, snakes, or almost any tasty morsel that comes their way. Pickerels are solitary fishes and prefer sluggish streams. Chain pickerel, *Esox niger,* move well down into the estuarine regions of Chesapeake Bay streams. They are large fish, up to two feet long. Their name comes from the distinct reticulated chainlike pattern along their sides. The redfin pickerel, *Esox americanus,* is a smaller pike, only a foot or so long

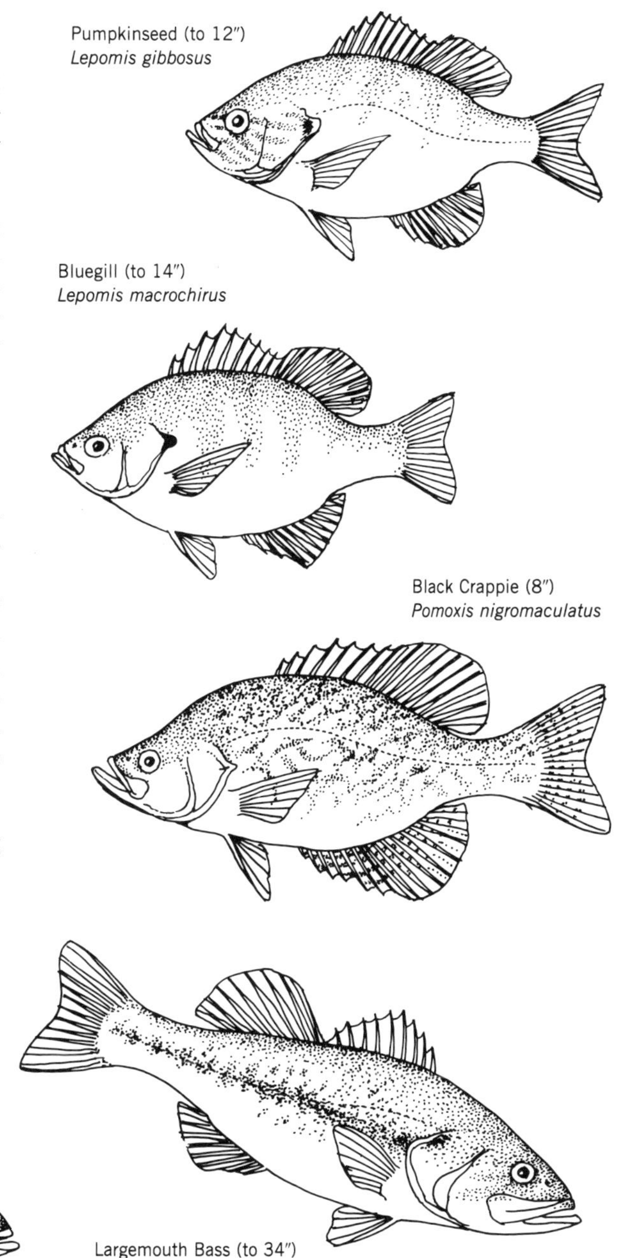

Sunfishes of the estuary. The white crappie, not illustrated, can be distinguished from the black crappie by the number of spines in the dorsal fin; the white crappie has six spines, the black crappie seven.

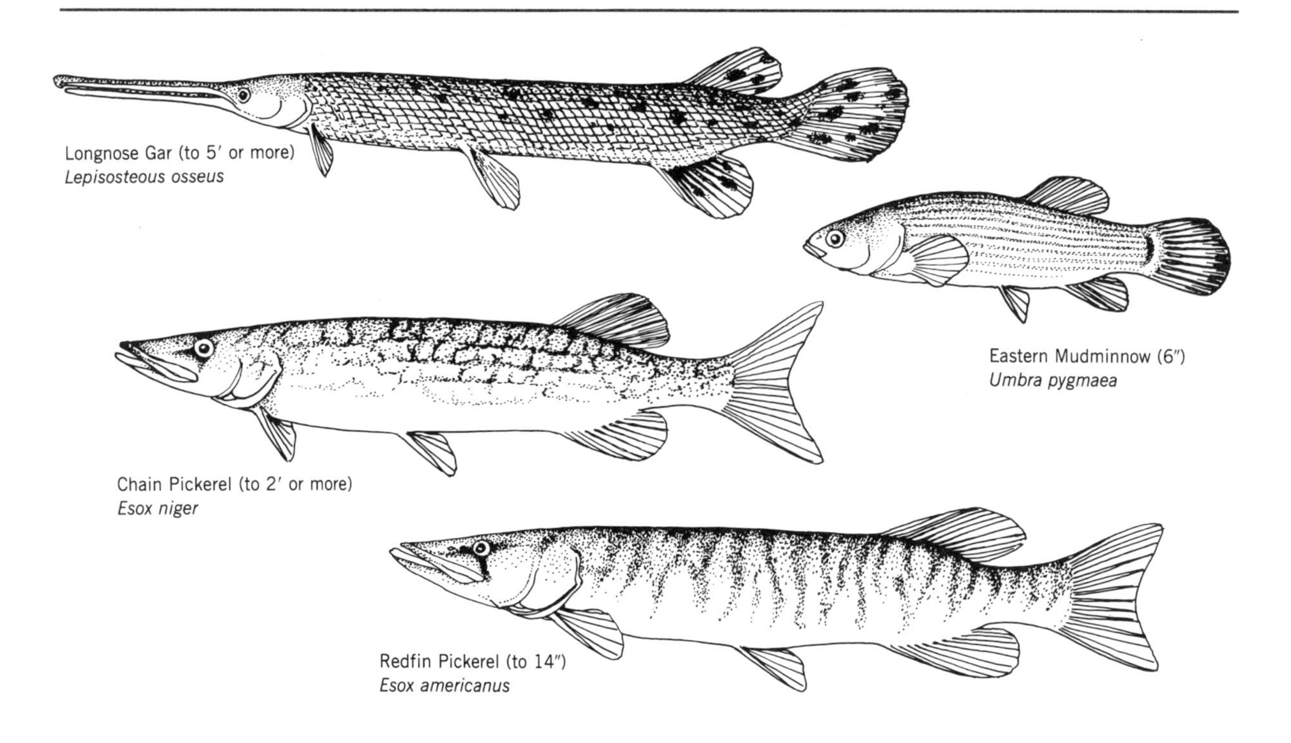

when fully grown, which lives in the same habitats as the chain pickerel and often interbreeds with it. Redfin pickerels have a barred pattern along their sides and can be confused with young chain pickerels, which have similar markings before the adult chain pattern develops.

The longnose gar, *Lepisosteus osseus,* is also an elongated fish of weedy freshwater and brackish streams. This species is a survivor of an ancient and primitive group that flourished during the age of the dinosaurs, before most modern fishes had evolved. Their scales are hard and square (*Lepisosteus* means "hard-scaled"), and their long-beaked mouths are armed with teeth, the top jaw having a pointed tip that overhangs the lower jaw. Gars frequent shallow mud flats as well as weedy areas.

The eastern mudminnow, *Umbra pygmaea,* is a small—bare six inches—unobtrusive inhabitant of muddy-swampy shallows of small streams and is often found downstream into slightly brackish waters. It has a flattened square head and a rounded tail fin. It buries itself in soft bottom silts, where its brown mottled coloration keeps it well hidden.

PLANKTON SOUP

Myriad forms of plants and animals float free in the waters of the shallow shore zones, composing the plankton, a soup bowl of tiny organisms on which the higher animals feed. It can be a broth or a stew depending on the number and size of the plankton species. There may be hundreds of thousands of individual organisms in a single quart of water, yet the water will appear clear. The organisms are so small as to be completely invisible to the naked eye. Another sample of water may be clouded with specks just large enough to be seen. Floating plants of the plankton are referred to collectively as the phytoplankton; floating animals, as the zooplankton. In addition, larger floating animals, such as jellyfishes, are included as part of the plankton.

Phytoplankton and Zooplankton

Phytoplankton (meaning "green-celled wanderers") are, typically, microscopic single-celled organisms called algae. There are many groups of phytoplankton in the Chesapeake Bay, such as green algae, blue-green algae, dinoflagellates, and diatoms. In tidal freshwater and in slightly brackish waters, blue-green algae can become so thick that they color the water green. Some blue-green species congregate thickly near the surface, and as the cells die they aggregate, along with the minuscule organisms that feed on them, to form what is called water flowers. Billions of these tiny water flowers entangle at the surface to form large floating mats. In higher-salinity regions, dinoflagellates are the more important component of the plankton. Dinoflagellates have whiplike flagella with which they can propel themselves in a vertical spiral path. Heavy blooms of certain dinoflagellates cause the water to turn a mahogany color, a condition referred to as "mahogany tide." Phytoplankton, like all green plants, convert sunlight into living organic material. They, in turn, are consumed by small animals, the zooplankton, that are themselves consumed by other animals; thus they are a basic foundation of life in the estuary. The greatest proportion of animals feeding on microscopic algae are minuscule forms themselves, including many, many types of planktonic larval stages of invertebrate animals. Water fleas are an abundant component of plankton, although most species are too small to be seen. An exception is what can only be de-

MICROSCOPIC PLANTS: THE PHYTOPLANKTON

1. Dinoflagellate
2. Dinoflagellate
3. Diatom
4. Green Alga
5. Golden-Brown Alga

MICROSCOPIC ANIMALS: THE ZOOPLANKTON

6. Barnacle Nauplius Larva
7. Barnacle Cypris Larva
8. Polychaete Worm Larva
9. Snail Larva
10. Oyster Larvae
11. Crab Zoea Larva
12. Herring Egg
13. Striped Bass Larva
14. Sea Nettle Ephyra
15. Copepod
16. Copepod
17. Giant Water Flea
18. Hydromedusa

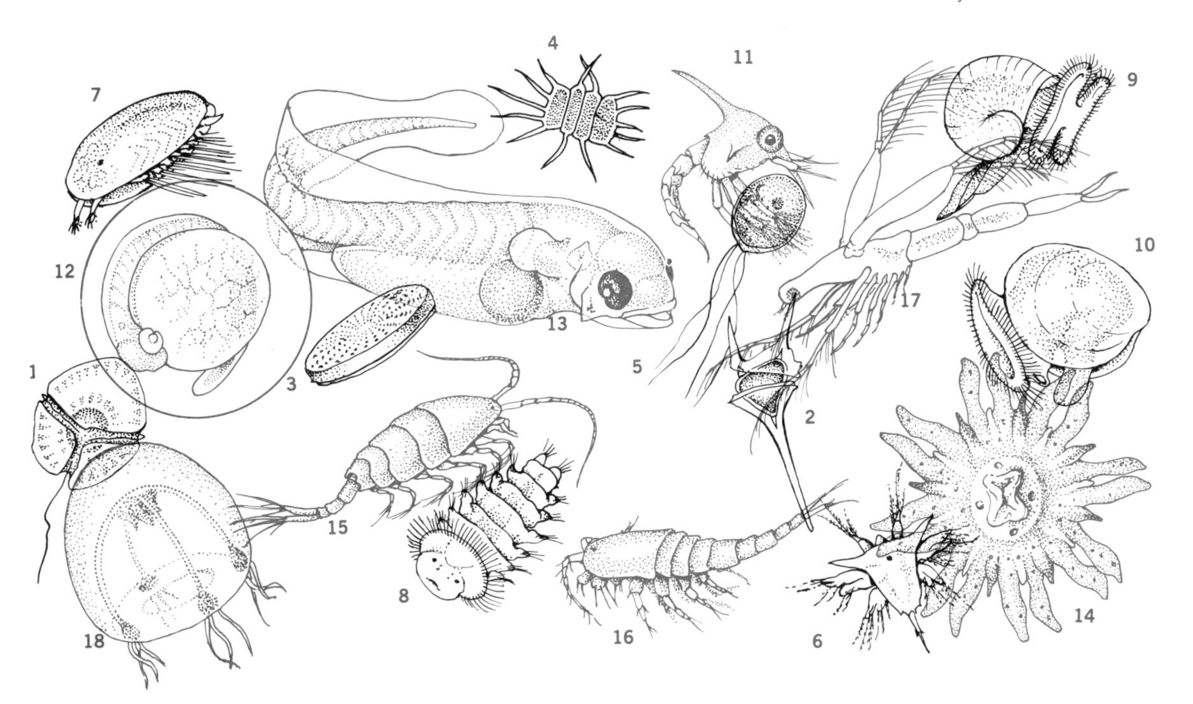

The planktonic organisms in black cannot be seen individually without the aid of a microscope. Those in blue are larger and can just be discerned by the unaided eye.

scribed as the weird, giant water flea, *Leptodora kindtii,* which looks like an extraterrestrial visitor with its single eye and two fringed "wings" splaying from its back. Giant water fleas are almost three-fourths of an inch long, but they are transparent and are easily overlooked. They are found primarily in tidal freshwater and slightly brackish water zones.

Copepods are considered by many scientists to be the most abundant group of multicelled animals on earth. They proliferate throughout the Chesapeake, in shallows and in open deep waters. There are 50 or more species of copepods in the Chesapeake Bay, but only a few species make up 95 percent or more of an average population. Fully grown copepods can just be seen in a jar of Bay water, jerkily propelling themselves about. As they move through the water, they create vortices that draw small food particles toward them. Some copepods feed on phytoplankton, others on detritus (decaying organic matter), bacteria, or protozoans (microscopic single-celled animals). Copepods, in turn, are a major food source for many larger invertebrates and fishes. Most larval fish would never survive without the dense populations of copepods that occur in the nursery grounds.

Sea Nettle
Chrysaora quinquecirrha

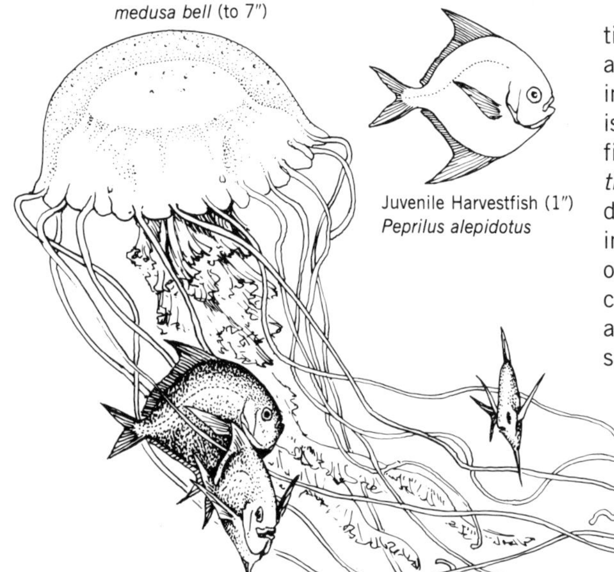

Juvenile Harvestfish (1")
Peprilus alepidotus

Juvenile Butterfish (1")
Peprilus triacanthus

Jellyfishes

Jellyfishes are the largest and best known of the floating plankton. The infamous stinging sea nettle, *Chrysaora quinquecirrha,* is the best known of Bay jellyfishes. Populations of sea nettles become so thick and widespread during most summer months that swimming, wading, waterskiing, and other water sports become a trial rather than a pleasure. Sea nettles are distributed throughout the mid and lower Bay and all its tributaries. They do not survive in fresh or low-salinity waters, so residents of the upper Bay and tributaries and the uppermost regions of lower Bay tributaries do not have to contend with them. Jellyfishes are capable of some propulsion accomplished by rhythmic contraction and expansion of the medusa bell. Nevertheless, they are not strong swimmers and are at the mercy of the currents and tides, which may transport them far and wide, concentrating them in some areas, clearing them from others. Sea nettles are carnivorous and opportunistic, entangling and capturing any prey that comes their way. Their tentacles (like those of their cousins, anemones and hydroids) are covered with stinging cells. After the prey is stunned it is passed through long ruffled mouth lappets up into the mouth, which lies under the center of the bell. An undigested small fish or shrimp can often be seen inside a jellyfish.

Sea nettles are almost 90 percent water—not a particularly hearty meal, yet a number of fishes, sea turtles, and crustaceans feed on them. Certain fishes seem to be immune to their poison. A curious symbiotic relationship is often found between the sea nettle and young harvestfish, *Peprilus alepidotus,* or butterfish, *Peprilus triacanthus.* Many times these flat little silver fish can be seen darting in and out of the long, floating tentacles like flashing half dollars. Apparently the young fishes have developed an immunity to the toxins in the stinging cells. Other common symbionts are small spider crabs, which travel along on top of the medusae and sometimes hollow out a space in the bell.

LIFE CYCLE OF THE SEA NETTLE

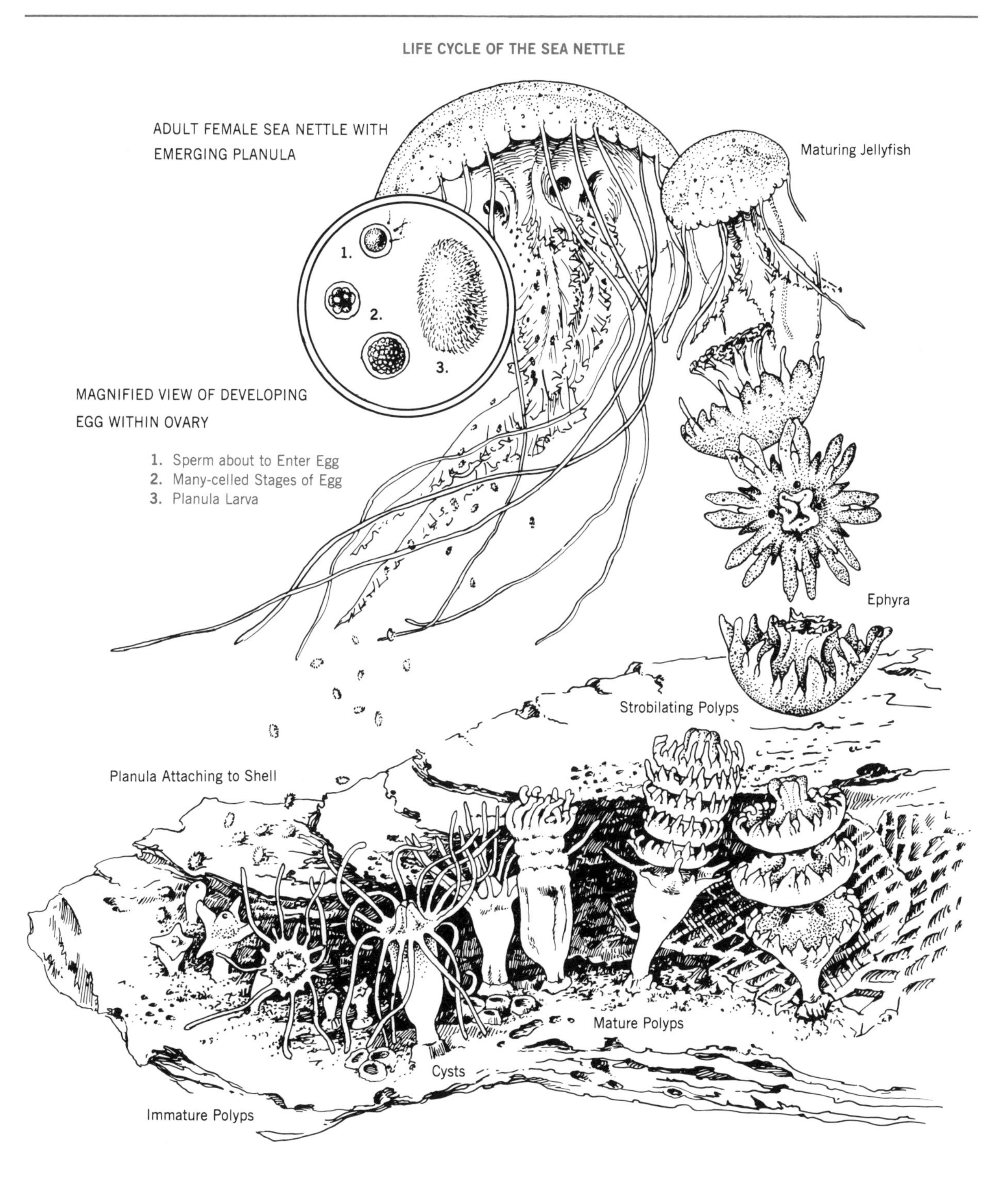

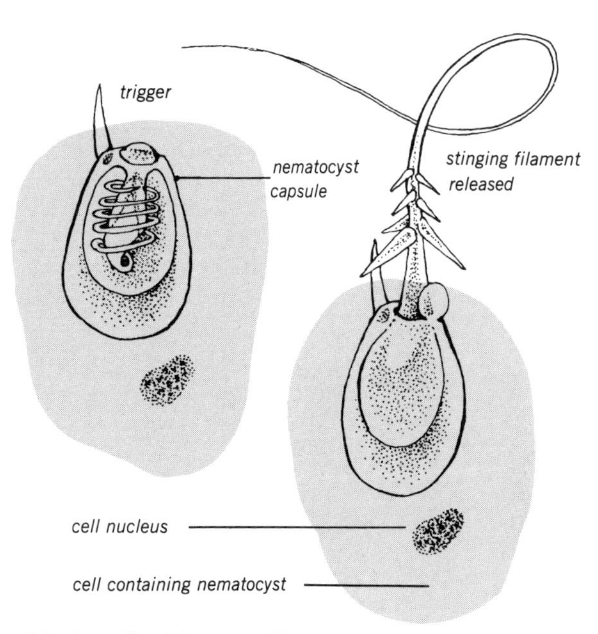

Stinging cells of the sea nettle.

When a swimmer touches a jellyfish tentacle numerous stinging cells (nematocysts) release microscopic darts, which can easily penetrate thin-skinned regions of the body such as the arms or legs. The skin of the palm of the hand is thicker, so if you care to impress the crowd, carefully pick up a sea nettle by the top of the bell, where there are no nematocysts, taking care to expose only the palm to any stray tentacles. Each nematocyst is a capsule encasing a long, coiled thread coated with venom and equipped with a "trap door," and a hairlike "trigger" on the surface. The threadlike dart is quickly everted as soon as the trigger is touched. Each tentacle is armed with hundreds of nematocysts. The venom produces a stinging sensation and inflammation in humans, which in most cases soon disappears.

No matter how many sea nettles there may have been during the summer, by late fall they are gone. Sea nettles grow and mature throughout the spring and summer. By mid-summer, ripening gonads, shaped like a four-leaf clover, can clearly be seen through the medusa bell. Male gonads are mostly white to pink; those of females, olive to dusky-colored. Mature males release sperm into the water, which is pumped into the females' bell, where the eggs are fertilized. Each egg inside the bell eventually develops into a tiny, ciliated, planula larva, smaller than a grain of salt. Planulae escape into the water and become part of the plankton. Mature sea nettles die after spawning. Planulae settle to the bottom after a few days and attach to a hard substrate. A preferred location is the underside of oyster shells, but they will attach to other shells and to bottles, cans, and other suitable substrates as well.

After attaching, the center of the planula soon bulges out and develops into a flowerlike polyp (much like a miniature sea anemone). Sea nettle polyps are able to reproduce many times over by budding off bits of tissue, each of which grows into a new polyp. Polyps can also form cysts, which is a protective stage that allows them to endure periods of environmental stress.

With warming spring temperatures, sea nettle polyps produce tiny floating medusae through a process called strobilization. The outer end of the polyp forms into a series of discs that looks like a stack of saucers. The saucers are eventually released and float free in the water as tiny medusae, called ephyra, only one-sixteenth inch in diameter. When ephyra grow to about one-fourth inch, they develop tentacles and mouth lappets and soon begin to look like adult sea nettles. In the Chesapeake, sea nettle ephyra appear in smaller creeks and streams by late April and May and move out into the larger tributaries and the open Bay by early summer.

The lion's mane or winter jellyfish, *Cyanea capillata,* is sometimes as abundant in the Bay as sea nettles, but it occurs only during winter and spring months and is thus not nearly as well known to the summer-oriented visitor to tidewater. Lion's mane jellyfish ephyra appear in the Bay by late November or December and grow to adulthood through the winter. Lion's mane jellyfish disappear by the end of May or the beginning of June.

Lion's mane jellyfish are orange-brown with relatively short tentacles and mouth lappets. They are about the same size as sea nettles but are not nearly as formidable, having a comparatively weak sting. In the northern Atlantic, they grow to be much larger. They may be carried by winds and currents into low-salinity waters, but they reproduce only in the higher salinities of the lower Bay (Zone 3).

The moon jellyfish, *Aurelia aurita,* is the largest of the Chesapeake Bay jellyfishes, growing to more than a foot in diameter. The bell is nearly flat, white, with dark horseshoe-shaped pink gonads at the center and with a mere fringe of short tentacles at its margin. Moon jellyfish are

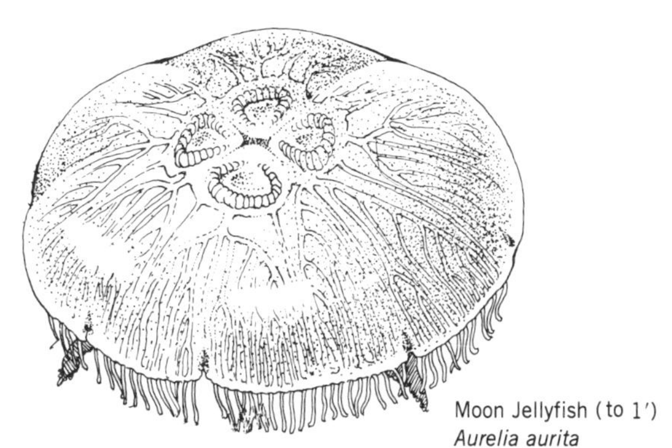

Moon Jellyfish (to 1′)
Aurelia aurita

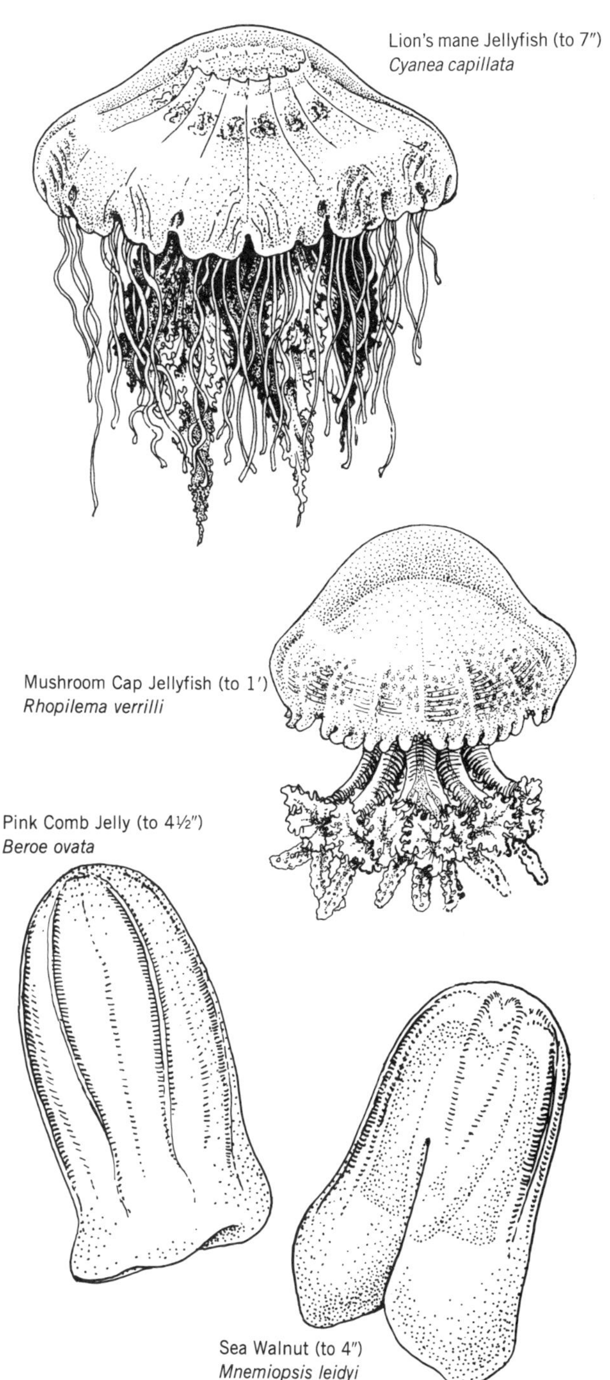

Lion's mane Jellyfish (to 7″)
Cyanea capillata

Mushroom Cap Jellyfish (to 1′)
Rhopilema verrilli

Pink Comb Jelly (to 4½″)
Beroe ovata

Sea Walnut (to 4″)
Mnemiopsis leidyi

sometimes abundant in the lower Bay, less so up-Bay, and they generally appear later in the summer than sea nettles.

The mushroom cap jellyfish, *Rhopilema verrilli,* is an occasional ocean intruder into the Bay in autumn and early winter. It has a thick, deep bell, without tentacles, over a series of strong mouth-arms. The mushroom cap jellyfish is creamy white with darker brown or yellow markings on its mouth-arms.

Comb Jellies

Comb jellies far surpass sea nettles and other stinging jellyfishes in numbers in the Chesapeake Bay. Comb jellies—or ctenophores, as they are labeled by scientists—are omnipresent throughout the year but occur in greatest numbers in spring and summer. Comb jellies are transparent and hard to see during the day, but at night, if you disturb the water just slightly, they will luminesce with a soft green light. They have no nematocysts and therefore do not sting. They feed voraciously on planktonic organisms by continuously pumping sea water into the cavity within their globular bodies; they have been known to consume almost five hundred copepods per hour. Comb jellies break apart easily when removed from the water. To see their true shape and graceful beauty, capture one by dipping it out with a clear glass container and observe it through the glass. Brightly iridescent lines of color will be seen to run in bands along its body. These are rows of cilia, which propel the animal by beating against the water within the body cavity. Two species of comb jellies are found in the Chesapeake Bay: the sea walnut, *Mnemiopsis leidyi,* and the pink comb jelly, *Beroe ovata.* Sea walnuts, as you might suspect, are shaped like walnuts and are colorless. They are the more widely distributed species, found from upper Zone 2 and throughout Zone 3. The pink comb jelly is shaped more like a simple sac and is limited to higher salinities (lower Zone 2 and Zone 3).

INSECTS OF THE SHALLOWS

There are many species of aquatic insects that spend all or part of their lives in the water, some only as larvae, such as certain beetles and gnats, and other species that are almost totally aquatic and develop into a specialized larval form from eggs deposited on the underside of a submerged leaf. They leave the water for a short time to undergo transformation to adulthood and then return to the water. Their many methods of reproduction mark the success of insects as a group.

When certain species of insects hatch from the egg, they greatly resemble the adult; entomologists call this method of reproduction simple metamorphosis. As usual in the sphere of living organisms, there are predictable exceptions. The dragonfly undergoes simple metamorphosis: it develops from the egg into an immature form and then into the adult; but the nymph, or immature form, hardly resembles the adult. Then there are those insect species that hatch from the egg into a soft-bodied, wormlike stage, called a larva. The larvae, variously called caterpillars, grubs, and maggots, increase in size and complexity as they molt. After several molts, depending on the species, the larva transforms into a nonfeeding stage called the pupa, and finally into the adult. The process of development from the egg into a larva and pupa and, ultimately, to an adult is called complete metamorphosis.

Some insects require brackish to fairly salty water to carry out their complete life cycle from egg to adult, and some can tolerate only slightly brackish water, as they are actually freshwater species. The familiar mosquitoes and no-see-ums spend their early life stages in the water; however, we know them best when they fly through the smallest holes in a screen and make their presence only too well known. There are also many species of insects which remain closely tied to the water, where they feed on minute plants or other creatures. The tidal freshwater reaches of many of the Chesapeake Bay tributaries harbor many aquatic insects. Some species can tolerate a relatively swift current; others inhabit the quiet backwaters of a small creek. Certain species spend most of their time underwater carrying their air supply in bubbles attached to hairs on their bodies; other insects are surface dwellers and seldom spend time below the surface of the water.

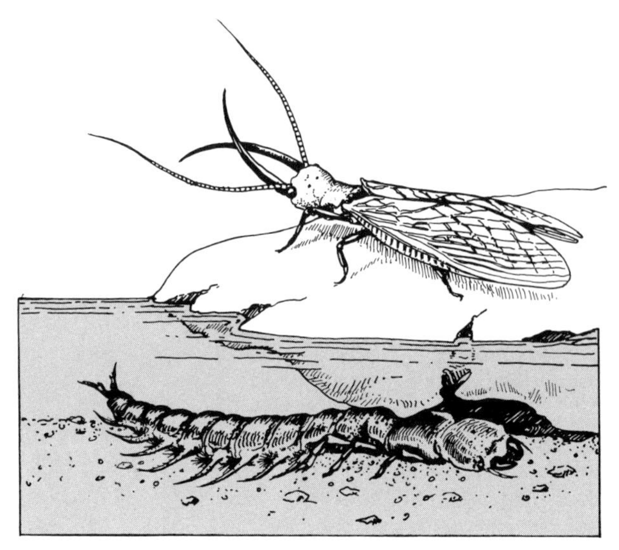

A male dobsonfly, *Corydalus cornutus* (to 2 inches), rests on an overhanging rock harboring a dobsonfly hellgrammite larva (to 4 inches) lurking below.

The Dobsonfly, *Corydalus cornutus,* is not really a fly at all, but a very large winged insect with enormous mandibles (tusklike jaws), especially in the male. Dobsonflies belong to a group of insects called Megaloptera, which means "great winged." Ironically, these impressively winged creatures, some of which have a wingspread of four and one-half inches, are fluttery, weak flyers. The tusklike mandibles are quite impressive and look formidable, but they are little used: the nonfeeding adults are short-lived, and after mating and depositing their eggs they perish.

The female lays eggs on overhanging branches, on stones, or on the underside of bridges and piers in fresh and brackish waters. Hatching occurs in about five to six days, and the larvae then drop into the water. These fiercely predacious larvae, known to freshwater fishermen as hellgrammites, begin to prey on almost any organism they can subdue, including their own kind. Hellgrammites have powerful jaws, paired lateral filaments extending from each abdominal segment, and gill-like tufts at the base of each filament. The dobsonfly larva is long-lived, sometimes existing as a hellgrammite as long as three years. The larvae live under rocks and are often found where there is swift-flowing water such as around bridge abutments and other structures. The mature hellgrammite crawls out of the water and constructs a pupal chamber in the bank or in a rotting log; a short time later, the handsome dobsonfly emerges and prepares for the next generation.

The predacious diving beetle, *Dytiscus* spp., is another fierce underwater denizen. The female lays individual eggs in plant stems; they may hatch in 5–8 days, or perhaps not for months, depending on temperatures. The larvae, aptly called water tigers, emerge and begin feeding by grasping prey in their sicklelike jaws. They will eat anything they can overwhelm, including fish and tadpoles larger than themselves. Their sharp jaws are channeled; when their prey is pierced by the jaws, a toxic digestive juice is injected through the channels, which kills the organism and initiates digestion. The water tiger then sucks up their prey's tissue in its mandibles. The larvae cannot store oxygen or breathe underwater, so that from time to time they come to the surface of the water and break the surface film with their abdominal tip or terminal hairs and take in air through specialized pores called spiracles. When mature, the larva crawls out of the water and constructs a pupal chamber in a moist area. After several weeks, the adult predacious diving beetle emerges and enters the water, where it spends most of its time. However, this beetle is also attracted to lights and is often seen clinging to the side of a building.

An adult predacious diving beetle, *Dytiscus* sp. (to 1⅝ inch), swims with its oarlike legs. A ferocious water tiger (to 2½ inches), the larval stage of the predacious diving beetle, comes to the surface to breathe through spiracle pores on its abdominal tip.

Mature predacious diving beetles have specialized hind legs, which are used like oars rather than being moved alternately as in most swimming beetles. The adult, like the larva, obtains a new supply of air by breaking through the surface film or from bubbles on submerged aquatic plants. Predacious diving beetles are found in saltmarsh areas, as well as in freshwater zones of upper tributaries throughout the Bay.

In shallow waters of the upper tributaries, where the water is fresh, or almost so, there are insects that propel themselves across the surface of the water in varying ways—scuttling here, whirling there. They are almost always gregarious. Some of these insects belong to the Order Hemiptera, which means "half wing." Their first pair of wings is leathery at the base and membranous towards the tip; they also have sucking mouth parts, usually in the form of a slender beak. They are known as "true bugs."

The water boatman, *Corixa* spp., is one of the true bugs that can live in both fresh and brackish waters. It is an elongate, oval bug; its head and eyes are as wide as its body, and it is mottled gray and brown with yellowish cross bands. Water boatmen's short forelegs are modified into scoops, which enables them to feed on algae and plant particles as well as on minute organisms. Their middle and hind legs are flattened and fringed with hairs, a design that supports them on the water surface as they scull about. In the spring the adult female attaches her eggs to various water plants and sticks. The eggs hatch in one or two weeks; after several intermediate stages, a new generation of water boatmen are propelling themselves across the water's surface. The adults can fly and often end up in swimming pools and birdbaths. They are relatively defenseless and are often preyed upon by more predacious insects and fish. Water boatmen are a favored food for birds and fishes, and their eggs are considered a delicacy by humans in Mexico, where they are gathered from lakes by the tons.

The water strider, *Gerris* spp., is another true bug; it is also known as a pond skater. Some species of water strid-

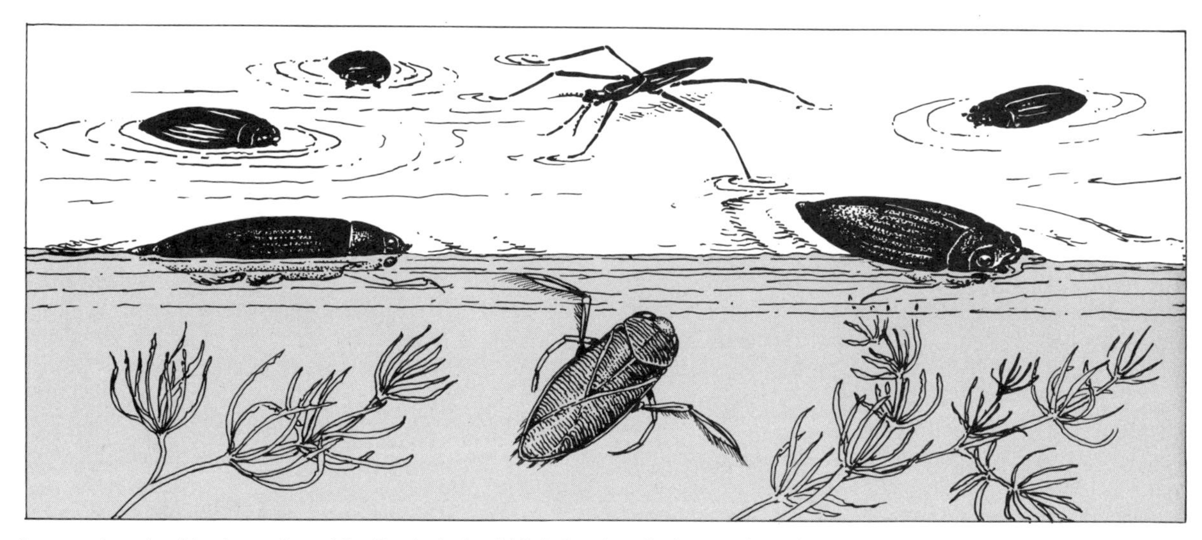

Common insects of fresher waters of the Bay include whirligig beetles, *Gyrinus* sp. (to ¾ inch), twirling on the surface; a water strider, *Gerris* sp. (to ¾ inch), walking on the surface film; and a water boatman, *Corixa* sp. (to ½ inch), rising to the surface.

ers prefer ponds and lakes or quiet backwaters of streams; others inhabit swiftly flowing water, and one species even lives on the surface of the open ocean. Water striders have very long legs which are adapted in such a way that their legs do not break through the surface film of the water—actually, the film is bent or depressed, but not pierced. They are very agile and quick and dart about in large numbers preying on all sorts of organisms, including members of their own species. The female cements her eggs in parallel rows to objects floating along the shore; after several weeks and several stages, new adult water striders emerge.

Whirligig beetles, *Gyrinus* spp., also known as waltzing beetles and scuttle bugs, actively whirl and gyrate on the surface of the water. Their scientific family name, Gyrinidae, means "to whirl and turn about in a circle." Whirligigs are oval, black beetles with compound eyes which are divided into two parts, allowing them to see below and above the surface while swimming. These beetles congregate in large swarms or clusters: often there are three or more species of whirligig beetles mixed together. They are commonly found in fresh or slightly brackish water of the upper Bay and tributaries and are generally quite active during the day.

The adults can dive readily and are good flyers. Unlike water striders, whirligig beetles break through the surface film of the water but are supported by their bodies. When a whirligig beetle dives below the surface, it takes in air and stores the quicksilver-like bubble under its wing covers, where the respiratory pore is located. The female lays eggs attached to submerged plants and sticks, sometimes in rather deep water. Predacious larvae develop from the eggs in one to three weeks and feed on other insect larvae, worms, and even fish, which they kill by injecting a poison through a canal in their curved mandibles. The larvae have feathery gills attached to their abdominal segments, which means that they are not dependent on surface air. The mature larva emerges from the water and crawls up the stem of a cattail or threesquare sedge and constructs a pupal chamber in which it remains until it emerges as a whirligig beetle adult.

CRUSTACEANS OF THE SHALLOWS

Crustaceans come into their own in the shallow-water habitats. Numerous species of amphipods and isopods, shrimps, and crabs occur here in great abundance. Most small crustaceans of the intertidal flats are quiescent when the tide is out and become active when the water returns. Other crustaceans will rarely, if ever, be found intertidally but may be quite abundant close to shore in very shallow waters.

Isopods

The slender isopod, *Cyathura polita,* is a typical, subtidal estuarine isopod frequently seen over sandy-bottom shallows. It occurs throughout the Bay but proliferates in upper regions of the estuary. It is a brownish cylindrical isopod, unlike the more typically flattened isopod, about an inch or so long with a fanned-out tail. Look for it over debris-covered bottoms.

Often young shore fishes are caught with "bugs" just visible under their gill flaps. Small striped bass, white perch, bluefish, and silver perch are frequently infested with the parasitic fish gill isopod, *Lironeca ovalis,* which clings tenaciously with its many hooked feet to the gill filaments. These parasites apparently do not feed directly on the host, but they became so large—perhaps an inch or so on a four-inch fish—that they destroy much of the gill tissue, stressing the fish and stunting its growth. Another parasitic isopod, the fish mouth isopod, *Olencira praegustator,* invades the mouth cavities of menhaden—*praegustator* means "before the taster." It grows to over an inch long, surely a bothersome lump to its unlucky host.

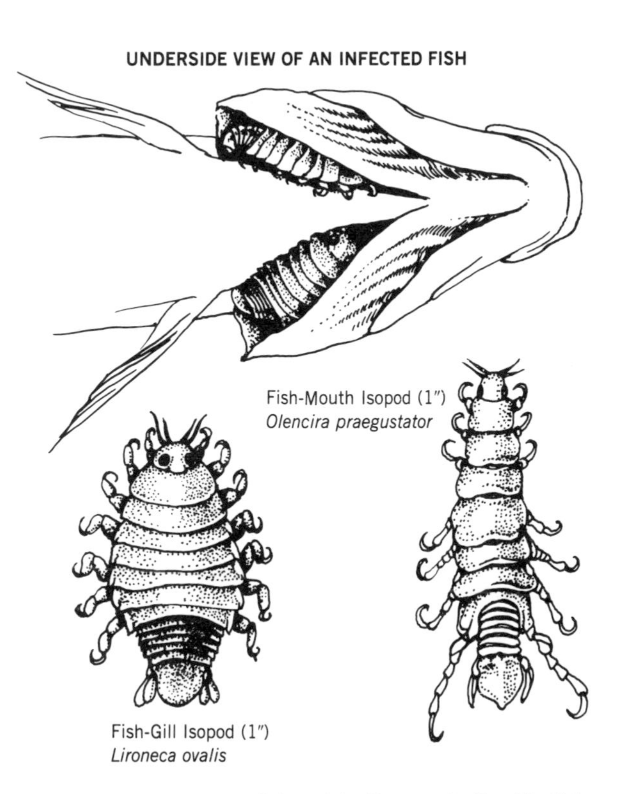

Parasitic isopods found on fishes of the Chesapeake Bay. The fish-gill isopod can easily be detected clinging to fish gills. The fish-mouth isopod attaches tenaciously to tissues in the cavity of a fish's mouth.

Shrimps of Many Sorts

Bay opossum shrimp are shrimplike creatures that carry their young in a specially adapted marsupium, or pouch, which gives them their common name. When released from the pouch, the young are well developed and able to fend for themselves. Opossum shrimp belong to the order Mysidacea and are, accordingly, also referred to as mysid shrimp. Bay opossum shrimp, *Neomysis ameri-*

cana, the common Bay species, are often abundant in shallow waters over sandy bottoms. They are only half an inch long and are transparent, with stalked eyes and fan-shaped tails. Opossum shrimp live on the bottom during daylight hours and rise into the water column at night to feed. Because of their nocturnal habits and small size, they are easily overlooked, yet they occur in dense populations throughout the estuary (Zones 2 and 3) and are an important food for many fish. Populations tend to be heavier in the tributaries than in the open Bay. Opossum shrimp, like many creatures of the estuary, move to deeper channels in the winter.

Few people suspect that market shrimps occasionally occur in the Chesapeake Bay. Though not found in quantities suitable for sustaining a commercial fishery, surprising numbers inhabit surprising places in the Bay. The pink, white, and brown shrimps, which form the basis of one of the most important fisheries in the United States, are harvested by the familiar shrimp trawlers, with their stubby bows, squat pilothouses, and flaring net booms, from North Carolina south to the Gulf of Mexico.

For various reasons—successful spawning and consequent high production, errant ocean currents, or salubrious climate—these large, highly valued crustaceans occasionally enter the Bay and, depending on the species, are sometimes found in low-salinity areas of Zone 2 as well as in saltier open waters of the lower Bay.

Penaeid species, the pink, white, and brown shrimps, spawn in the offshore waters of the ocean, where the eggs hatch and develop through a series of larval stages. The juveniles enter the estuary in spring and ascend tidal creeks, whose muddy habitat they seem to prefer. As fall ap-

SHRIMPS OF THE CHESAPEAKE BAY

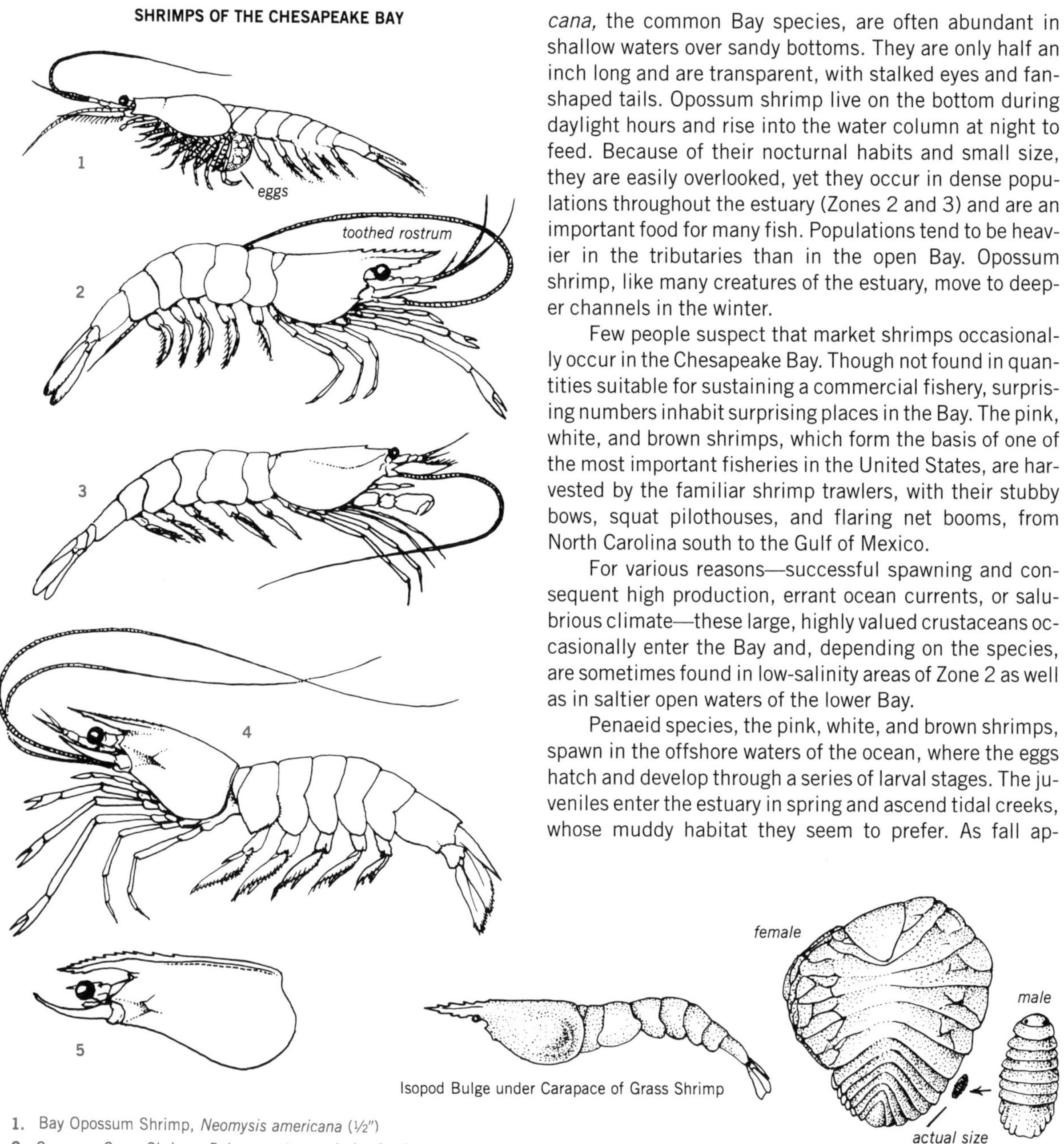

Isopod Bulge under Carapace of Grass Shrimp

Shrimp Parasite Isopod, *Probopyrus pandalicola*

1. Bay Opossum Shrimp, *Neomysis americana* (½″)
2. Common Grass Shrimp, *Palaemonetes pugio* (to 1½″)
3. Sand Shrimp, *Crangon septemspinosa* (to 2″)
4. White Shrimp, *Penaeus setiferus* (5″–8″)
5. Rostrum of Pink Shrimp, *Penaeus duorarum* (to 8″), and Brown Shrimp, *Penaeus aztecus* (to 8″)

proaches, the juvenile shrimps, now approaching sexual maturity, move into the deeper waters of the lower Bay and finally into the open ocean, where the spawning cycle is once again set in motion.

Penaeid shrimps are burrowers, to a greater or lesser extent, depending on the species and the bottom type. They are rapid-growing and attain sexual maturity and a length of approximately six inches in about a year. Like most crustaceans, shrimps are opportunistic feeders, preying on small organisms when available, on microbial flora on sea grasses, and on detrital organic matter.

The penaeid shrimps are somewhat difficult to determine to species without some knowledge and study of the genital structures. The first three pairs of legs are clawed. The white shrimp, *Penaeus setiferus,* may be separated from its relatives, the pink shrimp, *Penaeus duorarum,* and the brown shrimp, *Penaeus aztecus,* by noting that the lateral grooves lying along the dorsal toothed comb, the rostrum, do not extend beyond the rostrum. It is equally difficult to separate species based on their coloration. If you have a number of live specimens of each species it may be possible to separate the species. Coloration, however, varies from one specimen to the next owing to their stage of maturation—juvenile or adult—the area in which they were collected—ocean or estuary—and the immediate habitat from which they were taken—vegetated or sand and mud.

In contrast to the penaeid shrimps, which have an obligatory oceanic phase, the sand shrimp, *Crangon septemspinosa,* and grass shrimps, *Palaemonetes* spp., are not tied to the ocean but are ubiquitous, lifelong inhabitants of the estuary. The sand shrimp, however, can and does inhabit the ocean and is widely distributed on the Continental Shelf from Cape Cod south to the Chesapeake Bay.

The sand shrimp, which has a maximum body length of two and one-half to three inches, has a heavy, robust body, which tapers to a narrow posterior. It is flattened from top to bottom rather than from side to side, as is the usual case. Only the first pair of legs terminates in a claw. The rostrum, or toothed comb, is inconsequential when compared to that of the penaeid or grass shrimps. The sand shrimp varies from a pale, undistinguished color to the more usual ash gray, peppered with black spots, which provides camouflage for protection from predators.

Sand shrimp are distributed throughout the Bay in lower Zone 1 and in Zones 2 and 3. During the warmer months of the year they are more abundant in the shallower parts of the estuary. The shadows of a school of shrimp on the shallow bottom are often mistaken for a school of small fish, the shadow being taken for the substance! A second look will distinguish the almost transparent bodies of the shrimp above. They hover slowly as they graze, swimming in a manner quite different from that of fish. As the temperatures decrease they tend to concentrate in deeper waters, and have been collected in the deepest parts of the Bay.

Egg-bearing females (the egg mass will be obvious on the underside of the central body) are common from fall through spring.

This little shrimp, along with the grass shrimps, can be readily collected with a small, easily constructed seine made of three to four feet of aluminum window screen tacked to a pair of broom handles. Draw the seine through aquatic vegetation and over sandy areas where the shrimp burrow.

Sand shrimp apparently eat other small decapods and invertebrate eggs as well as organic debris. They are fed upon in turn by the many fish species that frequent their habitat, including flounder, striped bass, and seatrout.

The common grass shrimp, *Palaemonetes pugio,* is a delicate, almost transparent little shrimp with a strongly toothed rostrum extending beyond the eyes. The first two pairs of walking legs are distinctly clawed and the body is compressed from side to side, in contrast to the sand shrimp. This species is the most abundant of the four species of *Palaemonetes* known to occur in the Bay and is found throughout most of the Chesapeake system from Zone 1 through Zone 3.

The eggs of the grass shrimp can be clearly seen through its transparent body. This little shrimp spawns in the summer and carries its eggs in a brood pouch, where they hatch and are released as the larval zoeal stage.

Grass shrimps are shallow-water inhabitants of the estuary and abound in aquatic vegetation. They are often parasitized by a parasitic isopod, *Probopyrus pandalicola,* which causes an easily observed bulge, generally in the gill area.

Crabs

The blue crab, *Callinectes sapidus,* is a pugnacious, aggressive crab widely distributed along the Atlantic and Gulf coasts. It is most abundant and perhaps best known

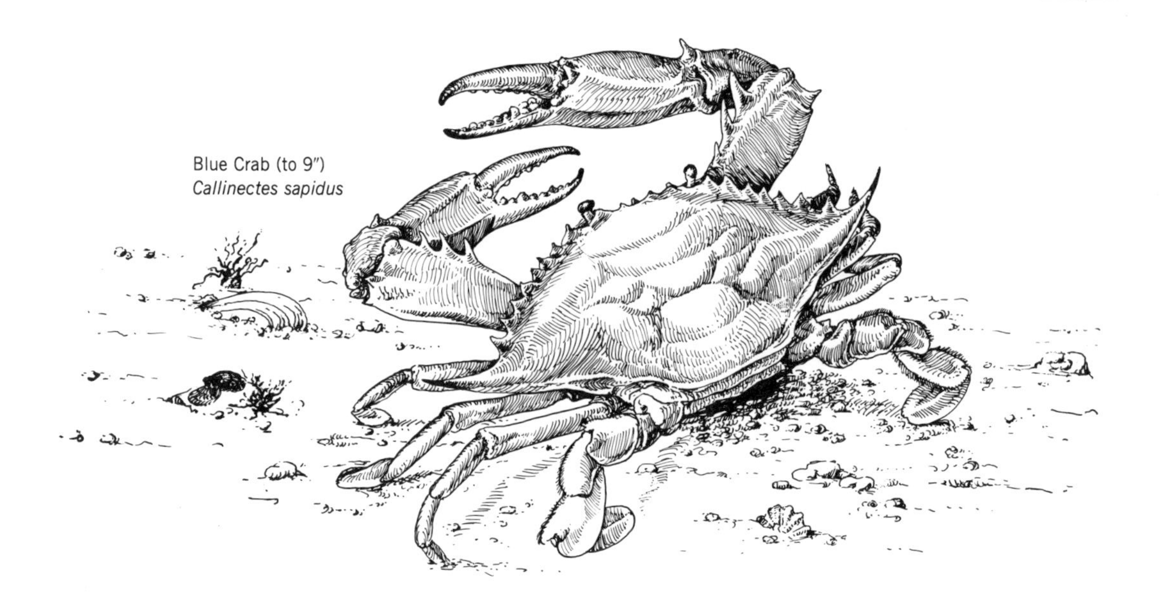

Blue Crab (to 9″)
Callinectes sapidus

from the Chesapeake Bay. Known as "the crab" in Chesapeake country, the blue crab is an active and proficient swimmer owing to its fifth pair of legs, which are modified into paddles. The musculature associated with the swimming paddles provides succulent morsels of seafood known as lump meat or back-fin.

The blue crab is a beautifully colored crustacean with bright blue claws—the mature female's fingers are tipped in red—and an olive to bluish green carapace. An important commercial and recreational species, "the crab" is caught with a variety of gear. An astounding 100 million pounds or more are caught in a good year. The blue crab ranges from the low-salinity waters of Zone 1 to waters of full-strength ocean salinity. The male ranges much farther into the upper Bay system than the female. Females congregate further downstream and down-Bay, where the water is saltier.

Mating occurs from June through October in the mid-Bay salinities. The male mates with a female that has just undergone her final molt and is in the soft-shell stage. After the female develops her new hard shell she begins a directed migration toward the salty waters of the lower Bay. It is common in the waning months of summer to see large numbers of individual females making their way down Bay, swimming along the surface with the aid of the ebbing tide. The majority of the males remain behind in the less saline

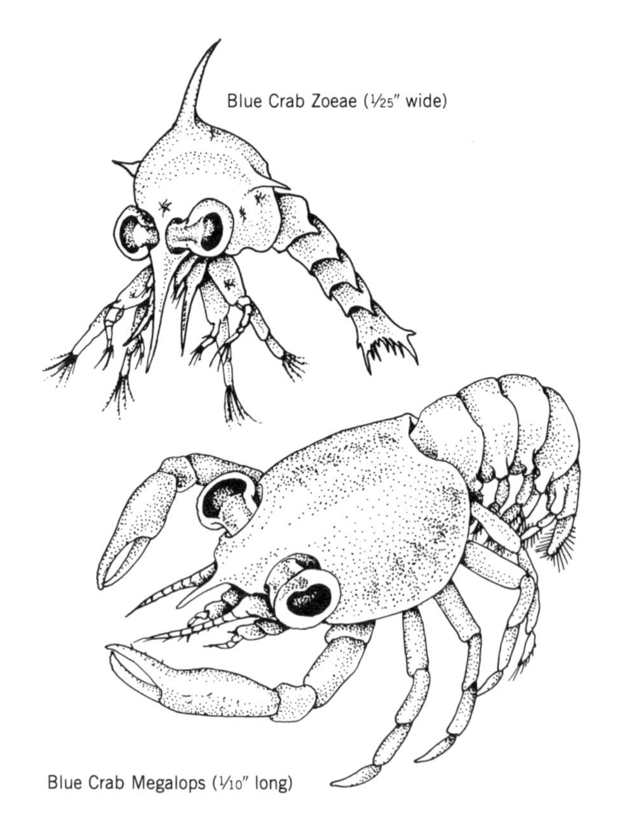

Blue Crab Zoeae (1/25″ wide)

Blue Crab Megalops (1/10″ long)

waters and overwinter in the muddy bottoms of deeper channel waters.

Blue crab females spawn near the mouth of the Bay from May to October. The sponge, or egg mass, which may contain three-quarters of a million to 2 million eggs, adheres to the undersurface of the crab. The color of the sponge, at first golden to orange, changes to black as the eggs near hatching. After a few weeks, small semitransparent zoeae larvae are released. Many of them are swept out into the ocean, where they begin a precarious, planktonic existence. There they mix with blue crab zoeae from other regions of the coast; eventually, onshore winds move the zoeae into estuaries such as the Delaware and Chesapeake bays. After a series of molts, the megalops, a second larval form, is produced. The megalops, easily mistaken for a tiny lobster or crayfish, moves along the bottom and begins to move up into the Bay system, where it molts into the tiny, but recognizable, blue crab. By the age of 12 to 16 months the crabs, which have molted several times, have reached an average size of five inches and have attained sexual maturity. The empty sheds of recently molted crabs washed up on the beach vividly demonstrate the abundance of crabs in the Bay.

Blue crabs feed on plant materials, clams, recently dead fish, oysters, and anything else to which they are attracted and can successfully attack, and are keen predators on the soft-shell stage of their own species. During the warmer months of the year they range the shallow waters, ghosting along the bottom and lurking in the cover of weed beds.

The lady crab, *Ovalipes ocellatus,* is a swimming crab related to the blue crab. It is a smaller species with a somewhat rounded carapace, yellowish gray, with small, reddish purple spots on its back. A species that inhabits high-salin-

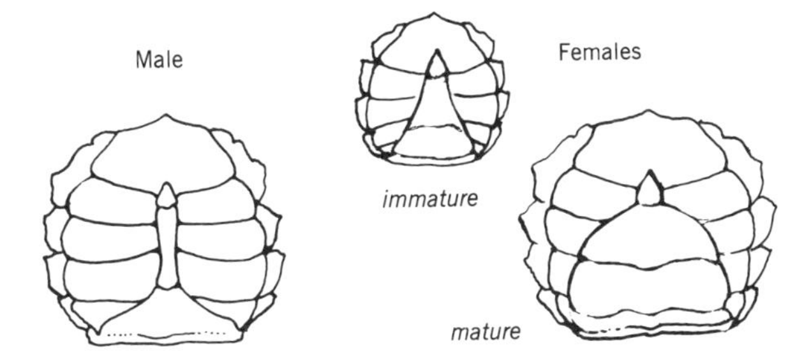

Sexual determination in a blue crab. The apron, or underside, of the crab easily reveals its sex.

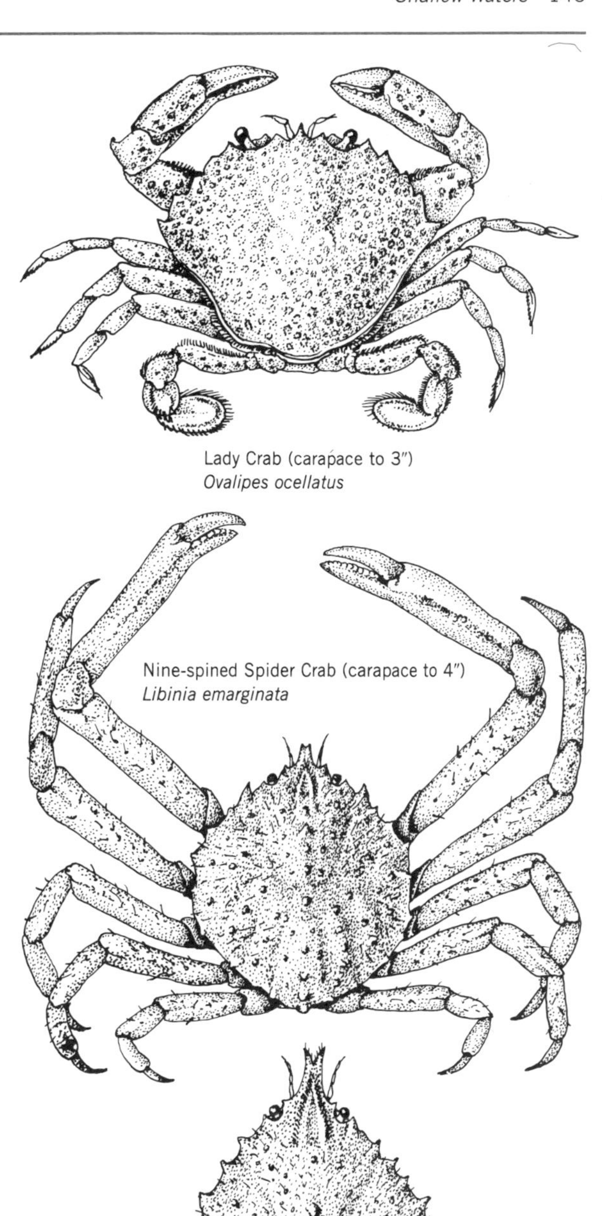

Lady Crab (carapace to 3")
Ovalipes ocellatus

Nine-spined Spider Crab (carapace to 4")
Libinia emarginata

Six-spined Spider Crab (carapace to 4")
Libinia dubia

ity waters in shallow subtidal areas on sandy bottoms, it is infrequently found in Zone 2 waters.

The nine-spined spider crab, *Libinia emarginata,* and the six-spined spider crab, *Libinia dubia,* are triangle-shaped, slow-moving crabs, found on a variety of bottoms from shallow water to depths exceeding 150 feet. They inhabit the waters of lower Zone 2 and are found throughout Zone 3. Their carapaces measure about four inches from front to back, and when their legs are stretched out they can reach a spread of one foot. The nine-spined spider crab has a median row of nine low spines on its carapace; the six-spined spider crab has only six. These sluggish, unaggressive crabs are khaki-colored and their carapaces are ornamented with various spines and tubercles. The younger crabs are often overgrown with sponges and hydroids.

CLAMS OF THE SHALLOWS

Most of the clams and snails of the Chesapeake Bay are well adapted to living out of water for periods of time; some, however, are not so tolerant and are rarely if ever found where they might be exposed. In the upper zones of the estuary, in tidal freshwater and in low-brackish waters, a number of bivalves are essentially subtidal animals, including two exotic species that have appeared in the Bay only in recent years and a variety of freshwater clams—or freshwater mussels, as they are also called. Most of the mollusks found on the intertidal flats are found subtidally, often in even greater abundance.

Freshwater mussels are large, yellow to dark brown or black bivalves, characterized by a heavy, hatchet-shaped foot and two short siphons. They bury themselves in the bottom, orienting their siphons to the current. Some species are buried only partially with most of the shell remaining out of the bottom, while others bury deeper, leaving only the siphon exposed. A number of freshwater mussels, particularly those belonging to the genera *Anodonta* and *Lampsilis,* occur within tidewaters. These species are large—up to four or five inches—so they are not easily overlooked. Freshwater mussels are able to move over the bottom surface by continually extending and retracting their foot horizontally, much in the same way that other clams burrow vertically. In this way they lurch across the bottom, leaving a distinctive trough-shaped trail behind. Their trails can sometimes be detected in clear, quiet shallows.

Unlike other mollusks of the Chesapeake, which produce veliger larvae, freshwater mussels produce a highly specialized larva called a glochidium, which look like microscopic two-valved clams or a set of false teeth. Glochidia are released from parent mussels and immediately sink to the bottom, where they lie on their backs with their valves open upward. The glochidium must come in contact with a fish within a few days or it will perish. When a fish brushes against the glochidium, its valves immediately close and clamp tightly to the gills, fins, or other parts of the fish's body. In a matter of days the fish tissue grows over and encapsulates the glochidium in a cyst, in which it remains and develops for two to four weeks. Extensive patches of small nodules of encysted mussel larvae are often discovered on fishes removed from freshwater, but they apparently do the fish no great harm. When a larva develops into a juvenile stage it looks very much like the adult mussel.

At one time freshwater mussels were highly prized for manufacturing into buttons, but the modern-day plastics industry has just about eliminated their use for that purpose. However, the Japanese pearl industry uses freshwater mussel nacre as seeds for producing pearls in Japanese oysters.

Two recent additions to Chesapeake Bay fauna have proliferated in the tidal fresh- and low-salinity waters of many of the major river systems. The Potomac River below Washington, D.C., the Rappahannock River below Fredericksburg, and the James River below Richmond have thick beds of both the brackish water clam, *Rangia cuneata,* and the Asian clam, *Corbicula fluminea.* The brackish water clam began to appear in Bay waters in the early 1960s. Originally found only in low salinities, it has spread upstream to tidal freshwater and downstream into somewhat higher salinities. The brackish water clam has become a favored food of waterfowl. Some attempts have been made to harvest it for food; however, it often takes on a musty algal taste, so is not too appetizing.

The Asian clam originally came to America's West Coast in 1938, brought there by Chinese immigrants working on the Columbia River, in Washington. They cultured this clam for food, and as it proliferated it spread across the United States, reaching the lower Chesapeake Bay tributaries by the late 1960s and the Potomac River in 1975. The Asian clam can either attach itself to hard substrates or bury itself in the muds. It is patchily distributed, but can grow rapidly into dense populations. In many locations baby clams have entered cooling water systems of power plants or other hydroelectric facilities. They attach to the inside of the pipes and create costly fouling problems.

SNAILS OF FRESHER WATERS

The pouch or tadpole snail, *Physa gyrina,* and the coolie hat snails, *Ferrissia* spp., are freshwater species found in Zone 1. The pouch snail is generally abundant in quiet, stagnant backwaters and on oozy mud, but it is sometimes found in the eddies of rapidly flowing water. The pouch snail is the snail often seen in freshwater aquaria. The shell is black while the animals live, but shining yellow to red after cleaning. This is an easy species to identify because the shell is sinistral, with the aperture on the left, rather than on the right as in most species.

The young, like the adults, are very active. They glide swiftly about and are often suspended upside down on the undersurface in calm water. The eggs of the pouch snail are laid in large, gelatinous masses, each mass containing upwards of 200 eggs. The eggs are attached to almost any submerged object—stones, logs, and aquatic vegetation.

Coolie hats are also called freshwater limpets. They are very small snails, about one-quarter inch long, light brown at the edges and darker at the summit. Because of their diminutive size they are easily overlooked; however, they can be quite abundant on the inside of dead freshwater clam shells and among stones and boulders and in fast-running water. The usual gelatinous egg case is roughly circular in shape and is divided into several capsules, usually no more than nine, with one egg in each capsule. Look for their egg cases on the underside of aquatic plants, such as water lilies.

Both the pouch snail and coolie hat snails are "air-breathing" snails and lack specialized gills and an operculum, the leathery seal attached to the foot which effectively closes the shell in other snails.

Mollusks in brackish and freshwater may form dense colonies. Asian clams, *Corbicula fluminea* (to 2 inches), clustered on the rocks at the upper left closely resemble the brackish water clams, *Rangia cuneata* (to 2½ inches), and identification requires careful examination of the shells (see Shell Guide). The large freshwater mussel, *Anodonta* sp. (to 5 inches), lies half-buried in the mud, its siphons aligned with the downstream water currents. In the foreground, pouch snails, *Physa gyrina* (to 1 inch), graze over some rocks along with tiny coolie-hat snails, *Ferrissia* sp. (to ¼ inch).

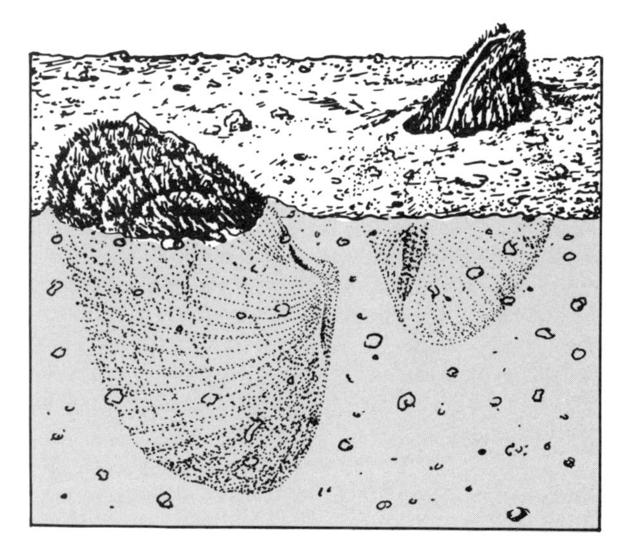

Ponderous arks, *Noetia ponderosa* (to 2½ inches), lie half-buried in the bottom.

ARKS OF THE LOWER BAY

Arks are boxlike bivalves that position themselves much like freshwater clams with their shells just at or protruding out of the bottom. They are found only subtidally in higher salinities of the lower Bay (Zone 3), both in the shallows and in deeper waters. Ark shells are often deposited on the beach, but they are usually bleached and eroded and bear little resemblance to live arks. Arks have a heavy foot, which anchors them firmly in sand or mud. They have no siphons but possess a figure-eight opening in the mantle which serves the same purpose. Arks have thick shells, which help them tolerate abrasion by currents and waves.

Three species of arks are found in the Chesapeake Bay—the transverse ark, *Anadara transversa,* the ponderous ark, *Noetia ponderosa,* and the blood ark, *Anadara ovalis* (see Shell Guide for identification). Arks are an ancient family of bivalves; a related species is a common fossil shell of the upper Bay Miocene deposits. The transverse ark is not always a burrower and may be positioned among the stems of sea grasses. It is a small ark, only an inch or so long, with a distinctive rhomboid-shaped shell. The ponderous ark is the largest and heaviest of the three Bay species. It is perfectly heart-shaped when viewed from the end. It has a furry brown periostracum often worn away from the beak. The blood ark gets its name from its red blood. Most bivalves have clear transparent blood, which goes unnoticed. It is somewhat startling to open a blood ark and see red fluid, especially since oysters, hard- and soft-shelled clams, and mussels, which are more commonly encountered, have clear or uncolored blood.

BIRDS OF THE SHALLOWS

Birds, of course, are not bound to a particular habitat, unlike aquatic plants or snails. They are free to range in search of food; they can roost in a tree upriver or bob on the open water at midday. The shallows, that huge, somewhat undefined and varied region throughout the Bay and its tributaries, offer diverse habitats and plentiful food for birds. The marshes, actually a landward extension of the shallows, attract many species of birds for the same reasons. Some birds have very specific requirements that tie them to a certain habitat, but most birds move freely between the shallows and the marshes, and occasionally deeper, open waters.

There are two species of swans that live in the shallows, the elegant mute swan, *Cygnus olor,* which lives year-round in the Bay, and the migratory tundra swan, *Cygnus columbianus.* The mute swan is an introduced species from Europe; the tundra swan is native to North America. In the 1950s there were only three "mutes" reported near Gibson Island, between Baltimore and Annapolis, and a pair on the upper Miles River on the Eastern Shore. Now firmly established in the Chesapeake and growing in numbers every year, the mute swan has become a common sight in the shallow-water habitat throughout the year.

The tundra swan, formerly known as the whistling swan, nests in the Arctic tundra, from which it gets its name. In late October and early November it flies in large flocks to the mid-Atlantic and Chesapeake Bay, where it spends the winter. It is the most widespread swan in North America.

Both swans are all-white. The mute swan has an orange bill and a black knob at the base of the bill; the tundra swan has a black bill which often has a yellow spot in front of the eye. The males are slightly larger and heavier than the females. The tundra swan is smaller than the mute swan, although it is still a large bird with a wingspan of six to seven feet. On arrival in the Bay in fall, immature tun-

dras are as large as the adults and will molt from their light gray plumage to white by late winter or early spring. Mute swans often carry their long necks in a graceful S curve, and sometimes their wings are arched over the back. They occasionally lift one black foot out of the water and carry it alongside the body. Tundra swans can usually be distinguished from mute swans because they most often hold their slender necks straight rather than in a gentle S curve. Mute swans are generally silent, but they do softly gabble as they gather on the water, and they hiss when threatened. Tundra swans are much more vocal and utter a high-pitched "who-who-who" that softly echoes across the water on a quiet, cold winter night.

Mute swans build ground nests of reeds and cattails, lined with soft grasses and down, along the marshy, low-lying shores of the Bay. The females lay 4 to 10 eggs shaded light gray to bluish green; the clutch size averages about six cygnets. Cygnets feed on aquatic invertebrates the first month or so of their lives and then switch to aquatic plants. When the cygnets are very young, the female will lower her tail so that they can scamper aboard her back for a ride. The young are light gray or brown and gradually change to white by the second year. The male is aggressive and fiercely protective during nesting season and when the cygnets are young. The male will hiss repeatedly and if aroused will not hesitate to half run, half fly with his long neck extended toward the intruder. Few people stand their ground when a swan is on the offensive. Mute swans display this same behavior toward tundra swans, sometimes continually harassing a group until they move on.

Both swan species are strong flyers and begin their flight from the water by running several feet across the water and slapping the surface with their wings before becoming airborne. Mute swans often fly in twos or threes or more with necks outstretched and wings making a whooshing sound overhead. Tundra swans tend to fly in larger flocks. Both species feed heavily on widgeon and redhead grasses, sago pondweed, and other aquatic vegetation in shallow areas, where they extend their long necks down to the vegetation or actually tip up like puddle ducks to reach the grasses, looking like marshmallow puffs on the water. They have been observed feeding on razor clams as well. They also feed on winter wheat and other grains, much to the consternation of farmers. There is some concern about the increasing numbers of mute swans because of heavy feeding on submerged aquatic vegetation and, particularly, because they are extremely aggressive toward the more docile, native tundra swan.

The early fall arrival of the Canada goose, *Branta canadensis,* heralds the oncoming winter like no other species in the Bay. By mid-September most of the ospreys have migrated south; the tundra swans will not arrive until November; and only a few small groups of ducks have appeared here and there. But there is nothing else like the primitive two-note "honk-a-lonk" or "haronk" of a V-shaped flock of high-flying geese that so symbolizes winter on the Chesapeake Bay. As the fall season progresses, more and more geese veer out of the migrating flocks, set their wings, and come to a skidding landing on the water. On the Chesapeake Bay and other mid-Atlantic coastal estuaries, geese gather in large flocks, drifting with the tide and loafing in the sun and then moving into the shallows to feed on aquatic vegetation as swans do. According to experts, there are 10–11 races (some consider them subspecies) of Canada geese. The geese along the Atlantic flyway are paler than their cousins on the Alaskan coast, and there are also some differences in the markings and in size between various races.

The Canada goose has a broad, round-tipped black bill, a long black neck, a white cheek and chin patch, a gray-brown upper body, and light gray to buff breast and underparts. Canada geese are plump birds with relatively short wings and short, wide-spread legs. They fly with

Canada geese, *Branta canadensis* (25 to 45 inches), rise from their feeding grounds.

deep, slow wingbeats and can exceed sustained speeds of over 35 miles per hour. When flying long distances they tend to fly in a V formation at an average altitude of 2,000 feet. Geese and other birds such as swans, cormorants, and ibises will often fly in a wedge or V with older birds at the head. Some believe that the older, thus stronger, birds decrease the wind resistance for the younger birds toward the rear of the flock. Another theory is that geese fly in a V so that all of the birds except those in the lead gain lift from the wing-tip air circulation of the bird in front. They are quite vocal: they honk when flying or when alarmed and continuously gabble or cackle when on the water. Like swans, Canada geese take off from water by running and flapping to gain altitude, but when alarmed they can literally jump from the water like puddle or dabbling ducks.

Canada geese are found in freshwater lakes and reservoirs and estuaries. They feed in the shallows and the uplands, where they thrive on corn and other grains; on golf courses; and in park areas. Typically, when feeding, one or more mature geese, called sentinels, stand guard over the others. There may well be more Canada geese now than in colonial times, probably because of increased agricultural production. In fact, many birds do not migrate back to their Arctic nesting grounds but linger year-round to nest and feed on grain and lush golf course grass. Canada geese usually breed in the third year and produce an average of four to seven goslings. They mate for life and live to an average age of 20–25 years.

The mallard duck, *Anas platyrhynchus,* is a duck that is familiar to almost everyone. It is widely distributed throughout North America any place where there is water—fresh and saltwater marshes, rivers, ponds, lakes, estuaries, and city parks. It is the largest species of dabbling or puddle duck, that group of ducks which feed in the shallows by tipping tail-up to reach aquatic vegetation, seeds, insects, and snails. They generally do not obtain their food by diving. Dabblers have small feet and their legs are positioned forward, in contrast to diving ducks, which have larger feet on short legs located to the rear. Mallards, like all dabblers, spring up into the air from the water and then fly horizontally with rapid wingbeats.

The male mallard, or drake, is so commonly seen that we often overlook the fact that it is one of the most strikingly beautiful of all the ducks. The male has a metallic green head and neck with a white neck ring, chestnut breast, yellow bill, and orange feet. The black feathers of the central part of the tail curl up. The speculum, a brightly colored area on the trailing edge of the upper wing, is purplish blue bordered in white in both sexes. The female is a mottled brown with an orange bill marked with brown or black. Juveniles and males in eclipse plumage, the dull, mottled plumage of the male after breeding, have markings similar to but duller than the female's; they also have an olive-yellow bill. Mallards are capable of readily interbreeding with the white domesticated Peking duck, muscovy duck, and the closely related native American black duck. This interbreeding produces an array of colors and patterns, regionally, between the mallard and the other species.

Mallards breed and nest in the northern pothole

Two male and two female mallards, *Anas platyrhynchos* (23 inches), rest along a bank edge.

prairies of the Dakotas and Canada, but they also breed in local populations throughout the country, including the Chesapeake Bay and mid-Atlantic. The premating behavior of mallards is active and somewhat frenetic: the males surround the female and chase her until she flies off, immediately followed by her potential mates. They dash and lunge at her, stand on the water and flap their wings, and chase her up onto land until she finally chooses a mate. The female is very vocal and is the one that voices the familiar "quack quack"; males call in soft, whistling or reedy notes. The female lays 8–10 pale greenish buff eggs in a shallow ground nest usually located near the shore or in marsh grass. The eggs hatch in 28 days, and almost immediately the ducklings are in the water and following the female. The young ducklings actively seek aquatic insects and other small invertebrates along bulkheads and the shore. They are able to fly about two months after hatching. Mallard populations are on the rise in the northern pothole prairie areas as well in the Chesapeake Bay.

Shallow waters teem with shrimp and small fish, from the oil-rich menhaden, anchovies, and silversides to killifish and the young of striped bass, yellow perch, and spot. The red-breasted merganser, *Mergus serrator,* is a winter visitor that feeds on the fishes and crustaceans of the shallows. Mergansers are diving ducks, also known as sea ducks, and are able to swim strongly above and below the surface of the water. They are equipped with long, slender bills with toothed or serrated edges, which enable them to grasp slippery fish underwater. The male red-breasted merganser has a lustrous green head with a shaggy or wispy crest, a white collar, and a speckled black and brown breast. The female has a shorter crest and a reddish brown head and neck merging into a lighter breast. They typically fly low over the water in single file. They are very active swimmers and constantly patrol the shallows. It is a joy to

Red-breasted Merganser (23″)
Mergus serrator

watch them quickly diving and then bobbing to the surface several yards away—never quite where you expect them. Mergansers usually hunt in mixed groups of males and females, and, like loons, herd fish and feed underwater. Red-breasted mergansers are common along the shorelines of the Bay and the Atlantic coast in late winter and early spring. They migrate north in mid-March to breed and nest from the Great Lakes to the Arctic.

Often you will see a small bird perching on a tree branch or shrub near the water that looks somewhat like a blue jay or a squat, stubby woodpecker with a crest. But perhaps the raucous, dry, rattling call of the belted kingfisher, *Ceryle alcyon,* is the first indication that one is nearby and patrolling its territory. Kingfishers live along the edges of freshwater lakes and marshes and on estuarine shorelines. Belted kingfishers range throughout most of North America. They are crow-sized birds and live primarily on fish, although they sometimes feed on small snakes, invertebrates, frogs, crayfish, and young birds. Belted kingfishers are stocky birds with large heads, large bills, and short necks. Both sexes have a ragged crest, a white neck, and a gray-blue breast band; the female also has a visible rust-colored belly band.

Belted kingfishers live year-round in the mid-Atlantic and the Chesapeake Bay and are often seen on electric power lines and on trees and shrubs near the water's edge.

Belted Kingfisher (13")
Ceryle alcyon

During breeding season, the male and female dig a horizontal burrow in a sandy bank along the water. Their strong, short legs and bills are well adapted for digging and removing the sand and debris from the tunnel. The burrow into the bank is generally three feet long or more. The end of the chamber is lined with grasses and leaves in which the female deposits six to seven white eggs. After the young birds have begun to fly, the parents will drop food into the water to tempt the fledglings to leave their perches and go after their meals.

Kingfishers are normally solitary birds except during breeding season. They will often hover over the water; when they sight a fish, they plunge into the water after their quarry. They are very wary of humans and dash from cover along the edge of the water, then fly with rapid, halting wingbeats to another perch a short distance away. Their dry, rattling, staccato call once heard is not easily forgotten.

Waders Along the Water's Edge

Many wading birds are common to the shallows, particularly when low tides reveal mud flats and shallow pools of stranded fish and shrimp. These waders are often found in the marshes as well. Some of these birds tolerate boaters, to a degree, and even allow homeowners and birdwatchers to approach, but only so close. It seems that each species and each individual sets up an invisible boundary which allows it to pursue one of its most fundamental activities, searching for food, while simultaneously existing in a tension zone that is rich with prey and trying not to become prey itself.

The hunched gray bird huddled on a piling suddenly squawks and unfolds. Slowly, it launches from its perch and begins its slow, deep-winged flight with its legs trailing behind and its head and neck drawn back in an S shape. This fluid flight is characteristic of the ubiquitous great blue heron, *Ardea herodias.* Great blue herons are shallow-water stalkers. When a great blue is hunting it will take slow, deliberate steps, sometimes stopping and then moving on. It points its head forward, which gives its body a long and angular shape. Suddenly it thrusts its long bill into the water and seizes one of the many fishes it will eat that day. Blue herons will also plunge into the water from a pier after food and occasionally will even be seen floating in deeper water.

Great blue herons belong to a family of birds that includes herons, egrets, and bitterns. Vernacular names

such as the great blue heron and great egret tend to obscure their actual close relationship. Herons and egrets resemble each other in generally having long bills, necks, and legs. They differ in coloration, size, behavior, and habitat preferences, but all are usually recognized immediately as belonging to the same family.

The great blue heron is the largest heron in the Bay and along the middle Atlantic. The head is largely white with a black stripe extending from the eye and terminating in a feathery crest; its long neck is streaked with black, and the back and wings are grayish blue. Breeding great blues have yellowish bills and plumes on the breast and back.

Great blue herons are one of six species of colonial nesting wading birds in the Chesapeake Bay and are year-round residents. They are gregarious during nesting season and solitary and territorial the rest of the year. They build nests of sticks and twigs in the tops of trees, often in the company of other birds such as ibises, night herons, and cormorants in nesting areas known as heronries. Some great blues, however, are solitary nesters. There are several large nesting areas in the Bay: the Nanjemoy Creek heronry on the Potomac River, with its more than 1,200 nesting pairs, may be the largest colony on the Atlantic coast. There are other large concentrations of nesting birds on Poplar Island, Romney Creek on the upper Bay, Tangier Island, and Mobjack Bay in Virginia. The female lays an average of four light bluish green eggs, which hatch in about a month; the young leave the nest in two months. The peak egg-laying season is between mid-March and mid-June. Great blue herons can live to be about 15 years old.

There are three white birds belonging to the heron family in the Bay area: the great egret, *Casmerodius albus,* the snowy egret, *Egretta thula,* and the cattle egret, *Bubulcus ibis.* The great egret is the largest, followed by the snowy and then the cattle egret. The cattle egret has spread from Africa to Europe and South America and in the 1950s first established breeding populations in Florida and Texas. This is primarily a species of the uplands, commonly seen in newly plowed fields and pastures where horses and cattle feed. The bill of the cattle egret is shorter than that of the other egrets, and during the spring and summer breeding season the adult bird has short buff-orange plumes on its head, breast, and back.

The great egret and the smaller snowy egret were hunted by plume hunters to such a great extent in the late 1800s and early 1900s that populations plummeted to very low numbers. They are now protected—and plumes, fortunately, are not the fashion rage they once were, so that these beautiful birds have now recovered and have extended their range in some areas of the country. Both the great egret and the snowy egret, like the great blue heron, are creatures of fresh and saltwater marshes, ponds, mud flats, and tidal shallows. The great egret has a long, heavy yellow bill and black legs; the snowy egret has a more slender black bill, black legs, and very obvious yellow feet. The great egret can also be distinguished by a dark line that extends from the gape of the bill to beyond the eye. The adults develop beautiful plumes called aigrets during breeding season—the great egret has long plumes trailing from the back and breast. The snowy egret has a corona of plumes on its head, and also on its neck and breast. The great egret, usually solitary, has a slow, almost ponderous wingbeat, and its flight pattern is somewhat sinuous because the heavy, deep strokes of the wings first push the bird up, then down. Snowy egrets have much more rapid wingbeats than great egrets; they also frequently fly in flocks in the evening as they return to their roosting areas. Cattle egrets also fly in flocks, and they too have rather rapid wingbeats; however, their bills and yellow legs and feet are shorter than the snowy egret's. The great egret hunts much like the great blue heron—slowly, methodically, and deliberately. The snowy egret is much more busy in its search for a meal, dashing here and there and using its strikingly colored feet to stir up the bottom muds and then stabbing repeatedly for fish and aquatic invertebrates.

Both egrets are colonial nesters and are often found in association with other herons, egrets, and ibises. The great egret arrives in the Bay area breeding colonies in mid-March; one of the largest breeding colonies on the Bay is located in Canoe Neck Creek in St. Marys County, Maryland. The snowy arrives a bit later at breeding colonies on Poplar Island, Tangier Island, and Fishermans Island near the mouth of the Bay. All three species, the great egret, the snowy egret, and the cattle egret, depart the Bay in the fall and spend the winter from the Carolinas south.

There are many shorebirds such as the ruddy turnstone and the dunlin that visit the Bay for a short time in the spring and fall as they pass through on their annual migrations. The greater yellowlegs, *Tringa melanoleuca,* and the closely related lesser yellowlegs, *Tringa flavipes,* are two such shorebirds that wade in the shallows and mud flats of the Bay area in the spring and fall.

The two species of yellowlegs are difficult to differentiate unless they are seen together, which is not usually the case. The greater yellowlegs is the slightly larger of the two; as the name implies, it is about 14 inches long as compared to 10 inches for the lesser yellowlegs. The greater yellowlegs has a longer bill, which may be slightly upturned. The plumage of both species is similar, with brownish gray backs marked with white speckles, white below, and brownish gray streaks on the breast and neck. The lesser yellowlegs is not as strongly marked and is somewhat paler. Both have long yellow legs, sometimes tending to orange. Their call is a "tew-tew," but the greater's call has more notes which usually descend the scale, while the lesser's call has fewer notes and is higher pitched. These are very vocal shorebirds that often gather in large flocks. The greater yellowlegs is more common in the Bay than the lesser yellowlegs. They feed in the shallows and mud flats and along the margins of fresh and saltwater marshes, often moving their heads in a side-to-side or sweeping motion while seeking insect larvae, worms, snails, shrimp, and small fish. They are quite commonly seen on the tidal flats of the Blackwater Wildlife Refuge on the Eastern Shore of Maryland.

TURTLES AND SNAKES

There are other animals of the shallows and the connecting uplands that feed along the edge and in the water. Some emerge to dig shallow nests in the sand and lay their eggs; others carry and nurture their eggs internally and ultimately bear their young live. These are the turtles and snakes, evolutionarily the primitive relatives of the birds, for they share certain characteristics. They lay eggs in one form or another, and the scales of the snake are represented in a bird's scaly legs as well as its bill. Birds have made the evolutionary leap from reliance on an external heat source to maintain body temperature to dependence on frequent meals to fuel their internal engines. Birds regulate their temperatures principally through the efficient insulating quality of their feathers. Snakes and turtles, while not exactly cold blooded, depend on external warm temperatures to get them up and running, or crawling. Constant feeding does not increase their metabolic rate or maintain a regulated internal temperature as in birds and mammals, so they feed less often and bask in the sun a great deal.

Turtles are great baskers. Some species prefer to repose on an old tree snag that long ago fell in the water; oth-

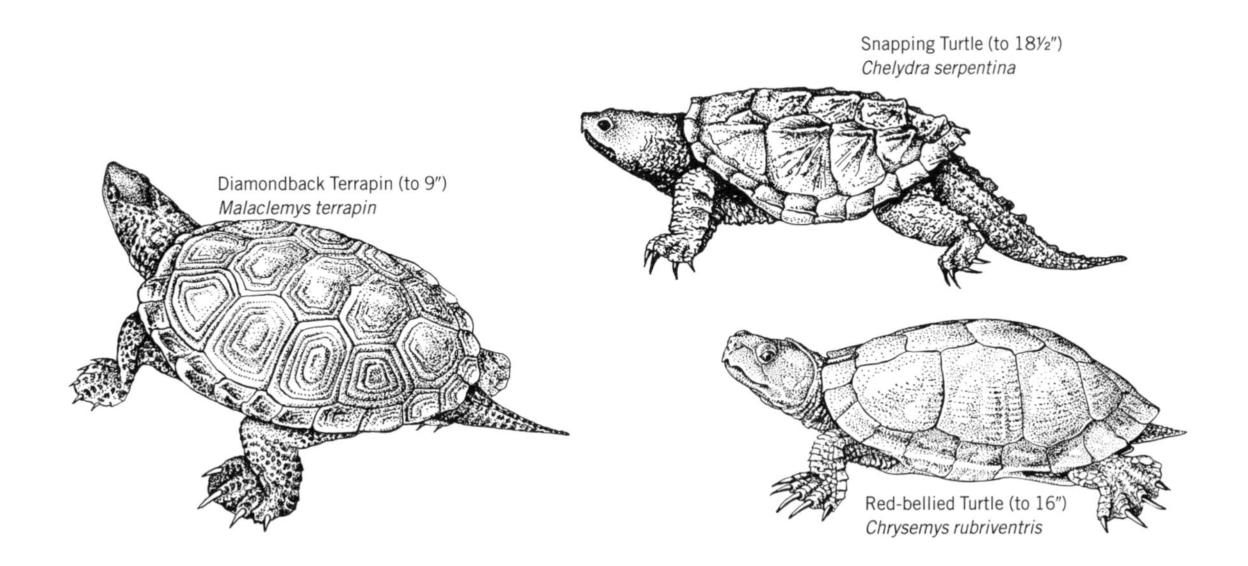

ers will select a rock, which not only gives them a dry perch on which to capture the sun's welcome rays but also radiates warmth—sort of a top and bottom tanning booth at the same time. Other turtles will sometimes bask along the shore, but more frequently they simply come to the surface of the water and laze along. Snakes also sun themselves along the shore and on overhanging branches and, depending on the species, frequently swim both above and below the surface. All snakes can swim. Some species prefer the water at certain times; others are more definitely creatures of the drier uplands, but they can and do swim.

The diamondback terrapin, *Malaclemys terrapin,* is a typical inhabitant of mid- and lower estuarine embayments and is found, as one of several subspecies, from southern New England to the coast of the Gulf of Mexico. The Chesapeake Bay has always had numbers of diamondbacks in the marshes and along the shallows. At one time, diamondback terrapins were a favorite of gourmets in restaurants and clubs from Baltimore to New York and beyond: a bowl of steaming rich terrapin soup laced with sherry was, and still is, a very flavorful dish. As was the case with ducks in the day of market hunters, terrapins were caught in large numbers in earlier times and the predictable decline set in. Regulations were imposed to protect the terrapin; appetites changed; and now the terrapin has increased in abundance. Nowadays, the terrapin is best known in this area as the symbol of the University of Maryland.

The diamondback terrapin is a medium-sized turtle: the female grows to a maximum of about nine inches and the male is smaller. The upper shell, or carapace, is sculpted with geometrically-shaped growth rings and ranges in color from a grayish green to almost black. The head and neck are light gray or brownish and have fine black speckles. The bottom shell, the plastron, is yellow to green-gray. The female lays an average of 10–15 eggs in shallow nests dug from sandy areas along marshes and upper beach areas; the young hatch in about three months. The female may nest several times in one season and, all told, lay 35 eggs or more. Diamondback terrapins feed on clams, worms, snails, crabs, carrion, and some vegetation. Terrapins are attracted to bait in the ever proliferating crab traps set out in the Chesapeake Bay, and some of them drown once they enter the trap; if they escape entrapment and predation, they live to be 25 years or older.

The upper tributary streams, marshy ponds, and brackish water areas are home to the red-bellied turtle, *Chrysemys rubriventris.* They bask along the shore, sometimes singly, sometimes in groups. Wary and shy, they quickly slip into the water when alarmed. They are somewhat larger than the diamondback terrapin and can reach a length of 16 inches and a weight of three to four pounds. Their notched upper jaw is heavy and strong; the carapace is brownish or black, sometimes tinged with red; and the marginal plates, or scutes, have a central red stripe. The plastron is red, hence the name red-bellied. The female lays an average of 10–15 eggs in an earthen nest near the water; hatching occurs in two and one-half to four months. The red-bellied turtle is an omnivore and feeds on crayfish, insect larvae, snails, and vegetation.

The largest freshwater turtle in the Chesapeake Bay is the snapping turtle, *Chelydra serpentina,* which can weigh 35 pounds or more. Snappers have a fearsome reputation and, particularly when out of the water, they can strike swiftly and inflict a nasty wound. They are distributed widely throughout North America from southern Canada to the Gulf of Mexico, and from the coastal Atlantic states westward to the Rocky Mountains. They live in lakes, in slow-moving rivers, and in marshy estuarine rivers. The snapper is easy to identify with its large head, long, thick neck, and a tail as long as its keeled carapace. The tail has three rows of hard knobs down its length. The light yellow to cream-colored, cross-shaped plastron is quite small—and, oh yes, to complete the snapper's repertoire of characteristics, it can release a foul-smelling anal musk when alarmed. Apparently, snapping turtles are mainly carnivorous in the spring and will eat just about anything—insects; crustaceans and amphibians; ducklings and mammals. However, later in the spring and summer, when aquatic vegetation appears, they depend more on plants for sustenance. Snappers are very aquatic and are not usually found out of the water, but they do bask on floating logs, generally some distance offshore. If their pond is semipermanent and begins to dry up, snappers are capable of migrating substantial distances over land to reach another suitable body of water. The female lays approximately 30–50 eggs during a protracted season extending from April to November; she may lay two clutches per season. Depending on the seasonal temperatures, the eggs may hatch in 80–90 days; however, the eggs can survive winter temperatures in the nest, and hatching will then take place the following spring. Snappers hibernate during the winter in muskrat houses, burrows, and under logs; they emerge in March. Snapping turtles make excellent turtle soup, and there are small regional fisheries here and there to supply turtle meat to restaurants.

Every snake seen in the water by an uninformed observer is potentially a poisonous "water moccasin," and the immediate impulse is to run for cover or to kill the snake and save someone's pet or child from certain death. All snakes can swim, and some are even venomous; but the odds are overwhelming that the snake in the water, or even the one in the grass, is harmless. "Water moccasins," by

Northern Water Snake (to 53")
Nerodia sipedon

the way, are actually cottonmouths and are extremely dangerous, but they are found only in the extreme southeast corner of Virginia and farther south. There are approximately 40 species of snakes in Maryland and Virginia, only 4 of which are poisonous: the cottonmouth, the copperhead, and the timber and canebrake rattlesnakes.

The snake that is most often seen in the water and along the shallows is the northern water snake, *Nerodia sipedon. Sipedon* is derived from Greek and means "swimmer." Northern water snakes are highly variable in color; they may be brown, tan, gray, or reddish. Older individuals are generally darker and may even be uniformly black. They usually have dark crossbands alternating with lighter patches on the neck and anterior portion of the body; the hinder parts of the back are marked with dark irregular spots and lateral bars. The under surface is variably marked with red, brown, or yellow spots. The overall pattern is generally more prominent in younger snakes. Northern water snakes are active swimmers, on the surface and below. During both day and night they move along the shallows, where they prey on fish, small rodents, and turtles. They bask along sea walls, on rocks, and in marshy areas; they may curl up under a shrub on the lawn or under a parked car. These are very aggressive snakes when threatened and will not hesitate to bite, which is painful but not poisonous. They also exude a malodorous anal secretion, which is convincing to any predator as well.

The copperhead, a venomous snake found in many areas of Maryland and Virginia, appears to have some resemblance to the northern water snake, but actually its markings are hourglass-shaped and its coloration is much brighter. Typically, copperheads are snakes of drier areas away from moist shores; they prefer wood and sawdust piles, old sheds and barns, and rocky outcrops.

● The ever-changing shallow water habitat affords the observer an opportunity to discover the diverse fauna that feed and live within this richly productive area. Without need of a boat or of sophisticated collecting gear, you can literally mingle with schools of flashing fish or closely inspect a beautiful, pulsating ephyra, the immature stage of a soon to be fully grown sea nettle. Freshwater and brackish shallows are home to aquatic insects, snakes, and turtles. And everywhere there are birds—elegant swans, loafing waterfowl, solitary herons, and flighty shorebirds. Life in the shallows is abundant and busy.

SEAGRASS MEADOWS AND WEED BEDS

Large stands of rooted aquatic plants grow lushly in shallow shoreline areas throughout the Chesapeake Bay and its tributaries. These marine plants are similar to our familiar land plants, with green leaves, flowering buds, seeds, and roots firmly anchored in sandy to muddy bottoms. Aquatic plants grow below the low-tide line out to depths of nine feet. Lack of light inhibits their growth in deeper waters. Flower buds can often be seen just ruffling the surface of the water. Plant beds may be exposed for short periods during extreme low tides, but for the most part they flourish only when entirely submerged. Stands of aquatic vegetation create a special habitat quite different from clear unvegetated shallows. They grow rapidly, normally in dense growth, and provide food and shelter for many animals of the estuary.

The plants are eaten directly by only a few aquatic organisms; more importantly, they serve as an indirect food source by providing great amounts of detrital material, which is formed as the leaves die and decompose. Detritus is utilized by snails, fish, amphipods, isopods, and many other creatures and is flushed out of the weed beds to enrich adjacent open waters. Plant leaves provide a base for a lush covering of microorganisms, which are, in turn, grazed on by larger animals. The root systems stabilize bottom sediments and the mass of leaves dampens the effects of waves, creating a more stable environment. Fish, crabs, and other animals find a haven and a rich food supply among the leaves. Submerged plants constitute the principal source of food for many species of waterfowl and small mammals such as muskrats, nutria, and others. In most instances, the number and variety of organisms is greater within an aquatic plant bed than in an adjoining unvegetated area. Many of the same species live in both habitats, but certain species are typical residents of vegetated areas and are not generally found elsewhere. In winter many of the aquatic plants die back leaving the surface

6. SEAGRASS MEADOWS AND WEED BEDS

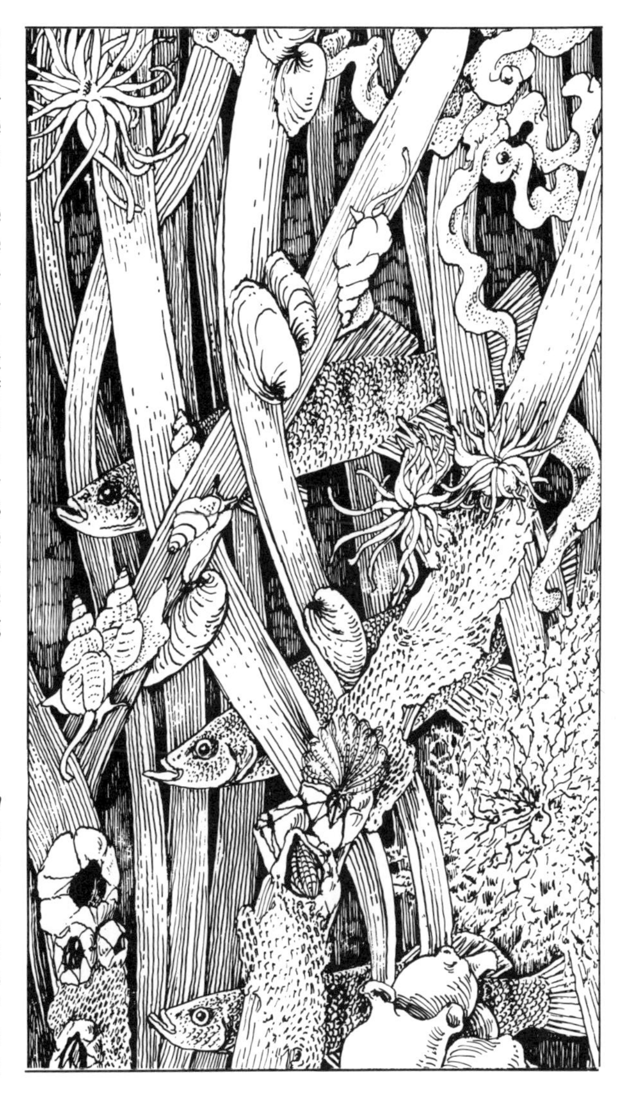

bare, and the habitat essentially disappears. Fish and shrimp and crabs move away to deeper waters, snails and worms bury in bottom muds, and the area becomes a continuation of the unvegetated shallow habitat. In spring, the roots or rhizomes sprout and seeds germinate to replenish the plant beds with green growth.

There has been a decline in recent decades in the amount and variety of submerged aquatic plants in the Chesapeake Bay and its tributaries. The loss of this important habitat can have drastic secondary effects on waterfowl, fish, and other animals that depend on it for food, shelter, or protection. A number of reasons have been put forth for the decline: chemical additions from the runoff of agricultural fields, changes in normal temperatures and salinities, destruction by cownose rays, siltation, disease, and boat traffic are some of the factors that may have contributed to the decline. Since the early 1990s there has been some recovery of aquatic vegetation here and there throughout the Bay. The vast underwater meadows have not yet returned as they were in the mid-1900s, but the decline in submerged aquatic vegetation may be turning around.

UNDERWATER PLANTS

Eelgrass, *Zostera marina,* and widgeon grass, *Ruppia maritima,* are the only two species of sea grasses common to the higher salinities of the Chesapeake Bay (lower Zone 2 and Zone 3). They form thick, expansive stands often referred to as seagrass meadows. Eelgrass (or grass wrack, as it is sometimes called) has historically been the more dominant and widespread plant in these regions. Eelgrass has long, narrow, ribbonlike leaves (*zostera* means "belt" in Greek), which vary in shape depending on depth of the water, salinity, and other environmental characteristics. In shallow, sandier, more turbulent areas, the leaves tend to be shorter and narrower; in deeper, muddy-bottom waters less exposed to wave turbulence, leaves tend to be longer and wider. In the Bay, eelgrass is commonly found over sandy-mud bottoms. Widgeon grass has more delicate narrow branching leaves than eelgrass. Because it tolerates lower-salinity waters than eelgrass, it is found farther upstream, even into tidal freshwater. Both eelgrass and widgeon grass are probably the most important aquatic plants for waterfowl in the Bay.

As the water becomes fresher, the variety of submerged plants in aquatic weed beds increases. In the upper mid-Bay regions of Zone 1 and upper Zone 2, redhead grass, *Potamogeton perfoliatus,* sago pondweed, *Potamogeton pectinatus,* and horned pondweed, *Zannichellia palustris,* are common water plants. They are often found along with widgeon grass. Although redhead grass and sago pondweed are closely related, they look nothing alike. Redhead grass has delicate, small oval leaves, which "clasp" the plant stem where they connect. The stems are slender, branching more toward the upper part of the plant. Sago pondweed has long, narrow, tapering leaves, which characteristically spread like a fan at the water's surface. Horned pondweed has short, narrow leaves and looks much like widgeon grass, but its leaves rise in pairs from each joint of the stem, whereas leaves of widgeon grass arise from alternate joints. Horned pondweed seems to have two spurts of growth, one in late spring and the other in late summer. It often disappears or is overgrown by other weeds in mid-summer months.

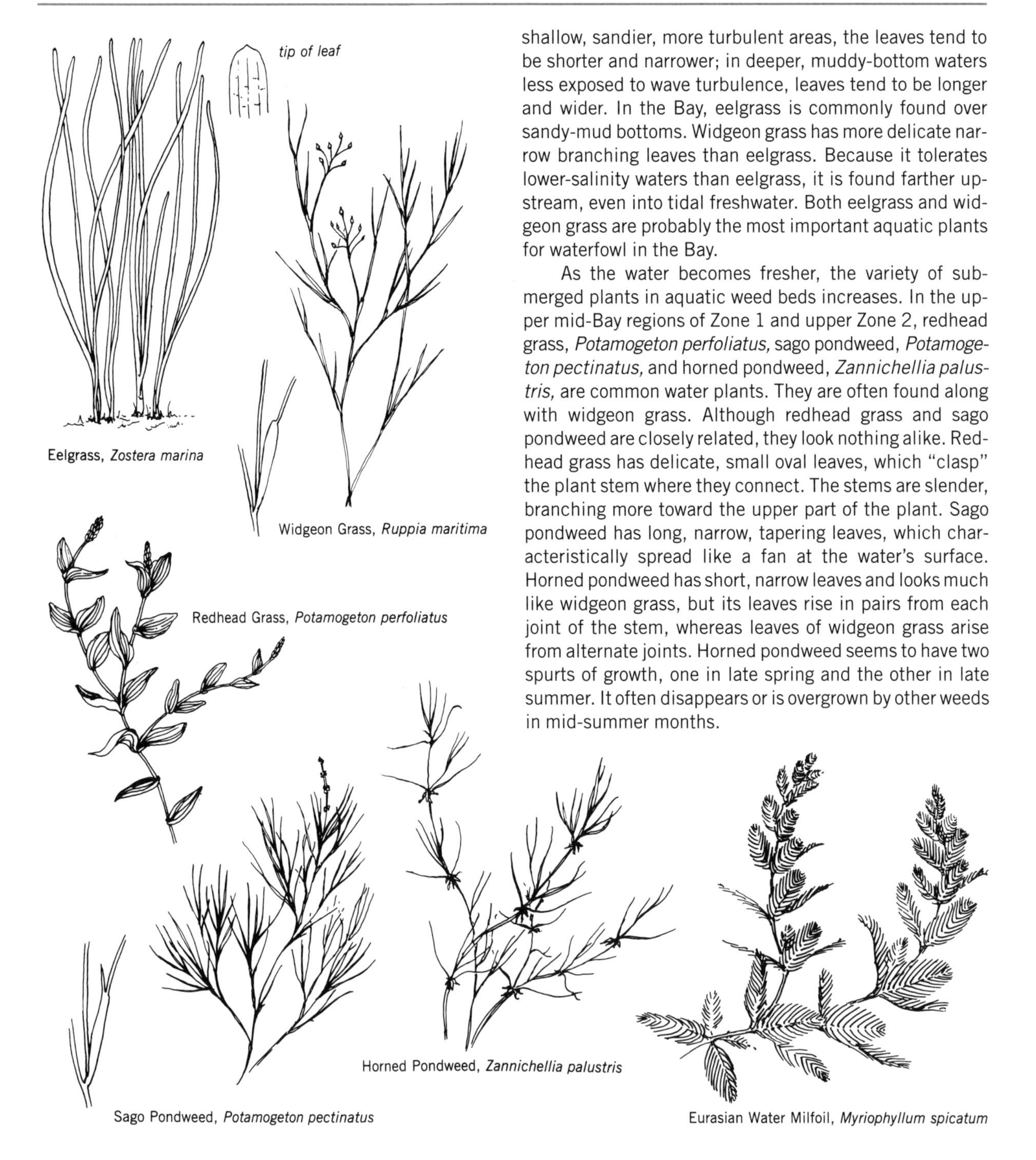

Eelgrass, *Zostera marina*

Widgeon Grass, *Ruppia maritima*

Redhead Grass, *Potamogeton perfoliatus*

Horned Pondweed, *Zannichellia palustris*

Sago Pondweed, *Potamogeton pectinatus*

Eurasian Water Milfoil, *Myriophyllum spicatum*

Eurasian water milfoil, *Myriophyllum spicatum,* also occurs in lower-salinity zones of the mid-Bay. This soft, feathery plant was introduced into the United States from Europe and Asia at the turn of the century. Each leaf is divided into 20 or more pairs of upward-curving threads, which resemble wet feathers. It is a prolific and persistent plant, and by the early 1960s, dense tangled mats carpeted one hundred thousand acres of Bay waters. Some observers claimed that the mats of milfoil were so thick that birds could stroll about on top of them. The milfoil became a nuisance, creating anaerobic conditions, becoming odoriferous as the plants rotted, causing hazards for small boat traffic, and restricting fishing and crabbing. Fortunately, water milfoil declined as quickly as it had proliferated, probably because of disease, and now only scattered stands occur in localized areas.

The number of plant species is even greater in the upper tidal freshwater and low-salinity regions of the Bay and its tributaries. The more common plants here include wild celery, *Vallisneria americana,* common waterweed, *Elodea canadensis,* coontail, *Ceratophyllum demersum,* and bushy pondweed, *Najas quadalupensis.* Wild celery is freshwater "eelgrass," with long, ribbonlike leaves, each with a light-colored central stripe, which can be seen when it is held to the light. Wild celery is tolerant of muddy, roiled waters. Common waterweed is recognizable as a popular aquarium plant. Its narrow oval leaves cluster two or three to each stem node. Common waterweed can become a pest by invading and choking small embayments or ponds. Coontail, or hornwort, as it is also called, is a rootless plant often seen floating in large masses that drift just below the surface in still or slow-moving waters. It is a bushy, feathery plant much like milfoil but with whorls of 9 to 10 sparsely branched filamentous leaves along the branches. Bushy pondweed, also called water nymph, varies in appearance but is generally bushy with branching stems and short, slender leaves. In the 1960s hydrilla, *Hydrilla verticilata,* native to Africa and perhaps southeast Asia, became established in the United States, and in 1982 scientists found it in the upper Potomac River near Washington, D.C. It now covers thousands of acres of shallow water downriver from the capital city. Some consider it a nuisance plant because it grows profusely, chokes channels into marinas, and can limit boating and other water sports. On the other hand, the dense mats of hydrilla are an important habitat and provide a source of food for insect larvae and small fishes. Largemouth bass lurk under the mats, and

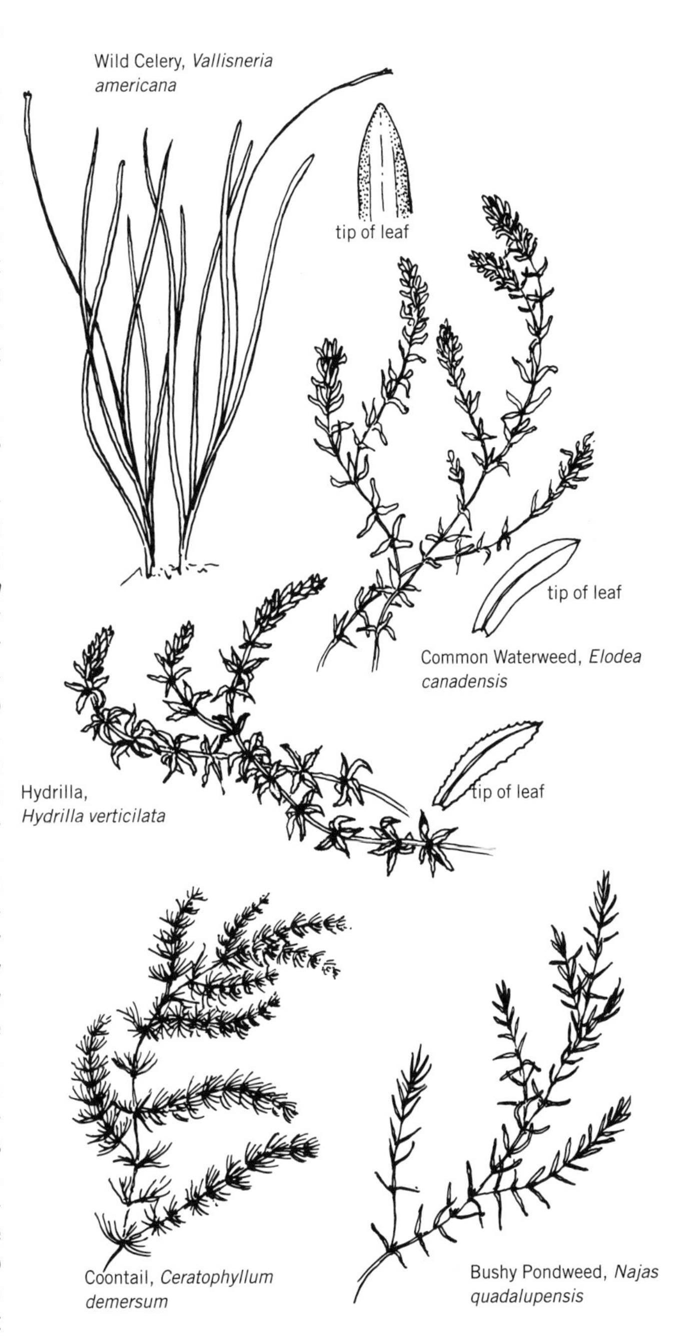

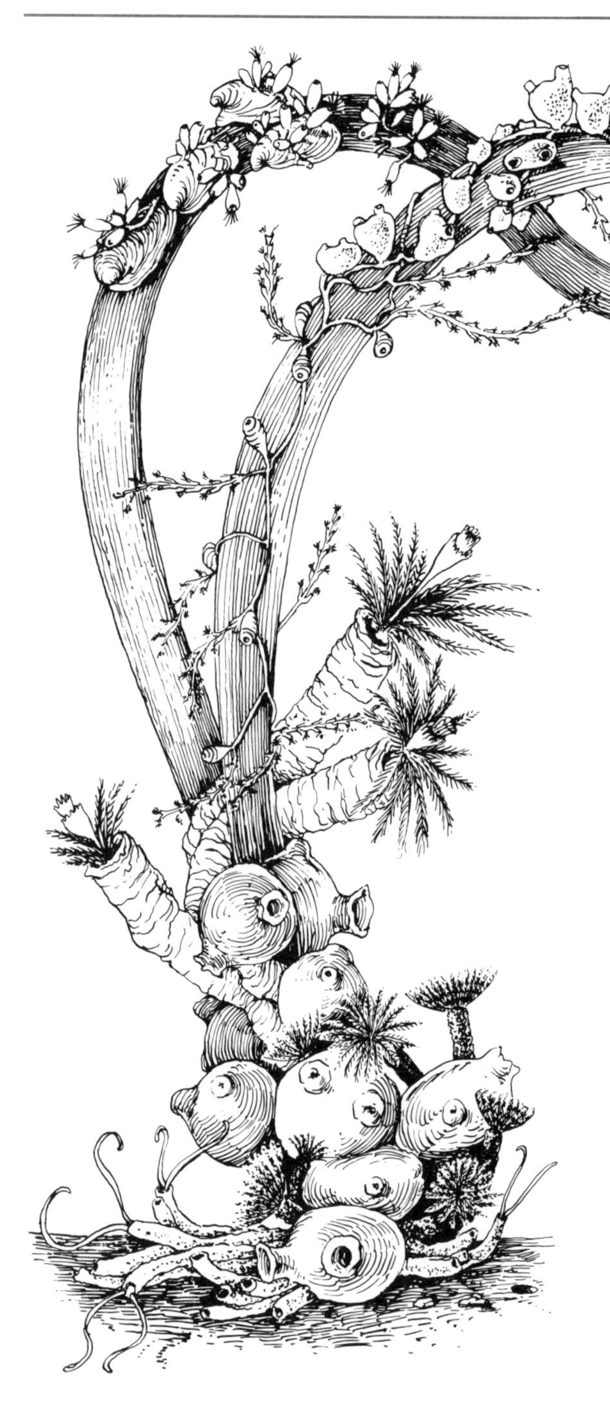

mallards, black ducks, coots, and Canada geese feed on the vegetation.

Hydrilla has long, branching stems with three to five toothed leaves growing in a whorl at each stem node. They do flower and can reproduce by pollination and eventual seed production, but more often hydrilla reproduces by growth from its rhizome system or from tubers or tuberlike turions that the plant produces. Hydrilla grows well in fresh water and tolerates brackish water to a limited extent. It does not require high levels of light to flourish, so exists well in turbid water.

SEAGRASS MEADOWS

The seagrass meadows of the mid and lower Chesapeake Bay regions are composed of stands of eelgrass or widgeon grass or, sometimes, a mixed bed of both. Typical fauna inhabiting seagrass meadows are a varied lot, some living on the blades and stems or around the base of the plants, others living in the bottom substrates. Some motile or swimming animals are permanent members of the

The horn garland hydroid, *Dynamena disticha* (branches to ¼ inch long), coils upward around the eelgrass blade in the foreground to intermingle with a colony of green beads tunicate. The creeping bryozoan, *Bowerbankia gracilis* (zooid shown enlarged, 1/32 inch), finds a place for attachment on small convex slipper shells, *Crepidula convexa,* on the other eelgrass blade. Attached to the base of the plant are clusters of sea squirts and the hard white cases of the limy tube worm, *Hydroides dianthus* (to 3 inches), and the mud-encrusted cases of the fan worm, *Sabella micropthalma* (to 1⅜ inches). The club-shaped operculum of the limy tube worm seals the opening of the tube when the feathery red tentacles are withdrawn. Whip mud worms form soft clusters of tubes at the base of the plant.

meadow community and move in and out of the area periodically in search of food or protection.

Animals Attached to the Plants

The grass blades provide a firm substrate for many sessile invertebrates and plants. In a sense, the attached life on an eelgrass frond is a piling or jetty community in miniature, colonized by many of the same organisms. The surface is covered with a scum of microscopic algae and protozoans. Seaweeds such as sea lettuce, hollow-tubed seaweed, and the like attach or float among the waving fronds of the rooted plants. Two of the most abundant animals of the piling community, the bay barnacle and the sea squirt, are frequently attached to eelgrasses. Colonies of sea squirts can grow so heavy through the summer that they weigh down and kill the eelgrass, creating bare spots in the meadows. Small patches of the volcano sponge or the yellow sun sponge appear on the blades or around the base of the plants. Redbeard sponge is also found within seagrass meadows, but it is usually attached to bits of shell or pebbles on the bottom rather than to the grass blades. Feathery hydroids find attachment on the blades. Seagrass meadows are the typical habitats of the creeping stolons of the horn garland hydroid, *Dynamena disticha.* This low, creeping hydroid encircles the blades with a stolon from which tiny branches lined with hydranths extend into the water. The branches look like delicate, quarter-inch-long strings fringing the leaf edge and are just discernible to the naked eye. Similarly, the creeping stolons of the little, green beads tunicate can be seen along the blades, and thick clumps of golden star tunicate may encircle the stems. Sea anemones attached to the leaves wave their graceful tentacles in the currents. Seagrass meadows are perhaps the best habitats in which to search for the green-striped anemone. Lacy crust and coffin box bryozoans often decorate the leaves with their delicate, spreading colonies.

Crowded colonies of the fragile convex slipper shell stacked one on top of the other along the blades of eelgrass, are a common sight. This slipper shell is cousin to the larger slipper shells found attached to moon snail shells, whelks, and other mollusks. The convex slipper shells on eelgrass are small (only one-third inch long) and elongated. Their shape is adapted to the narrow width of the leaves. Convex slipper shells, which attach to other shells, such as those occupied by hermit crabs, are wider and more robust. Slipper shells, unlike other gastropod snails, are sessile plankton feeders and do not move about the plant to graze. If a shell is carefully removed from the stem, a swirl of transparent egg capsules can often be seen attached to the stem by a sticky pad. Each capsule contains a half dozen or so embryonic slipper shells, which are brooded under the female's shell until ready to hatch to a well-developed stage. Baby slipper shells do not experience the rigors of a planktonic larval stage as so many other mollusks do.

A mound of slipper shells may be completely covered by the creeping bryozoan, *Bowerbankia gracilis,* barely visible as a soft, furry coating of minute, closely packed, vase-shaped zooecia.

Toward the base of the plants, where flocculent sediments accumulate among the many attached organisms, whip mud worms build thick, entangled mats of tube cases, as they do on almost any hard surface throughout the Chesapeake Bay. In the higher-salinity regions of the lower Bay, the tough, mud-encrusted cases of the fan worm, *Sabella micropthalma,* and the long, hard, white cases of the limy tube worm, *Hydroides dianthus,* can often be spotted at the base of the plants. Both worms are typical residents of the oyster bar communities of high-salinity regions.

Grazers over the Plants

The ever-present store of living food attached to sea grasses attracts a host of grazer organisms, which wander over the surface of every leaf and stem. Many species are

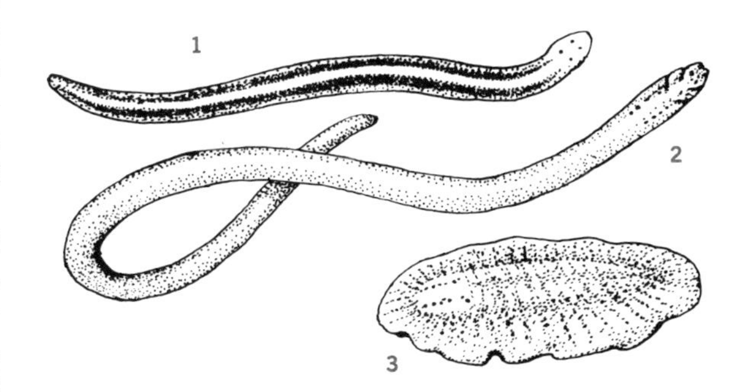

1. Four-eyed Ribbon Worm, *Tetrastemma elegans* (to ¾″)
2. Green Ribbon Worm, *Zygonemertes virescens* (to 1½″)
3. Slender Flatworm, *Euplana gracilis* (to ⅜″)

the same as those found among the fouling organisms of a pier, jetty, or other hard substrate. Some species favor the plant community over any other habitat.

A wide variety of errant worms move about the plants. The omnipresent common clamworm is here, as well as a number of other diminutive polychaete worms, most of them too tiny to identify without a microscope. Two small ribbon worms are common in seagrass meadows. In spite of their size, they are avid predators prowling the blades of grass for amphipods and isopods. The smaller of the two, the four-eyed ribbon worm, *Tetrastemma elegans,* is barely three-fourths of an inch long, more cylindrical than flat, black, but sporting a white stripe down the center of its back. It has four distinct eyespots, which form a square on its head with an eye at each corner. The green ribbon worm, *Zygonemertes virescens,* is somewhat longer at one and one-half inches. It is evenly greenish with numerous scattered eyespots and two oblique grooves on the side of its head.

Small flatworms graze over the surface as well, not only the little black oyster flatworm, the common grazer of barnacles, but also the yellowish gray slender flatworm, *Euplana gracilis.* The slender flatworm has eyespots in two rows along the head. It is without the short tentacles of the oyster flatworm.

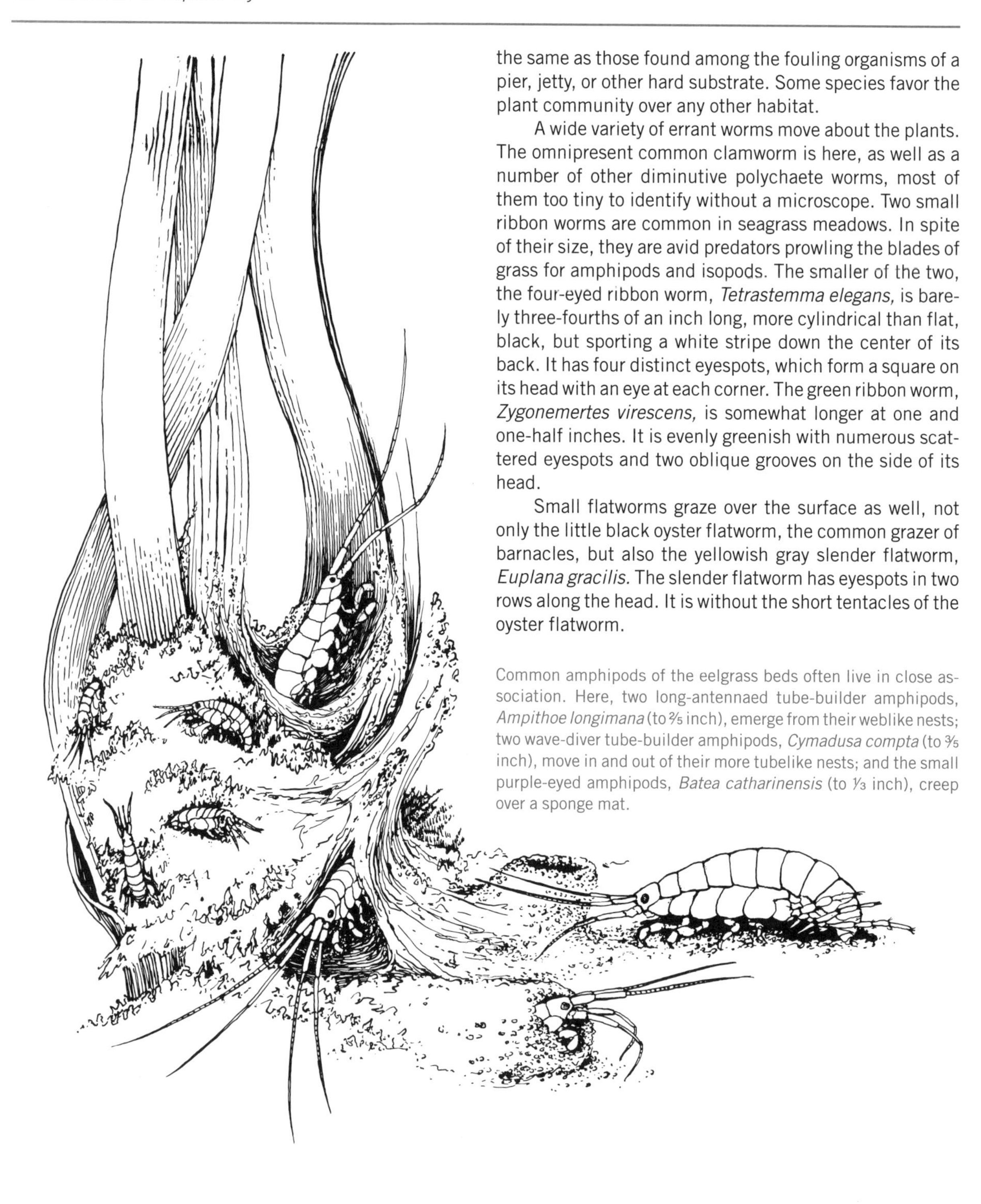

Common amphipods of the eelgrass beds often live in close association. Here, two long-antennaed tube-builder amphipods, *Ampithoe longimana* (to ⅖ inch), emerge from their weblike nests; two wave-diver tube-builder amphipods, *Cymadusa compta* (to ⅗ inch), move in and out of their more tubelike nests; and the small purple-eyed amphipods, *Batea catharinensis* (to ⅓ inch), creep over a sponge mat.

Isopods and amphipods are major predators of the seagrass community, easily observed as they scoot in and out of the faunal growths. The eelgrass pill bug, *Paracerceis caudata,* is one of the most numerous isopod residents, at least in higher salinities. The female, almost half an inch long, looks remarkably like the intertidal sea pill bug, its eyes set wide at the corners of the head. Scientists have long mistaken one species for the other. The males are somewhat smaller, with elaborate processes curving out from their tails. The elongated eelgrass isopod, *Erichsonella attenuata,* is also about half an inch long, with two long antennae. The mounded-back isopod, *Edotea triloba,* is frequently found with the others. It is small (one-fourth inch) and oval-shaped with two indented ridges down its back. In higher salinities, a somewhat larger and thus more noticeable isopod, the Baltic isopod, *Idotea baltica,* often appears in numbers on the plants, as well as floating on passing algae or loose seagrass strands. A single population may be composed of individuals of different patterns and colors.

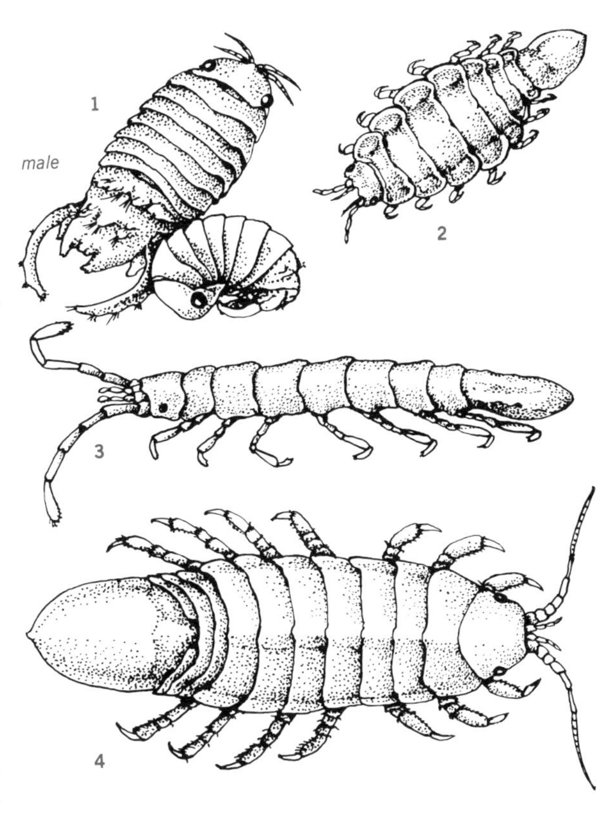

1. Eelgrass Pill Bug, *Paracerceis caudata* (male to ¼", female to ½")
2. Mounded-back Isopod, *Edotea triloba* (to ¼")
3. Elongated Eelgrass Isopod, *Erichsonella attenuata* (to ½")
4. Baltic Isopod, *Idotea baltica* (to 1")

Both tube-building amphipods and free-living roaming amphipods abound in seagrass meadows. The leaves, particularly those at the base of the plants, may be covered with soft mats of the slender tube-builder amphipod with its flattened body and stout front appendages—the same species that seems to cover almost any substrate where soft silts have collected. However, the most common tube-making amphipods with dense populations in seagrass meadows are two closely related species: the long-antennaed tube builder, *Ampithoe longimana,* and the wave-diver tube builder, *Cymadusa compta.* The long-antennaed amphipod constructs weblike tube nests of bits of plant material bound together with secretions. It feeds on the algal coating of the leaves but returns to its nest when not feeding. It is small, less than one-half inch, with long, large, black eyes and slender antennae almost as long as its body. The wave-diver also lives in a tube, wandering out to feed, preferably on detrital material rather than algae. It is somewhat larger, with antennae not quite as long as those of the long-antennaed tube builder, and is distinguished by large, light reddish, encapsulated eyes, which almost cover the front of the head. The wave-diver is abundant in unvegetated bottoms as well.

Many of the non-tube-building amphipods in the seagrass meadows are the same species found in abundance elsewhere. Spine-backed scuds, found so often under seaweed drifts on the intertidal flats, are plentiful here. Skeleton shrimps attach to hydroids and other epifaunal growths. The purple-eyed amphipod, *Batea catharinensis,* often creeps among hydroids, sponges, and bryozoans. It is a rather ordinary looking amphipod, marked, principally, by large, square-shaped purple-brown eyes.

Snails. Hundreds of tiny snails no more than a quarter-inch long when full grown may be observed crawling over almost any plant of the seagrass meadows. The most abundant species is the grass cerith, *Bittium varium,* a highly gregarious little snail found in dense concentrations which moves slowly over the grass blades feeding on detritus. Its shell is a slender, grayish brown spire covered with a criss-cross pattern of ribs and whorls. The shells of some of the older bittiums of the colony may be completely covered by soft snail fur, *Stylactaria arge.* This hydroid closely resem-

An Atlantic oyster drill, *Urosalpinx cineria* (to 2 inches), looms large against other snails of an eelgrass bed. Two black-lined triphoras, *Triphora nigrocincta* (to 1/4 inch), feed on a sponge covering the eelgrass blade on the left. A lunar dove shell, *Mitrella lunata* (to 1/4 inch), glides over a blade in the center; grass ceriths, *Bittium varium* (to 1/8 inch), some covered with a fuzzy coating of snail fur hydroids, graze on other blades. One impressed odostome, *Boonea impressa* (to 1/4 inch), probes into a sea squirt while another attacks a convex slipper shell.

bles the snail fur so common on hermit crab shells, but without the small spines of the latter. In some studies of eelgrass communities of the lower Bay, the grass cerith has been found to be more abundant than any other single animal if one excludes the microscopic organisms. The next most numerous snail is the ectoparasitic snail, the impressed odostome, *Boonea impressa,* which feeds on the ceriths as well as on slipper shells and other mollusks. It also attacks sea squirts. Odostomes feed on their victims by sticking a long proboscis inside their shell and sucking out the flesh. It is about the same size, perhaps slightly larger, than the ceriths but it does not hesitate to attack even larger snails. The shell is a milky white, elongated cone with flat, indented whorls.

Two other small snails, the black-lined triphora, *Triphora nigrocincta,* and the lunar dove shell, *Mitrella lunata,* are frequent visitors to seagrass beds. Black-lined triphoras have slender, elongated shells, dark brown with a "glassy beaded" surface and a dark band between the whorls; but the best clue for recognition is its left, or sinistral, aperture. Lunar dove shells are smooth and glossy, somewhat fusiform and quite beautifully marked with reddish brown or fawn-colored zigzag lines.

A much larger carnivorous snail, the Atlantic oyster drill, *Urosalpinx cinerea,* a typical resident of the oyster bar community (and discussed in Chap. 8), frequently moves into the plant beds to prey on slipper shells, barnacles, tube worms, and bryozoans.

Sea Slugs. Sea slugs, unusual and seldom-noticed inhabitants of seagrass meadows, are closely related to garden slugs but are far more diverse in form and graceful in shape and movement. Sea slugs are shell-less mollusks, the shells having disappeared gradually over the ages. Most slugs have vestigial coiled shells in their veliger larva stage. The fragile bubble shell snails of the intertidal flats and subtidal bottoms are closely related to sea slugs and look much like them when sliding about with their open mantle completely enveloping the shell. There are two basic types of sea slugs: those that feed on plants, the sea slugs known as herbivorous sacoglossans; and those that feed on animals, the carnivorous nudibranchs. Both types occur in the Chesapeake Bay.

Sea slugs slide over the substrate in a way that suggests the locomotion of snails. They are characterized by having two pairs of tentacles: a pair of oral tentacles at the front of their head and a second pair of tentacles, the rhinophores, somewhat behind and on top. They may have elaborate gill

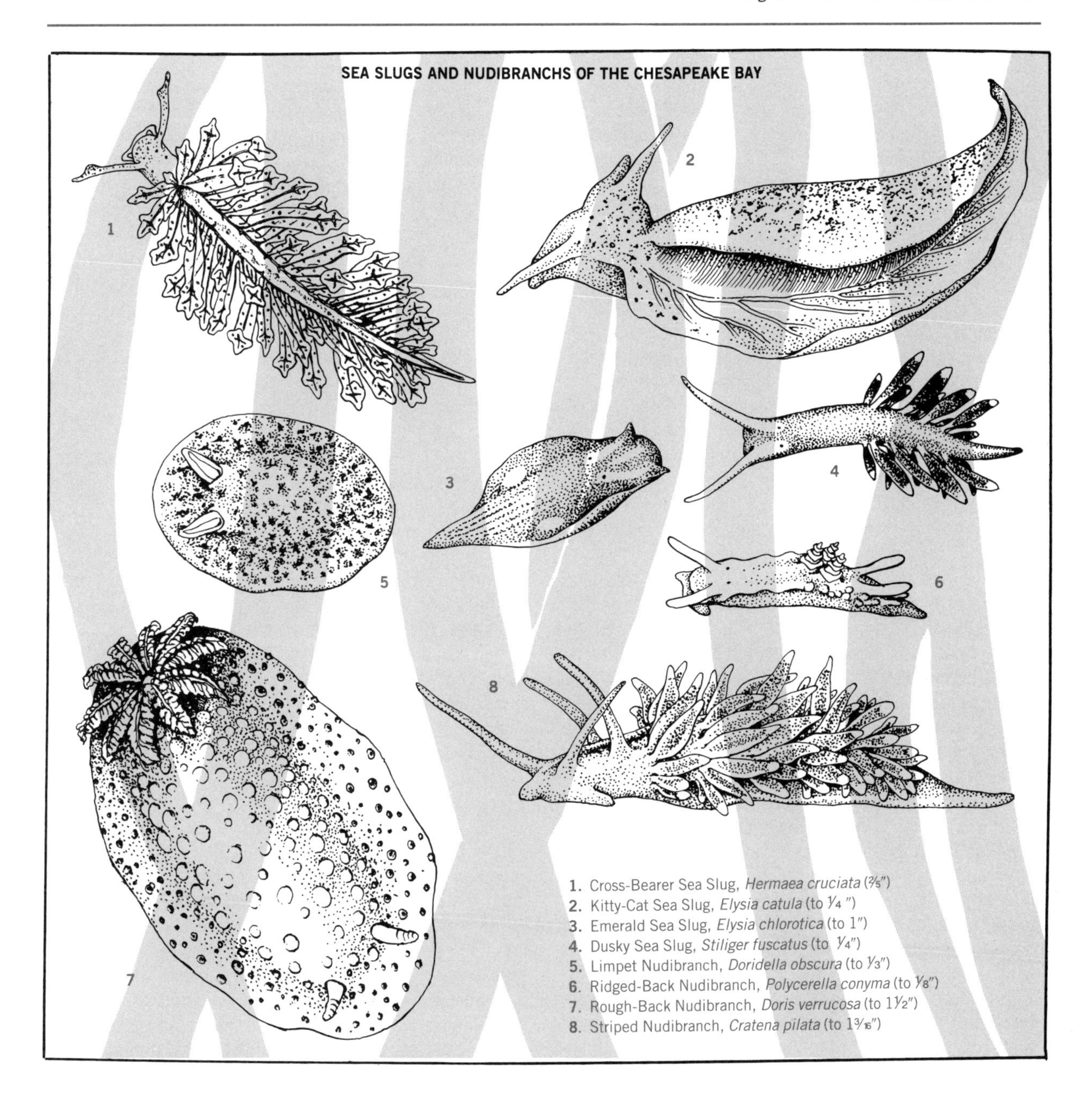

SEA SLUGS AND NUDIBRANCHS OF THE CHESAPEAKE BAY

1. Cross-Bearer Sea Slug, *Hermaea cruciata* (2/5")
2. Kitty-Cat Sea Slug, *Elysia catula* (to 1/4 ")
3. Emerald Sea Slug, *Elysia chlorotica* (to 1")
4. Dusky Sea Slug, *Stiliger fuscatus* (to 1/4")
5. Limpet Nudibranch, *Doridella obscura* (to 1/3")
6. Ridged-Back Nudibranch, *Polycerella conyma* (to 1/8")
7. Rough-Back Nudibranch, *Doris verrucosa* (to 1 1/2")
8. Striped Nudibranch, *Cratena pilata* (to 1 3/16")

structures, tubercules, or bulbous, club-shaped protrusions called cerata, which aid in their respiration. Sea slugs of the Chesapeake are much smaller than most garden slugs, most species being less than half an inch long. They are easily missed, even though there may be dozens on a single plant. Not only are they small, but they tend to take on the general coloration of the food they are grazing on, making it doubly difficult to notice them. Most species are soft-bodied and should be placed in water for viewing. They collapse into shapeless lumps when removed from water.

Two of the most common species of sea slugs on the seagrass beds are tiny herbivorous sea slugs, both about one-fourth inch long and well camouflaged against the plants. The kitty-cat slug, *Elysia catula,* is so named because its rhinophores are shaped like a cat's ears. It is a smooth-backed slug, variably brownish but with a definite greenish cast from the plant material it ingests. It has a whitish spot between its cat's ears and lightish streaks obliquely angled backward from the rhinophores. It has flaring "wings" along each side which give it a graceful, flowing appearance. Kitty-cat sea slugs are apparently confined to seagrass meadows of the lower Bay and do not extend into Maryland waters. The dusky sea slug, *Stiliger fuscatus,* is a dark slate gray with a series of sharply patterned black and white cerata on the back half of its body. There are four large cerata along each side with five or six smaller ones alternating below. The rhinophores are long and light-colored, the light color extending back from each rhinophore to behind the tiny, black eyes. The foot is pale yellow. The dusky sea slug apparently feeds on filamentous algae growing on the plants, rather than on the plants themselves.

The emerald sea slug, *Elysia chlorotica,* has been found only in Maryland waters of the Bay. It is a beautiful, graceful creature, an inch or so long, colored a brilliant green, mottled with white, and dotted with minute red spots. It, too, has winglike extensions along each side. Both *Elysia* sea slugs have not been found within the same general area. It is likely that they have merely escaped collection up until now.

One of the most beautifully delicate creatures in the Chesapeake Bay is the cross-bearer sea slug, *Hermaea cruciata.* It is not very common, but once seen, is instantly recognized. Its numerous long, translucent cerata are highlighted by brownish extensions of the liver which divide into cruciform shapes at the tip. The whole animal seems to sparkle with tiny white spots. It is a sacoglossan and has been found among red weed seaweeds as well as among eelgrass.

The two most common nudibranchs are even smaller than the sacoglossans. The limpet nudibranch, *Doridella obscura,* only one-third inch long, is well concealed as it passes over the colonies of encrusting bryozoans on which it feeds. Limpet nudibranchs are variably mottled with black, brown, and yellow pigment spots; some are almost colorless, others nearly black. They are round, with a caplike back that completely covers the head and foot. Two short rhinophores are located on the back toward one end. Limpet nudibranchs are common, not only in seagrass meadows but also in other habitats where encrusting bryozoans proliferate, as on pilings and oyster bars. The ridged-head nudibranch, *Polycerella conyma,* is a pale, translucent yellow-green, a color that blends with its favored food, the creeping bryozoan, *Bowerbankia gracilis.* It is totally inconspicuous as it grazes over the bryozoan's zooecia. Three-branched gills arise from the center of its wart-covered back. Two somewhat scalloped ridges extend along each side of the head back to the area of the gills.

The largest sea slug in the Bay is the rough-back nudibranch, *Doris verrucosa.* It is a flat, oval-shaped sea slug about one and one-half inches long with a dull orange color and a warty back. It is equipped with two small rhinophores at one end and a circle of feathery gills at the other. Rough-back nudibranchs lay long, sinuous ribbons of egg cases on the blades of eelgrass. The adults, however, are more often discovered on large sponges dredged from deeper waters. It is restricted to the higher salinities of Zone 3.

The striped nudibranch, *Cratena pilata,* has numerous club-shaped cerata along its back. It is pale grayish to greenish with irregular, broken russet-colored stripes on the head and along the body. Both pairs of tentacles are elongated. Striped nudibranchs are found among hydroids of the seagrass meadows. Jellyfish polyps are a favored food, so striped nudibranchs are also abundant on oyster shells, where polyps attach. They are apparently able to ingest the stinging cells of the polyp's tentacles without triggering them.

Animals in the Muds around the Plants

Much of the infauna of adjacent unvegetated areas surrounding the seagrass meadows live buried among the plant roots in the bottom muds and sands. Many species are more numerous here than they are in adjacent open waters. The great abundance of food available within the meadows supports this rich buried population. The common bristle worms of the intertidal flats and shallow habitats abound here—capitellid thread worms, glassy tube worms, clamworms, oligochaetes, and various mud worms. Tube-builder mud worms and fan worms are also found in the bottom sediments, as well as on the plants. Soft-shelled clams and Baltic macoma clams burrow in the bottoms, although populations are lower in the seagrass meadows near the mouth of the Bay. In winter, when the plants die off, the grass ceriths and lunar dove shell snails overwinter in the bottom muds. Most of the isopods and

A colony of long-antennaed four-eyed amphipods amongst eelgrass plants. One tube is shown semitransparently to indicate the position of the amphipod within.

amphipods found among the leaves during warmer months also find refuge in the bottom during the coldest seasons.

Three species of four-eyed amphipods construct tough parchment tubes in the bottom among the bases of the plants. Discrete colonies are formed with many tubes clustered one upon the other. The species are closely related to one another and are similar in appearance. Each bright pink eye is split into two separated eyes, giving them a somewhat cockeyed appearance. Four-eyed amphipods are pale-colored, with flattened heads and a distinctive hump on their telson (tail) segment. The long-antennaed four-eyed amphipod, *Ampelisca vadorum,* is about one-half inch long and builds a short, broad tube about an inch long. The small four-eyed amphipod, *Ampelisca abdita,* is half that size, yet it builds a longer, narrow tube of about one and three-fourths inches. The narrow-headed four-eyed amphipod, *Ampelisca verrilli,* builds tubes as long, but more robust. Tubes are constructed of gray, parchmentlike material with a smooth, whitish tube lining. The tubes are embedded with bottom sediments of grains of various sizes, depending on the species: those of the narrow-headed with coarse sand grains, of the long-antennaed with small sand grains, and of the small four-eyed with finer silts. The tubes are flattened from side to side and may project almost half an inch above the surface of the bottom. The bottoms of the tubes are constructed of a lighter membranous sheath.

The amphipods lie on their backs across the mouth of the tube and feed on planktonic matter brought to them by currents set up by their beating appendages, much in the feeding manner of the sand-digger amphipods of the sand beach habitat. They also feed on particulate matter scraped from the bottom by their long second antennae. Their four eyes are positioned above the edge of the tube, alert for any danger; when startled, the amphipod quickly retreats into its tube and the soft edges fold over to close the opening.

The amphipods leave their tubes to breed in the water during spring and summer. At this time, they become part of the epifaunal community of the seagrasses. After breeding, females settle onto the bottom and release their young, which soon construct their own tubes and begin a whole new colony.

Shrimps and Crabs

Many crustaceans move into the seagrass meadows to feed on the rich supply of food on the plants or buried in the bottom. Juvenile crabs utilize the plant beds as prime nursery areas, finding protection from predators as well as abundant food among the thick stands of leaves. Schools

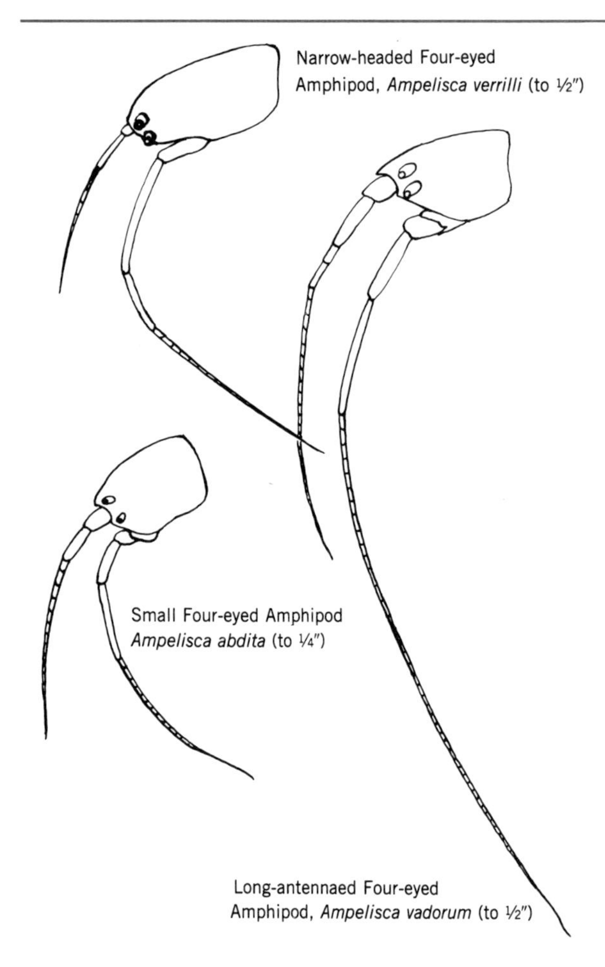

Differences in heads and antennae help identify the three species of four-eyed amphipods common to the Chesapeake Bay.

of juvenile grass shrimps are particularly abundant here. Sand shrimp and other bay shrimps also tend to congregate in vegetated areas. Shrimps are generally more active during nighttime hours, burying in bottom sediments during the day. Blue crabs move into the meadows not only to feed but also to find protection during the critical shedding period. Blue crabs are soft after molting and are unable to swim or crawl; they are, therefore, at the mercy of larger prey species. Opossum shrimp can be as numerous here as they are in more open waters. Occasionally mud crabs, hermit crabs, and spider crabs wander into seagrass areas, but shrimps and blue crabs are by far the dominant species of large crustaceans. The number of shrimps and crabs begins to increase dramatically in late spring and remains high through early summer. As the season progresses, the lushness of the sea grasses declines, juveniles grow, and as they begin to migrate they are fed upon by schools of predatory fishes, sharks, and rays. As cold weather approaches crabs and shrimps move to deeper channels, where they overwinter.

Fishes of the Seagrass Meadows

Fish populations within seagrass meadows are abundant and diverse. Some fishes are permanent residents within the beds and are rarely found in other habitats. Many larger predator fishes of deeper, open waters make frequent forays into the meadows to feed. Seagrass meadows in the saltier waters of the lower Bay have a more diverse assemblage of fishes than are found in seagrass meadows farther up in the Bay and its tributaries, owing to the greater influx of fishes from the ocean. Fishes begin to gather in the meadows in spring, as the plants begin vigorous growth, and remain plentiful throughout the summer and into the autumn. As winter approaches and the plants die back, they move to deeper Bay waters or leave the Bay entirely for the sea.

Fishes found almost exclusively among vegetation include pipefishes, seahorses, and sticklebacks. Most people are unaware that seahorses live in the Chesapeake Bay—the same familiar species brought home as a souvenir from a Florida trip. Pipefishes and seahorses are closely related. A seahorse is, in a sense, a pipefish with a curled tail and a cocked head. Both pipefishes and seahorses have elongated tubular jaws with small, toothless mouths at the end. Their bodies are encased in jointed, bony rings. Males have brood pouches, where the young are reared after the females have deposited the eggs inside. It is quite a sight to see a male seahorse spouting out a cloud of tiny seahorses from the top of his sacklike brood pouch! In pipefish, the brood pouch is formed by elongated lateral folds along the underside of the belly, the eggs inside lined up neatly in rows.

Small seahorses and pipefish are well developed when they leave the pouch and quickly assume the habits of adults. The Chesapeake Bay seahorse is the lined seahorse, *Hippocampus erectus.* It swims erect, frequently pausing to curl its tail around plant leaves while it feeds on minute organisms. The long, thin pipefish often align themselves vertically, melding perfectly with the long leaves of eelgrass as they sway gracefully in the water currents, imitating the movements of the leaves. Two species are found in the Bay. The dusky pipefish, *Syngnathus floridae,* and the northern pipefish, *Syngnathus fuscus,* are similar in appearance, both about six to eight inches long. The relative lengths of their snouts and dorsal fins are identifying characteristics: northern pipefish have a shorter snout and a longer dorsal fin extending along an equal number of body rings before and after the vent (the anal opening that divides the tail from the abdomen). The dusky pipefish has a longer snout and a shorter dorsal fin situated over one to three body rings before the vent and as many as six to seven behind the vent. The northern pipefish is by far the more common and widespread species, occurring well into the estuary as far as freshwater, although it is scarce in the upper regions of the Bay.

Sticklebacks are small, two-inch fishes with spindle-shaped bodies and very narrow tails. There are two species in the Chesapeake Bay, the fourspine stickleback, *Apeltes quadracus,* and the threespine stickleback, *Gasterosteus aculeatus.* The common names refer to the distinctive dorsal spines; however, a word of warning—fourspine sticklebacks appear to have only three spines and threespine sticklebacks appear to have only two because in both species the last spine is in close alignment with the soft-rayed dorsal fin. Fourspine sticklebacks are scaleless, whereas threespines have long, narrow plates (called scutes) along their sides. Sticklebacks live among vegetation and

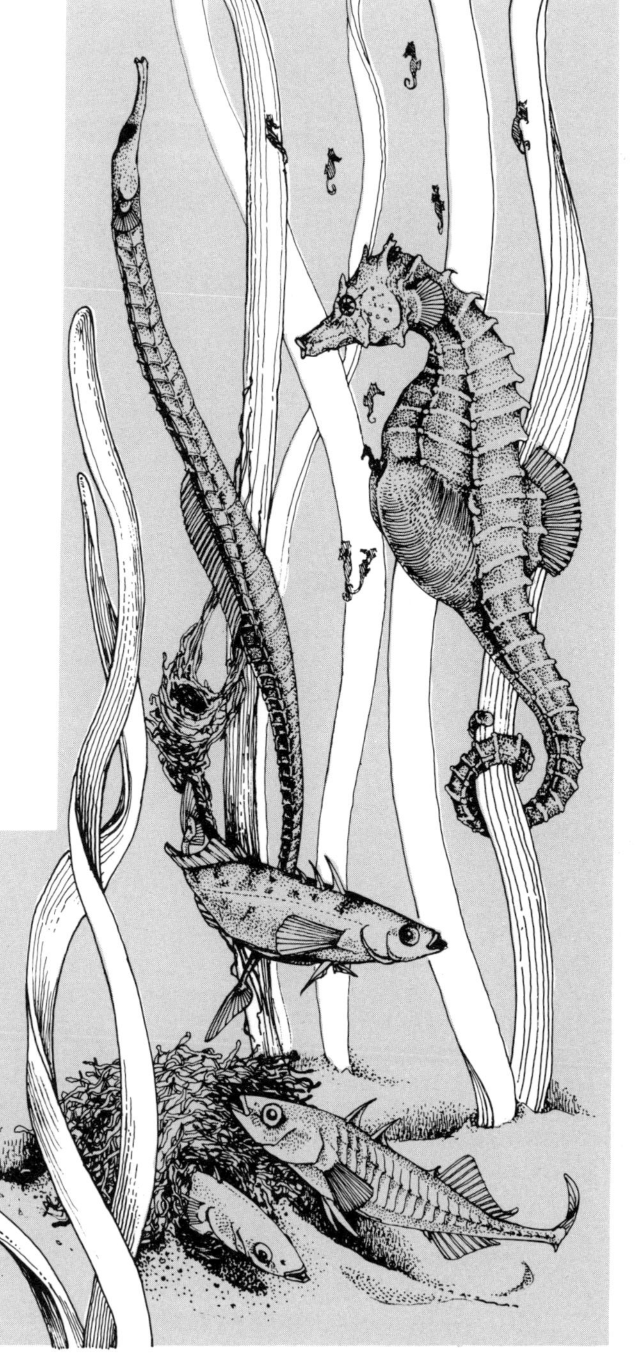

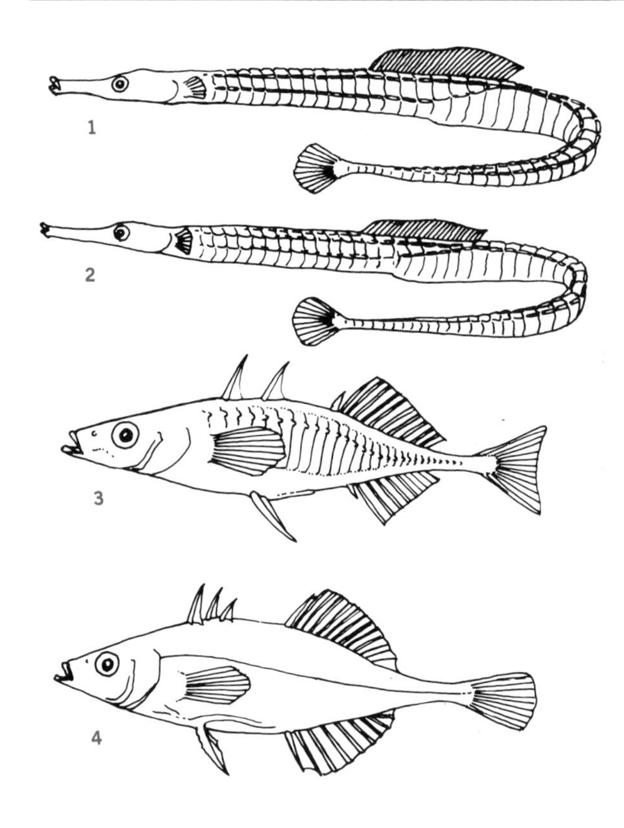

1. Northern Pipefish, *Syngnathus fuscus* (8″)
2. Dusky Pipefish, *Syngnathus floridae* (8″)
3. Threespine Stickleback, *Gasterosteus aculeatus* (2″)
4. Fourspine Stickleback, *Apeltes quadracus* (2″)

construct nests of leaves stuck together with mucus. Males carefully guard the nests, cleaning out any eggs that die and picking up and returning live eggs that fall from the nest. Fourspines build small, elongated (cone-shaped) nests off the bottom around the weeds; threespines build larger, circular (ball-shaped) nests on the bottom, sometimes embedding them in the soft muds.

Most of the small fishes schooling in the open shallows move into the seagrass meadows as well. Schools of anchovies and silversides and young menhaden can be found there. Some juvenile herrings and shads move down from upstream nursery areas; needlefish and halfbeaks, killifishes, menhaden, gobies, and blennies are all here at one time or another. These regions are important nursery areas for young of the drum family. Young spot are often the most abundant fish in seagrass meadows. Other juvenile members of the drum family, including croakers, weakfish, red drum, and silver perch, can be quite numerous in the seagrass meadows of the lower Bay regions. Adults of the same species as well as other predator fishes feed ravenously on the schools of smaller fishes. The larger fishes dig into the bottom in search of clams and worms or graze on the large variety of small invertebrates. Ravaging bluefish follow the schools of menhaden into the meadows. Large weakfish, spotted seatrout, and summer flounder are common predators.

Rays and Sharks

Rays and sharks can wreak havoc in the sea meadows. Cownose rays, *Rhinoptera bonasus,* have been known to denude large areas as they root up plants in search of soft-shelled clams—a favorite food. Cownose rays are also known as "double heads" because of an indentation around their snout which gives them the appearance of having an upper and lower head region. They are large, heavy rays with a wing span of up to three feet. As they swim in the shallows, their wing tips break the surface like a pair of shark fins. Cownose rays use their winglike fins to expose buried soft-shelled clams by flapping them against the bottom sediments, then crushing the shells with their powerful dental plates. Cownose rays are armed with a long, poisonous spine located at the base of the tail, so they should be handled carefully. There is an oft-repeated story of Captain John Smith's encounter with the spine of a cownose ray. Apparently he was stung severely and was in such pain that his crew despaired for his life. He recovered, but the incident was memorialized by his sailors, who named the site of this lamentable incident Stingray Point; located at the mouth of the Rappahannock River, it is still known by that name.

The bluntnose stingray, *Dasyatis say,* is another common visitor to seagrass meadows. It is a flat-nosed, round-shaped ray with a long, whiplike tail. The snouts of most stingrays are pointed, rather than rounded, as in this species. It, too, has dangerous spines, one or two of which are located back on its tail and often lie flat against it. This ray's most characteristic feature is a finlike fold of skin on the top and bottom of the tail. Its eyes are on the flat top of its head, its spiracles (respiratory openings) just below and behind them. The bluntnose ray is brownish to greenish on top and whitish below.

Schools of bluntnose stingrays frequently enter bays and estuaries along the Atlantic coast. As they migrate, they break the surface of the water with rapid splashes of their wings. They characteristically swim in spurts, between which they settle below the water for a time. An observer on shore can often see their splashing movements and their long tails breaking the surface. Bluntnose stingrays are bottom feeders that move into the seagrass meadows in search of worms, small clams, amphipods, and other invertebrates. When not feeding they bury themselves flat in soft sediment with only their eyes and spiracles exposed.

Many kinds of sharks enter the Chesapeake Bay in summer—most large sharks are solitary occasional visitors from deeper water habitats. However, schools of young sandbar sharks, *Carcharhinus plumbeus,* move en masse into the Bay in summer, and very young specimens are found in abundance in sea grasses. The large population of blue crabs are vulnerable prey to sandbar sharks. Adult sandbar sharks may reach a length of seven feet, but the young found regularly within Bay waters are generally from two to almost three feet long. The sandbar shark is a classic-looking shark with a long, round, pointed snout, a heavy first dorsal fin, and a long thrasher tail. It can be distinguished by the forward position of its first dorsal fin and by a thick ridge of skin along the midline of the back between their first and second dorsal fins. Sandbar sharks move well into the Bay and may even be found in Maryland waters.

Dabblers and Divers

Some dabblers, or puddle ducks as they are also known, can dive, and divers can dabble. Dabblers are ducks that tip up, stretch out their necks, and graze on submerged vegetation and aquatic invertebrates. They can dive to feed or to evade danger, but they usually dabble. Most ducks in this group have an iridescent or metallic speculum or wing patch on the trailing edge of their wings. When alarmed, dabblers literally spring out of the water and take flight. Divers typically dive for their food and to escape danger; they also sometimes dabble. As discussed in Chapter 5, divers have legs set far to the rear with very large feet; dabblers' legs are positioned farther forward, and they have smaller feet. Divers do not explode out of the water like dabblers; they must patter across the surface a number of yards before gathering altitude. Dabblers are able to land on the ground and graze on grass and grain; divers, at best, are ungainly on dry land because of the rearward position of their legs. Some divers such as the stiff-tailed ruddy duck will fall on their breasts after a few steps on land and must resort to scootching back to the water to take wing or dive. About eight species of dabbling ducks regularly occur in the Chesapeake Bay. All of them may be found, at one time or another, in the shallows, over seagrass meadows and weed beds, and in the marshes. Sometimes, like the divers, they will loaf in deeper waters as well.

Two dabblers that often feed over seagrass meadows and weed beds are the American wigeon, *Anas americana,* and the green-winged teal, *Anas crecca.* The American wigeon is perhaps better known as the baldpate or sometimes just as wigeon. Both birds are migratory and spend their winters in the Chesapeake Bay and along the mid-Atlantic coast. They are both found in fresh and brackish water where there is an abundant supply of aquatic vegetation including wild celery, redhead grass, sago pondweed, and widgeon grass. They also feed on grains and grasses in fields.

The baldpate is the larger of the two birds and is a saucy, alert puddle duck, seldom still and always quick to sound the alarm. The males have a soft three-note whistle, with the accent on the second note; the female's call is a croak. The drakes have iridescent green, commalike markings extending from around the eye and down the side and back of the neck. The forehead and the top of the head are all white, giving the appearance, even in a mixed flock, of a bird with a bald pate. The throat is a streaked gray-brown and the breast and back are brownish. The female has a gray-brown streaked head and a russet-brown breast and back. Both sexes have blue bills tipped with black. The American wigeon is a dabbler and not a proficient diver. To obtain some tasty bits of vegetation that otherwise may not be available, it has devised the habit of occasionally waiting for a deep diver such as a canvasback or redhead duck to bob to the surface with a succulent tidbit, at which point it quickly filches the food. Wigeons, like all dabblers, can leap vertically from the water. The flock reacts as one, grouping close together, wheeling, and veering much like a group of pigeons.

Green-winged teals are the smallest species of dabbling ducks. They are little, predominantly dark birds. The male has a russet head with a green patch extending from the eye to the back of the head; when at rest, a vertical white bar is obvious in front of the wing. Green-winged teals are named for the opalescent green speculum in both sexes. The male's call is a whistling twitter; the female quacks. They are swift fliers and resemble a flock of shorebirds as they fluidly leap from the water and quickly vanish.

Two divers that feed in vegetated areas are the canvasback duck, *Aythya valisineria,* and the ruddy duck, *Oxyura jamaicensis.* Canvasbacks are one of the largest ducks; ruddy ducks, one of the smallest. Historically, canvasbacks fed extensively on wild celery, from which they get their name, *valisineria.* Ruddy ducks also feed on plants, particularly in the winter and spring. As wild celery and other succulent plants began declining in the Bay, canvasbacks modified their feeding behavior and now feed on macoma clams and little surf clams, as well as other invertebrates, to supplement their preferred diet of aquatic vegetation. Canvasbacks, one of the most abundant of ducks and much sought after by hunters, used to winter in the Bay by the thousands. Up to the 1950s, about half of the North American population of "cans," as they are often called, migrated to the Bay every fall.

An estimated quarter of a million canvasbacks flocked to the Susquehanna Flats and other areas of the Bay to feed on the lush vegetation. Now only about 20 percent of the entire continent's "cans" winter on the Bay. The decline is due not only to the lack of vegetation; the loss of nesting areas in the northern prairie pothole country and other factors have also contributed to their reduced numbers. However, wildlife biologists and managers have observed an increase in reproduction of "cans" in the critical nesting areas. A great deal of work by various groups has been done to restore the prairie pothole region and improve the nesting habitat; this, coupled recently with excellent climatic conditions, provides some hope for a modest recovery in the canvasback population. The redhead duck, *Aythya*

Puddle ducks of the Bay. A male ruddy duck, *Oxyura jamaicensis* (15 inches), swims at left while his mate dives for her meal. A female green-winged teal, *Anas crecca* (14½ inches), dabbles for her food, her mate just above her. A male American wigeon, *Anas americana* (19 inches), another dabbling duck, looks on.

americana, a slightly smaller relative of the canvasback, was not able to make the switch from aquatic vegetation to clams and other invertebrates and has essentially deserted the Bay for areas to the south. Only about two thousand redheads winter in the Bay now.

Canvasbacks, typical of divers, have widespread legs set to the rear of the body, which makes them awkward walkers. The head of a canvasback slopes into its long bill in an almost unbroken line, giving it a silhouette like no other duck's. The male's head and neck are chestnut colored, the breast is black, and the back and sides are grayish to almost white. The female has a light brown head and neck and a paler gray-brown back and sides. Both sexes have dark bills and the typical short wings of divers, yet they are the swiftest flying duck. Canvasbacks often raft up in large numbers in deeper waters, where they loaf in the sun and preen their feathers. In recent years, "cans" in late winter have been gathering in large flocks on the Annapolis side of the Chesapeake Bay Bridge. They usually dive and feed in water depths of 12 feet or less, but they can dive deeper. Canvasbacks begin courting and pairing off in early spring prior to their migration north to the prairie pothole region. They return to the Bay in early to mid-December.

The perky little ruddy duck has also declined in the Bay area. It is an excellent diver and swimmer and, when alarmed, will generally dive underwater, using its stiff tail as a rudder, to escape the threat. Ruddy ducks are small, stubby birds with thick necks and large heads with very wide, broad bills. The male has distinctive white cheeks and a black cap and is grayish in the winter. The female is similar, but her cheeks are not as conspicuously white and she has a dark stripe running from the base of the bill, across her cheek, to the neck. The ruddy duck is one of the smallest ducks but, curiously enough, it lays one of duckdom's largest eggs—even larger than those of mallards and canvasbacks, which are much larger birds indeed. Shortly

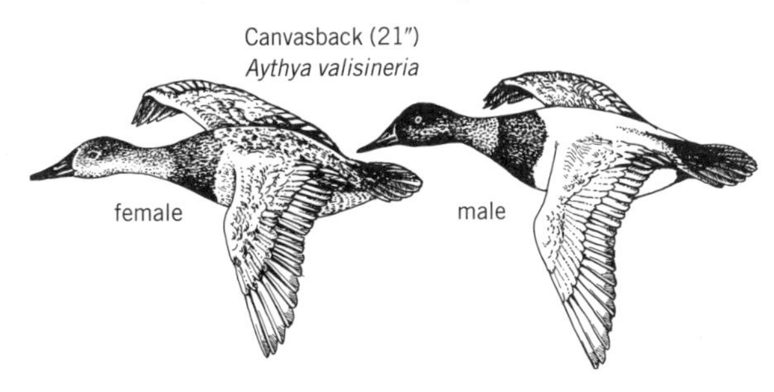

after the eggs hatch, the ducklings are capable of diving. Look for these sassy little birds to appear in the Bay about mid-October. They have very rapid wingbeats—in fact, some say they resemble large bumblebees; and with the male's conspicuous white cheek patches and its stiff tail cocked at an angle much like a wren's, they should not be difficult to identify.

WEED BEDS OF THE UPPER BAY

Where the water becomes less salty, eelgrass disappears and other aquatic plants begin to flourish. The typical seagrass meadows of the lower half of the Bay are replaced by the brackish-water weed beds of the upper Bay (upper Zone 2), typically vegetated by redhead grass, sago pondweed, and horned pondweed. Widgeon grass continues to thrive. The plants in these weed beds serve the same purposes as those in the seagrass meadows—they provide food, protection, and stability for a wide variety of organisms. Barnacles, encrusting bryozoans, whip mud worms, platform mussels, bloodworms, and clamworms are all found in these weed beds. Many of the same isopods and amphipods that occur in seagrass meadows also exist here—eelgrass isopods, Baltic isopods, and mounded-back isopods continue to thrive, as well as slender tube-builder amphipods and most of the same species of scuds. Some of the most numerous epifaunal species of seagrass meadows decline and eventually disappear upstream, including white and green-striped anemones, sponges, sea squirts, hydroids, and sea slugs. The typical snails of the seagrass meadows are not found in brackish-water weed beds. The convex slipper shell, oyster drill, lunar dove shell, and grass cerith snails are all high-salinity animals. However, the impressed odostome snail tolerates lower salinities and may be found here.

In brackish-water weed beds, the grass cerith is replaced by other minuscule snails, the seaweed snails, *Hydrobia* spp. These are thin, smooth, almost translucent brownish snails, less than one-fourth inch long.

Burrowers in the bottom substrates of these weed beds consist primarily of various mud worms, capitellid thread worms, soft-shelled clams, and macoma clams. Weed beds do attract a host of shrimps, crabs, and fishes. Grass shrimps, sand shrimp, and opossum shrimp are all found in these weed beds, as are molting blue crabs. All the small resident fishes of the estuary may be found here. There is a wider variety of killifishes in these lower-salinity areas than in eelgrass beds. Young spot and croakers utilize the weed beds, but young of the other drums, so plentiful in seagrass meadows, are not found in numbers. Young herrings and shads, on the other hand, are more plentiful. Sharks and rays do not often migrate upstream to brackish regions.

The tidal freshwater plant community has a wide variety of water plants including coontail, common waterweed, and wild celery. Here typical Chesapeake Bay animals are sparse and are replaced by a whole spectrum of freshwater creatures. Many different types of freshwater snails graze over the weed beds in tidal freshwaters, including the little pouch snail found also along the intertidal shorelines of these areas. The hornshell snail, *Goniobasis virginica,* is easily detected among the plant leaves. Hornshell snails have elongated, turreted shells, which are thin and smooth but not glossy. They are a brown to olive color and sometimes have reddish bands.

A number of species of small freshwater clams called fingernail clams are often found in weed beds in tidal freshwater areas. At full growth even the largest species are only half an inch long, the size of a fingernail. They have thin, fragile shells of a typical clam shape and are often mistaken for the young of freshwater mussels. Fingernail clams do not produce glochidium larvae, with a compulsory stage of encystation on fish, but brood their young in a special marsupial pouch formed within the gills. Fingernail clams are hermaphroditic, producing eggs that pass through a canal where sperm fertilizes them. The baby clams are not released until they are well able to fend for themselves. Adult fingernail clams generally lie buried in the bottom, but the young actively crawl about the plants. At times they suspend themselves from the plant strands

Seaweed snails, *Hydrobia* sp. (to 1/5 inch), on redhead grass.

Life amongst the thick plant growth of a waterweed community is diverse. Banded freshwater scuds, *Gammarus fasciatus* (to ½ inch), scamper over plant stems in search of food. Below them tiny pill clams, *Pisidium* sp. (to ⅛ inch), and the larger long-siphoned fingernail clams, *Musculium* sp. (to ½ inch), feed on detrital material. The hornshell snail, *Goniobasis virginica* (to 1 inch), glides along a plant stem while above, mosquitofish, *Gambusia holbrooki* (to 2 inches), feed on mosquito larvae.

with their byssal threads. The pill clams, *Pisidium* spp., are fingernail clams that, even as adults, continue to crawl over the plants. A handful of pill clams can often be collected by sweeping a fine mesh net through a stand of waterweeds. Pill clams are very tiny, less than one-eighth inch long, with rounded shells and a single small excurrent siphon. The proportionately large tongue-shaped foot is used both for burrowing and for crawling over vegetation. Incurrent water is pumped through a split in the mantle just below the single siphon. Two other genera of somewhat larger fingernail clams are the short-siphoned fingernail clams, *Sphaerium* spp., and the long-siphoned fingernail clams, *Musculium* spp.; the former have thicker, yellow to brownish shells, and the latter have thin, almost transparent, shells.

Freshwater scuds are easily discovered by day hiding among the leaves of the plants or under dying vegetation at the base of the plants. At night, they move from their hiding places in search of food. The banded freshwater scud, *Gammarus fasciatus,* is perhaps the most widespread and abundant member of this group in the Chesapeake Bay. It can be found all the way into slightly saline waters. It is about half an inch long, whitish, with large, pale, kidney-shaped eyes and is banded with brown or green along its body and the upper part of the appendages.

If you see a small fish nibbling at the surface of the water of freshwater weed beds, chances are it is a mosquitofish, *Gambusia holbrooki,* feeding on larval mosquitoes and other insects that are suspended just below the water's surface. Mosquitofish are also called topminnows, which refers to their surface-feeding habits. They are well adapted for this feeding mode, with flattened head, underslung lower jaw, and upward-opening mouth. They are small fish, up to two inches long, and look much like their close relatives, the well-known guppies of home aquariums. Mosquitofish are live-bearers, unique among Chesapeake Bay fishes. The male guides sperm directly into the female with its prong-shaped anal fin. The eggs develop and hatch within the female. The young, when born, are well developed and almost half an inch long. Mosquitofish are true live-bearers, unlike pipefishes and seahorses, whose eggs are fertilized outside the female, the males merely brooding them in external body pouches.

● Seagrass meadows and weed beds are a haven for fish and invertebrates. Home to organisms as varied as seahorses and sharks, scurrying shrimps, cleverly disguised nudibranchs, and dabbling and diving ducks, seagrass meadows and weed beds are also a rich source of nutrients that are circulated throughout the Bay system.

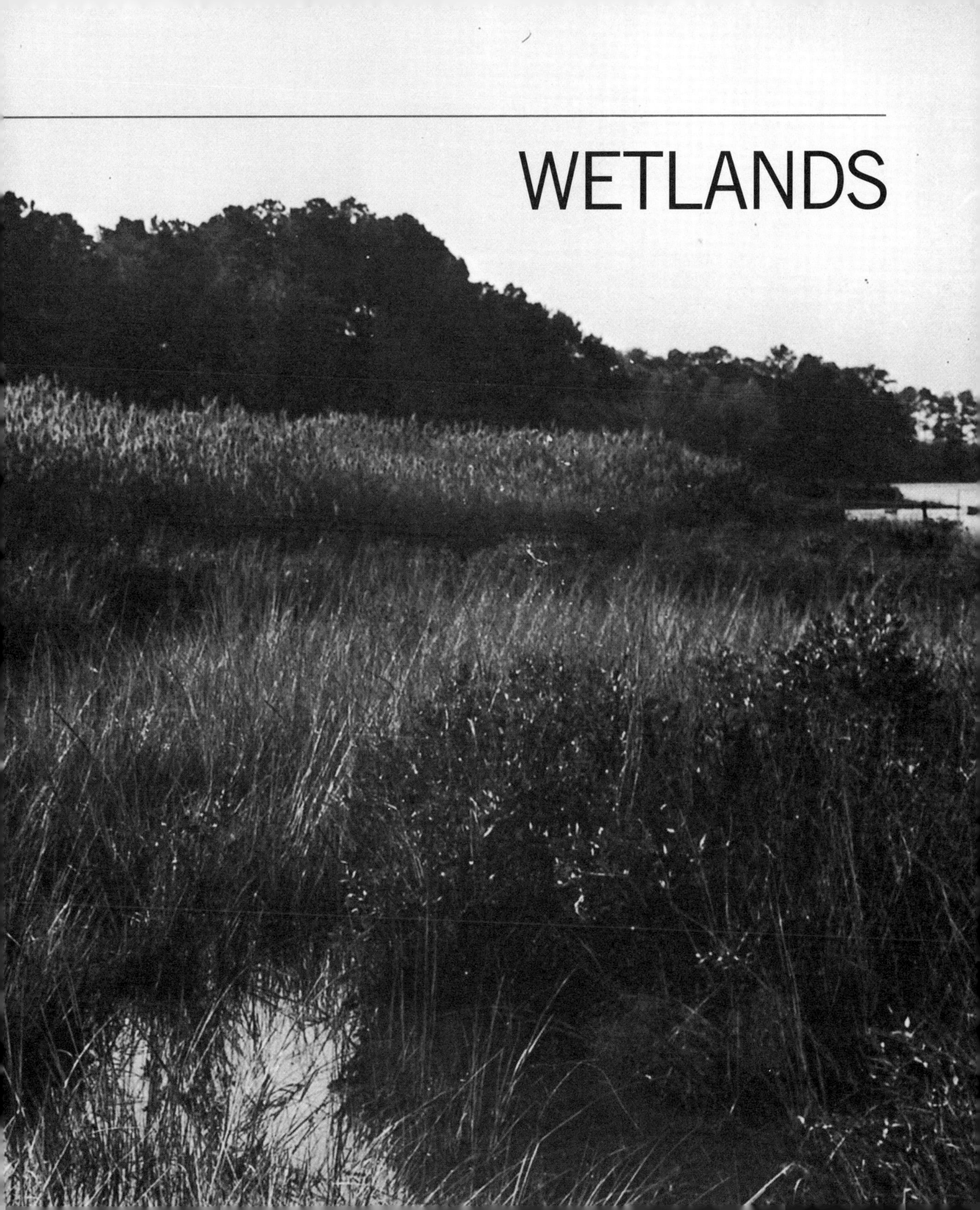

WETLANDS

The wetlands of the Chesapeake Bay form an important transitional zone between shallow water habitats and upland vegetation. There are upwards of a million acres of wetlands in the Bay system, extending from nontidal, freshwater, forested wetlands to the broad reaches of the lower estuary. Wetlands vary in size and type from flooded woodlands and riverine flood plains to small pocket marshes—often found at heads of tidal creeks—and narrow fringing marshes, to great meadowlands of green and gold-bladed plants that stretch, almost uninterruptedly, to the horizon. The diversity, character, and size of a wetland is dependent on salinity, frequency of tidal flooding, physical forces of waves and wind, hydrology, and land elevations.

Wetlands are vulnerable and fragile. Thousands of acres have been filled or altered, owing to man's lack of understanding of the ecological role wetlands play, coupled with his great desire for waterfront property. Much has been written in recent years about the fundamental importance of wetlands. The contributions wetlands make to the productivity of fisheries, as well as to birds and wildlife, are now better appreciated. Wetlands also provide effective erosion control and function as living filter systems for removing and utilizing organic chemicals and pollutants. Both Virginia and Maryland have taken steps to reduce the rate of wetland elimination attributable to man's activities.

Wetland plants, along with phytoplankton and submerged aquatic vegetation, provide a perpetual energy source that forms the basis for the immense annual production of estuarine life. Marshes and forested wetlands are periodically flooded by lunar and wind-driven tides or seasonal rainfall. Nutrients and minerals transported to the marshes by the increasing ebb and flow of rich estuarine waters, or to the wooded swamps from terrestrial sources, provide nourishment for the maintenance and continued existence of wetlands. But the tides and runoff carry more than nourishing chemicals; they provide a path-

7. WETLANDS

way for the larvae of crustaceans, mollusks, and other organisms to the rich organic surfaces of the wetlands. The muddy floors of these wet habitats, along with the roots, stems, and leaves of the plants, provide a fertile diet of microorganisms and decaying plant material. Snails glide up and down stems grazing on diatoms, insect larvae prey on small invertebrates, and fiddler crabs extract sustenance from the mud; plants and animals live and die in close relationships, continuously recycling nutrients. The flow of nutrients does not end in the wetlands. As the waters recede from the marshes and riverine flood plains, they transport nutrients and decaying plant detritus laden with bacteria that nurture anew the complicated food web of the estuary.

Tidal marshes and forested wetlands lie close to or border the water's edge, thereby allowing easy access to numerous and varied organisms when these rich areas are flooded. Frequently, a marsh has a single, usually sinuous, tidal stream, or it may be interlaced by small, natural channels or man-made mosquito ditches. Forested wetlands often have a connecting stream wending from the uplands through the wet woods and emptying into a larger tributary of the Bay system. The forested wetlands may also border

a broad flood plain adjacent to a river where, during times of high water, temporary seasonal ponds are created. These waterways and impoundments literally pulsate with life. Insect larvae, fishes, crabs, reptiles, birds, and mammals abound. The young of both freshwater and saltwater fishes often seek the shallow waters of wetland channels and pools, where their food is concentrated and where they have greater protection from predators. Marshes are havens for many birds. Herons and rails stealthily stalk their prey, and dabbling ducks congregate in the wetland shallows.

There are several types of wetlands in the Chesapeake region, all of which depend on water-soaked soil—the single unifying element in wetland habitats—for their existence. The particular wetland habitats discussed in this book are the tidal freshwater marshes of the upper Bay and rivers; the low marshes, or regularly flooded salt marshes; the high marshes, or irregularly flooded salt marshes; and the tidal and nontidal forested wetlands.

TIDAL FRESHWATER MARSHES

Many of the Chesapeake Bay rivers and subtributaries in Zone 1 have extensive tidal freshwater marshes. The variety of plant species found there is usually much greater than that found in salt marshes. A typical freshwater marsh may consist of various broad-leaved plants growing in a wide band in the river. Shoreward of the emergent plants are the rushes, mallows, cattails, and other species that are less tolerant of constant immersion.

One of the many plants that grows directly in the water, the so-called emergent species, is arrow arum, *Peltandra virginica,* which has arrow-shaped leaves up to three feet long. From May through July the flowering stalk bears white flowers, which later form a cluster of green berries. This species is often found with pickerelweed, *Pontederia cordata,* which is similar to arrow arum; however, the pickerelweed's leaves are heart-shaped and the

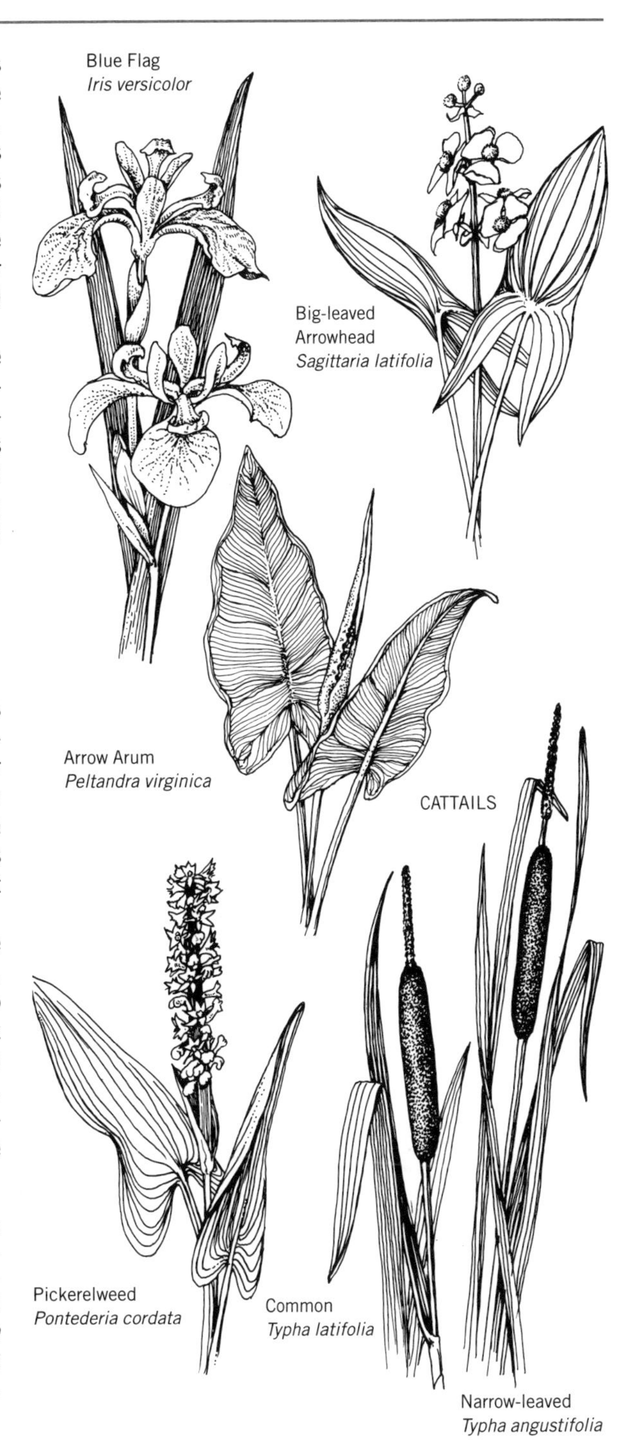

Feathery tufts of reed grass, *Phragmites australis,* bend in the breeze over the common cattail, *Typha latifolia;* the showy blossom of the marsh hibiscus, *Hibiscus moscheutos,* is in the foreground. The yellow pond lily, *Nuphar leutea,* with its floating round leaves, emerges from the water; nearby is arrow arum, *Peltandra virginica,* and pickerelweed, *Pontederia cordata.* Stands of big cordgrass, *Spartina cynosuroides,* and threesquare, *Scirpus* sp., line the far shore.

flowers are blue and bloom from May through October. The yellow pond lily, or spatterdock, *Nuphar lutea,* is a common plant that blooms in tidal freshwater marshes from late spring through early fall. Its heart-shaped leaves float flat on the water at high tide but stand rigid and upright at low tide. It has a solitary yellow flower. Big-leaved arrowhead, *Sagittaria latifolia,* is an herbaceous, nonwoody plant that grows to about four feet. The arrow-head shaped leaves originate from the basal clump, directly from the roots. The showy white, three-petaled flowers with yellow centers are borne on an elongate stalk. Big-leaved arrowhead grows in the shallow water zone and moist soils of tidal freshwater marshes, ditches, and swamps. It is also called duck potato for the starchy tubers growing on the underground stems. The seeds and tubers are eaten by many species of waterfowl, shorebirds, rails, and muskrats.

Blue flag, *Iris versicolor,* our beautiful native iris, which has a poisonous rootstock, or rhizome, also grows in shallow water and along moist shores. It arises from a thick, sprawling rhizome and grows to an erect plant, one to three feet tall, with long, flattened leaves. The blue or violet flowers are as beautiful as the cultivated ornamental irises. The yellow flag, a closely related species, also grows in this region.

As the water becomes shallower and the upland begins a gentle elevation, many other species, only a few of which can be described here, begin to appear. The common cattail, *Typha latifolia,* and the narrow-leaved cattail, *Typha angustifolia,* are common in this zone. The narrow-leaved cattail has leaves one-quarter to one-half inch wide. The familiar, sausagelike brown head, which is actually the female flower, is topped by a spike, the male flower. In this species there is a gap between the spike and the brown head. The common cattail has wider leaves and there is no separation between the brown head and the spike.

The sedge family has about 600 species worldwide with approximately 90 species in North America. There are a number of sedges found in various wetlands around the Bay; they superficially resemble grasses except that most sedges have triangular stems in cross section whereas grasses have round stems. The fruits of sedges are golden brown to russet nutlets which are quite large when compared to the small, grainy seeds of grasses. As a group,

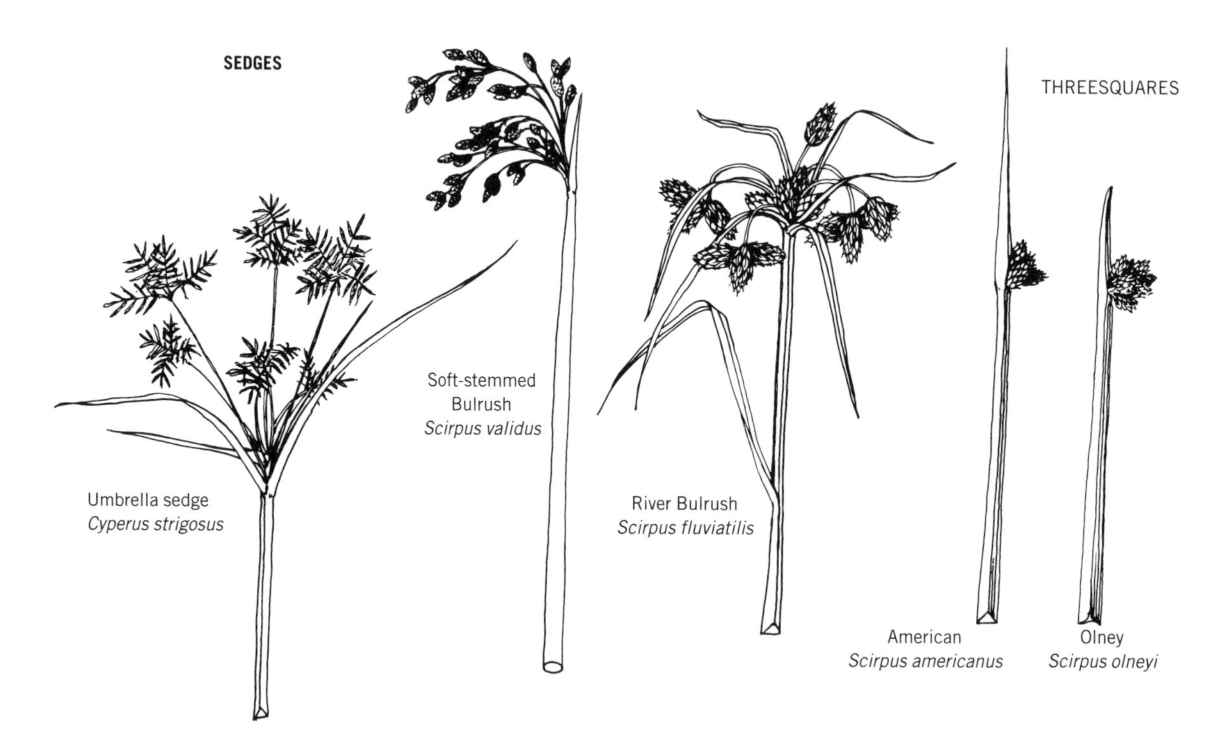

sedges are quite adaptive and range from moist swales and ditches to swamps and tidal salt marshes. Umbrella sedge, *Cyperus strigosus,* grows in moist fields, ditches, and swamps and on wet shores. River bulrush, *Scirpus fluviatilis,* is another sedge of the fresher waters; it is found in tidal fresh marshes and nontidal fresh waters along riverbanks and in marshes. The tall soft-stemmed bulrush, *Scirpus validus,* unlike most of the sedges, is round stemmed. It is distributed from shallow, fresh inland waters to brackish high and low marshes. The American threesquare, or chairmaker's rush, *Scirpus americanus,* and its close relative Olney threesquare, *Scirpus olneyi,* are found in brackish to salt high marshes. Olney threesquare is usually found in slightly saltier waters. The two species are similar except for position of the seed clusters, which are closer to the tip of the stem in Olney threesquare than in American threesquare.

Higher up in the tidal fresh and salt marshes grow two plants that belong to a large group of related species that are variously called smartweeds, tearthumbs, knotweeds, and docks. They share a common characteristic in that the leaf stems are sheathed and appear swollen where they join the main stem of the plant. Some of these species are aquatic or semiaquatic; others are terrestrial.

The halberd-leaved tearthumb, *Polygonum arifolium,* and water smartweed, *Polygonum punctatum,* grow along the edges of tidal freshwater marshes and slightly brackish areas as well as swamps and wet meadows, and occasionally shallow water. The halberd-leaved tearthumb,

Flowering plants of the higher marsh include the seashore mallow, *Kosteletzkya virginica,* with beautiful pink hibiscus-like blossoms; the creeping, prickly halberd-leaved tearthumb, *Polygonum arifolium,* with small pink flower clusters; and the tall upright water smartweed, *Polygonum punctatum,* with spikes of delicate pink to greenish white flowers.

probably named by the first person who grasped the angled, many-prickled stems with the downward-pointing spines, is a creeping perennial with arrowhead-shaped, triangular-based leaves. The flowers are small and pink; they grow in clusters from July through September. Water smartweed is an erect plant that grows to three feet or more. The stems are sheathed over each joint, and the eight-inch willowlike leaves taper at each end. The flowers are tiny, green, greenish white, or pinkish and are borne on long, flexible spikes. They are present from early summer to frost. This is the most abundant and widespread species of this group in the Bay.

Two of the most beautiful plants in tidal fresh and brackish marshes are the marsh hibiscus, or rose mallow, *Hibiscus moscheutos,* and the seashore mallow, or pink mallow, *Kosteletzkya virginica.* Both are members of the Mallow family, which includes okra, cotton, tropical hibiscus, hollyhock, and rose-of-Sharon. Their flowers are typical hibiscus blossoms with five petals surrounding a bright yellow, pollen-laden stamen. Marsh hibiscus have large, showy flowers up to six inches across. They may be white to pink and usually have red to pink centers. The pink flowers of the seashore mallow are generally more numerous but smaller than those of the marsh hibiscus, usually one to two inches in diameter. Both species have gray-green leaves, generally but not consistently three-lobed. The leaves of the marsh hibiscus are wider and have rounder bases than the seashore mallow's. Both the marsh hibiscus and the seashore mallow grow in the upper parts of marshes from tidal fresh to salt marshes; the seashore mallow, however, does not usually flourish as well in fresh water. The marsh hibiscus flowers from July through September, and the seashore mallow follows from August into October.

Water hemp has an interesting scientific name, *Amaranthus cannabinus,* which suggests that it is related to the marijuana plant, *Cannabis sativa,* but it is not. Water hemp has simple, narrow, willowlike leaves; marijuana has coarsely toothed, compound leaves and belongs to the true hemp family. Water hemp often grows in the wetter areas of tidal fresh, brackish, and salt marshes. In early summer, water hemp may be obscured by other marsh plants; however, as the summer presses on, water hemp seems to appear almost suddenly as a large, thick, red-stemmed, treelike plant. The small greenish yellow flowers are borne on spikes and often turn red in the fall.

Another beautiful, flowering perennial of the upper marsh, purple loosestrife, or spiked loosestrife, *Lythrum salicaria,* is related to the garden loosestrife frequently planted and admired in many flowerbeds. This species was introduced from Eurasia; it is a very vigorous and invasive plant in some areas and often crowds out other plants that are considered more valuable for wildlife food and cover. Purple loosestrife grows to about four feet. Its stems are fibrous, almost woody; the leaves are lance shaped and often clasp the stem. The flowers, which have four to six purple petals, grow in dense clusters on long spikes and bloom from June through September. A thick stand of these beau-

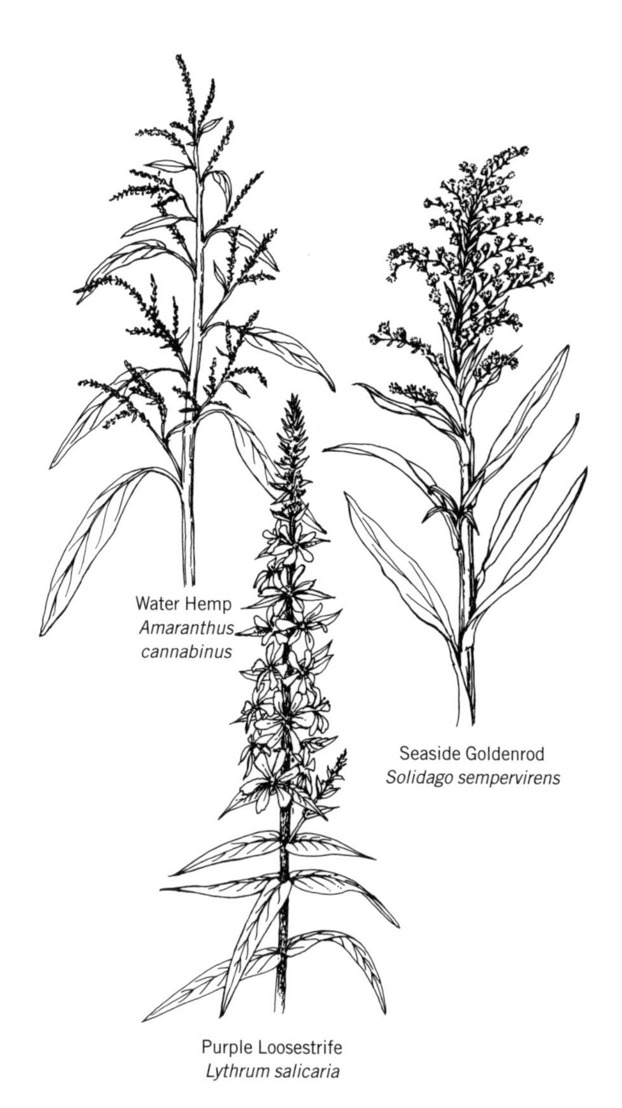

Water Hemp
Amaranthus cannabinus

Seaside Goldenrod
Solidago sempervirens

Purple Loosestrife
Lythrum salicaria

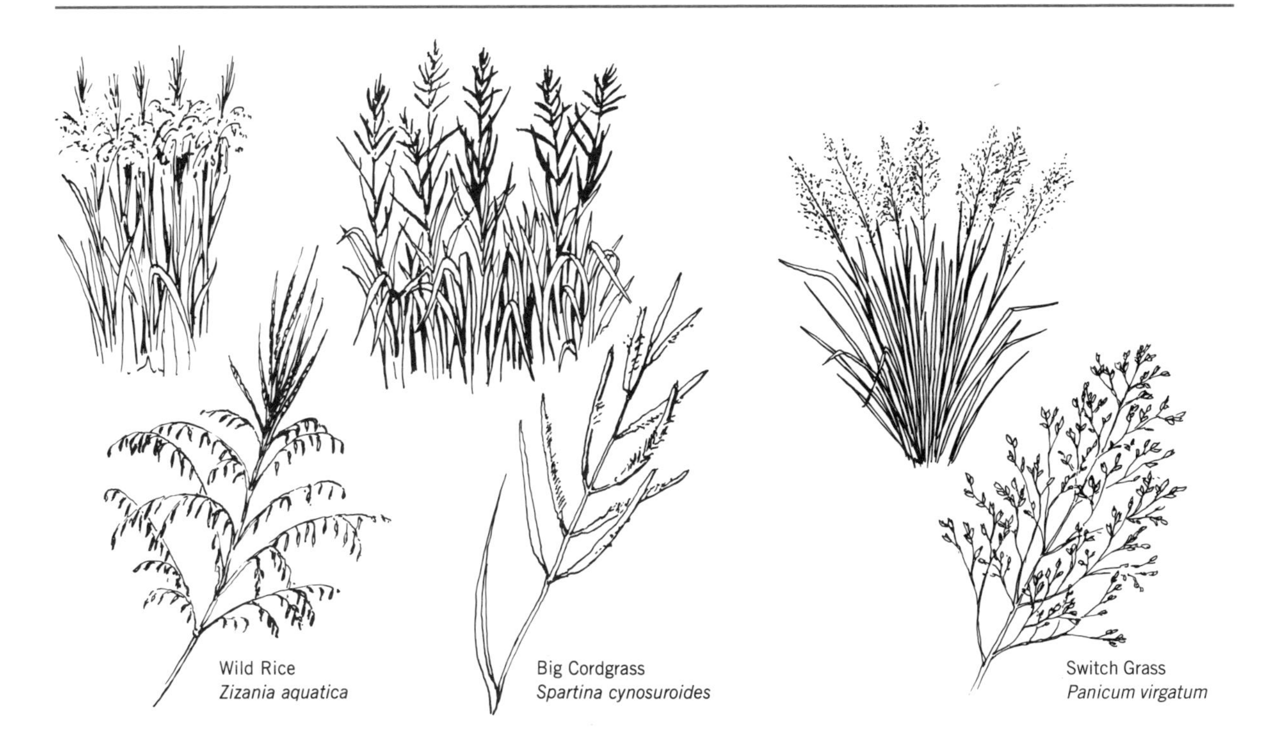

Wild Rice
Zizania aquatica

Big Cordgrass
Spartina cynosuroides

Switch Grass
Panicum virgatum

tiful purple flowering plants adorning a moist meadow or the fringes of tidal fresh, brackish, or salt marshes is such a lovely sight that it is almost possible to forget their ability to outcompete other, more valuable species.

The showy yellow blossoms of seaside goldenrod, *Solidago sempervirens,* paint a golden accent to the upper marsh. The flowers are clustered in large yellow heads on plants that grow to three or four feet or more. They bloom from late summer into autumn frost. The leaves are fleshy, toothless, lance shaped to oblong, and longer at the base; they decrease in length toward the top of the plant. The leaves are attached directly to the stalk of the plant and are alternately arranged. There are many species of goldenrod, and apparently some species often hybridize, making species identification very difficult. Seaside goldenrod is the only common species inhabiting the irregularly flooded areas of fresh, brackish, and salt marshes; beaches; and dunes.

Big cordgrass, *Spartina cynosuroides,* is the largest of the three *Spartina* species common to the Bay, often attaining heights of ten feet or more, and grows only in low-salinity or freshwater meadows. The leaves are flat and the edges are sharply armed with tiny teeth. The plant produces brown to purple grasslike blooms from August through October. Big cordgrass may stand well back from the regularly flooded tide line, forming a towering backdrop to the lower marsh community.

Switch grass, *Panicum virgatum,* related to big cordgrass and the other *Spartina* species, is a round-stemmed grass that grows in dense clumps at the upper edges of fresh, brackish, and salt marshes. Its flowers are inconspicuous and short-lived. It is often difficult to identify the species. Look for a feathery head and long, tapered leaves, 15–20 inches, arising from the base of the plant.

Wild rice, *Zizania aquatica,* is a grass that arises from soft mud and shallow water and may grow 10 feet tall. The stem is thickened at the base, the leaves are very long, 3 feet or more, and the flower clusters are divided into the male flower below and the female flower above. The feathery heads bear the blackish grains or seeds that are still harvested by the Menomini Indians in Minnesota and sold throughout the country as a gourmet addition to many dishes. Wild rice grows in the upper York River, tributaries of the Potomac, the upper Patuxent and Choptank rivers, and

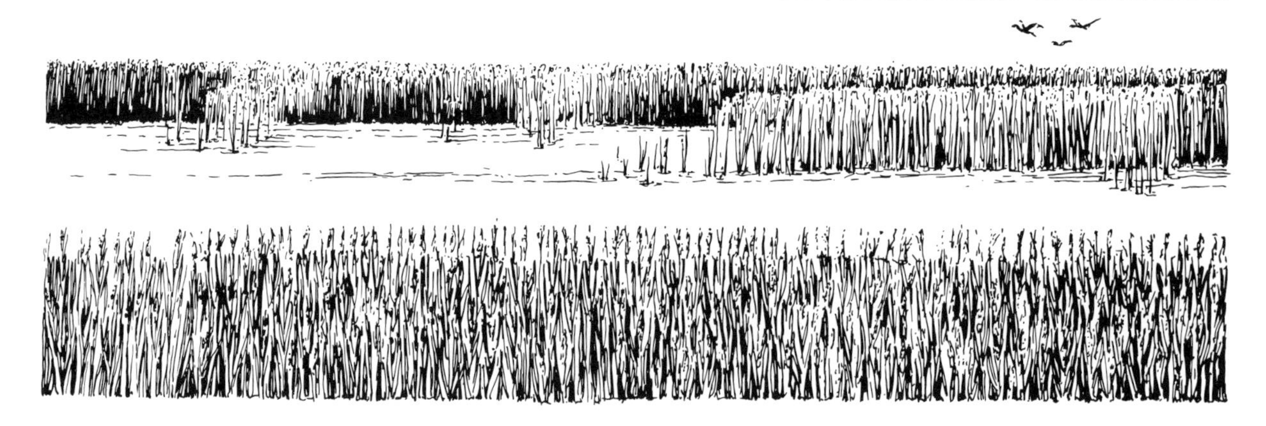

Pure stands of saltmarsh cordgrass, *Spartina alterniflora,* which is the dominant species of the regularly flooded or low marsh.

other areas of the Bay where the fresh water circulates across the shallow, muddy bottoms.

Reed grass, *Phragmites australis,* is a tall, coarse grass whose showy, feathery seed heads are purple when the plant is young and turn fluffy white as it ages. It borders not only freshwater marshes but also saltwater wetlands. This is an invasive species that outcompetes some of the more valuable wetland plants. It is found in great stands, particularly where the soil has been disturbed.

Tidal freshwater marshes are generally located adjacent to the very important spawning and nursery grounds of shads, river herrings, striped bass, sturgeons, killifishes, and many other estuarine and ocean species of fishes. In addition, typical freshwater fishes, such as largemouth bass, sunfishes, pickerels, bullheads, yellow perch, and catfishes, build spawning nests under the protective floating leaves of emergent plants. Immature and adult fishes dart about the underwater stems and roots evading predators and in turn preying on the insect larvae and other organisms so abundant in these areas. The pouch snail deposits its gelatinous eggs on the undersides of pond lilies and other emergent plants. Here, too, are found freshwater limpets, fingernail clams, and freshwater mussels.

SALT MARSHES

Salt marshes, to some, signify the very essence of an estuary such as the Chesapeake. Organisms, both plants and animals, exist because of the mixture of salt and fresh water that bathes these low-lying areas. Plants live and die; over the years, measured in centuries, the decayed vegetation transforms into carbon-rich deposits of peat. Marshes naturally take shape in response to wind intensity and direction, wave action, water depth, soil types, and land elevation. There are other factors, of course, such as shading by trees, siltation from runoff, and man-made changes in shorelines, which may cause erosion at one site and accretion at another. Marshes, where they exist, are shaped by these forces over time: they may take the form of broad meadows, thin fringe marshes along a shoreline, or compact, dense pocket marshes in a cove. In a marsh where the elevation is comparatively flat with few hillocks or meandering creeks, the vegetation may consist of just a few abundant species. In a marsh with a steeper elevation, low depressions, and other irregularities, the diversity of plant species will be increased because of the available microhabitats. Saltmarsh animals also respond in the same way, usually with fewer species and greater numbers in broad, uninterrupted marshes, and many species with relatively fewer numbers in marshes with varying habitats.

Low Marsh

Low, regularly flooded salt marshes are flooded twice daily and are dominated by saltmarsh cordgrass, *Spartina alterniflora.* This very important saltmarsh plant often occurs as only a narrow fringe seaward of the high marsh. It also grows in lush stands, referred to as pocket marshes, at the heads of tidal creeks. The most spectacular growth occurs in large, low-lying areas where the tidal waters sweep in and cover the marsh. These wetlands are frequently dissected by small tidal creeks, which carry the tidal flow deep into the marsh.

Saltmarsh cordgrass has a strong, interconnected root system and smooth, leathery, flat, narrow leaves. There are two recognized growth forms: the tall form, which grows along natural channels and tidal guts and in other areas where tidal flooding occurs daily; and the short form, which grows in slightly higher areas, near the upper limit of tidal inundation. Saltmarsh cordgrass is the single most important marsh plant species in the estuary. It is responsible for the growth of huge amounts of organic material and accompanying bacteria and algae that are eventually flushed out of the marsh and transported to the Bay waters. Saltmarsh cordgrass grows throughout Zones 2 and 3.

Sea lavenders, *Limonium* spp., and the glassworts, *Salicornia* spp., are some of the minor plant species that grow in the low marsh interspersed with saltmarsh cordgrass. Sea lavender has smooth, fleshy leaves, which grow in a circular pattern at the base of the stem. The small lavender flowers, used in dried floral arrangements, grow profusely on the branching stem. Look for blooms from midsummer through October. Glassworts have inconspicuous leaves and flowers and twigs that resemble asparagus spears. These plants generally grow in the drier salt pans of Zone 3 marshes.

Regularly flooded marshes are vital havens for young fishes and many species of invertebrates. The tidal guts connected to shallow ponds within the marsh provide feeding areas and protection from predators. The intertidal zone is home to shrimps and to fiddler and marsh crabs. Here, too, lives the marsh periwinkle, *Littorina irrorata,* and the Atlantic ribbed mussel, *Geukensia demissa.*

The marsh periwinkle, about one inch long, is a heavy, thick-shelled snail with whorls separated by shallow grooves. The color is usually grayish white to yellowish tan with minute dashes of reddish brown on the spiral ridges. This abundant and conspicuous snail is particularly common in marshes protected from heavy wave-action. Like saltmarsh snails, marsh periwinkles can usually be found on the stems of saltmarsh cordgrass and black needlerush in Zones 2 and 3. They are often so numerous on the grass stems that the plants appear to be bearing fruit. It has been long thought that these snails ascend the grass stems to avoid submergence and subsequent drowning. Recent research casts doubt on this assumption, and some scientists now believe that marsh periwinkles climb the stems to avoid predation by blue crabs.

The marsh periwinkle lays in the water individual eggs, which hatch into free-swimming veliger larvae. They develop and appear in the marsh in midsummer as small snails. Periwinkles feed on algae that grow on plants and on the marsh floor; they also consume plant detritus and bacteria. They have an operculum, found in most marine snails, which seals the snail against desiccation and serves as protection against cold weather.

The Atlantic ribbed mussel is two to four inches long

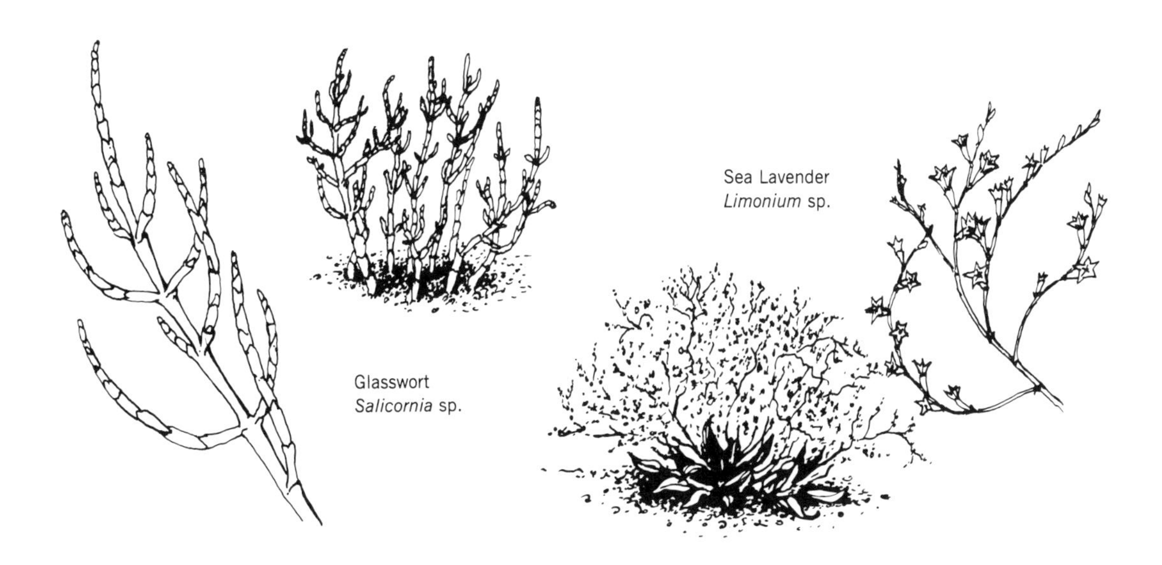

Mollusks of the low marsh. Marsh periwinkles, *Littorina irrorata* (to 1 inch), graze on organic material and algae on plant stems and the marsh floor; the Atlantic ribbed mussel, *Geukensia demissa* (to 4 inches), lies half-buried in the mud as tiny saltmarsh snails search for food in this productive habitat.

and black-brown in color. The interior is bluish white and the posterior end is tinged with purple or purplish red. As the name suggests, the shell is ornamented with strong, rough, radial ribs. Care should be taken not to confuse this species with the hooked mussel, which is strongly curved at the anterior end and is rosy-brown on the inside.

The ribbed mussel is an important representative of marsh fauna. This four-inch bivalve lives almost entirely buried in the mud in Zones 2 and 3, with only the anterior end protruding above the surface. It occurs in small clumps of individuals that frequently attach themselves to marsh cordgrass roots by means of their byssal threads. They occupy the intertidal zone and are well adapted to withstand fluctuating temperatures and dry periods.

Juvenile ribbed mussels move about freely over the surface of the marsh. When they come in contact with a clump of mussels, they are apparently stimulated to settle, thereby increasing the aggregation.

The ribbed mussel is a filter-feeder and probably feeds on detritus, algae, and zooplankton that are flushed in with the tides. This mussel plays an important role in cycling nutrients through the estuarine system. It concentrates organic phosphorus and deposits it on the marsh floor, where bacterial transformation to the essential nutrient, inorganic phosphate, takes place.

Small creeks meandering through marshes harbor dense populations of juvenile and adult fishes. Killifishes, white catfish, bluegill, largemouth bass, and golden shiners are closely associated with freshwater and low-salinity marshes.

In mid-salinity marshes the striped killifish, mummichog, bay anchovy, American eel, hogchoker, and many more species of fish feed on scuds, bristle worms, shrimps, and the rich plankton and organic material manufactured by the marsh plants.

The lower Bay high-salinity marshes support many of the invertebrate and fish species found in the mid-salinity areas. Young of the weakfish, spot, and striped mullet mingle with needlefish and menhaden, vying for food with the aggressive blue crab and the multitudinous silversides.

High Marsh

High marshes, also referred to as irregularly flooded salt marshes, are marshes that are flooded only by wind-driven or exceptionally high tides. These marshes are often a continuum of low, regularly flooded salt marshes. They are common in Zones 2 and 3, and the plant species reflect the salinity of the flooding water. A typical high salt marsh of the mid-Bay would be composed mainly of black needlerush, *Juncus roemerianus,* mixed with some threesquares in the lower elevations of the marsh. Black needlerush is a sharp-tipped, round-stemmed, stiff marsh rush that commonly grows in dense stands of uniform height. This species stands out against the other marsh plants because of its dark color, which appears as brown or gray-black. Interspersed among the slate gray, browns, and greens of black needlerush and threesquare are dashes of pink and lavender and blue and white, sure evidence that saltmarsh fleabane, *Pluchea purpurascens,* and saltmarsh aster, *Aster tenuifolius,* are in bloom. Saltmarsh fleabane is also called camphorweed for the aromatic odor of its crushed leaves. In fact, there are references that mention that fleabane, when applied to the skin, will repel fleas. Fleabane grows up to three feet tall. The long, oval-shaped leaves are fleshy, sharply toothed, and alternately arranged. The entire plant is somewhat hairy, particularly the leaves. The individual lavender to pink flowers are quite small, but they are densely crowded into slightly rounded, flat-topped clusters or heads making them easy to see. Fleabane grows in high marshes, swales, and ditches and blooms from late summer through early fall. Saltmarsh asters are small, daisylike flowers that dot the higher elevations of brackish and saltwater marshes. They are scraggly plants that grow to 30 inches in height with linear, fleshy leaves and pale purple, blue, or white flowers. Each plant has relatively few flower heads, but there are often many plants in the marsh to offer dashes of color. Saltmarsh aster is one of about 250 species of wild asters found in North America. Individual species are often difficult to identify.

Sea oxeye, *Borrichia frutescens,* is a small branched shrub that bears bright yellow asterlike blooms, for it also belongs to the aster family. The flower heads have twelve or more petals with large golden brown centers. The leaves are long to oval shaped, covered with dense, gray hairs; the margins may be smooth or toothed. Sea oxeyes arise from rhizomes and grow in dense colonies on the edges of salt marshes. They are seldom seen in Maryland, but are abundant in areas of Virginia and farther south.

In the higher elevations of irregularly flooded marshes, where flooding occurs even less frequently, saltmeadow hay, *Spartina patens,* and salt grass, *Distichlis spicata,* are the dominant species. These species are short, wiry grasses that form dense meadowland carpets. Salt grass tolerates higher salinities and wetter ground than does saltmeadow hay. It is a pale-green plant with short leaves and

Plants of the high marsh or irregularly flooded salt marsh live in close association. The matted plants in the foreground are saltmeadow hay, *Spartina patens,* intermixed with saltgrass, *Distichlis spicata;* marsh elder, *Iva frutescens,* is in the left foreground. Dark stands of black needlerush, *Juncus roemerianus,* interspersed with clumps of Olney threesquare, *Scirpus olneyi,* line the horizon.

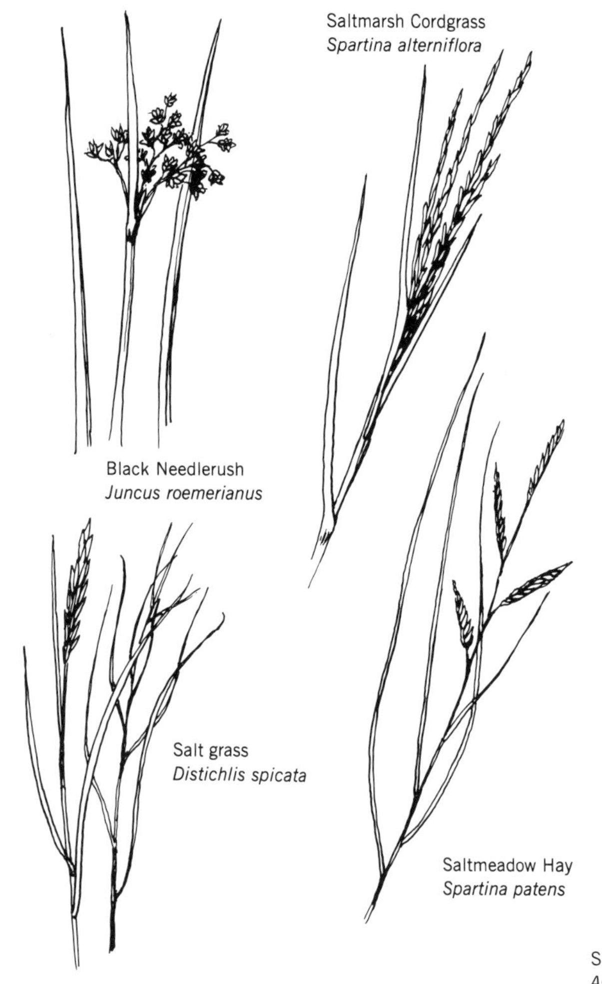

are thinner and are arranged in an alternate pattern on the stem. These shrubs flower in late summer through fall. The groundsel tree is a very showy shrub in the fall, particularly after most of the other wetland plants have bloomed. The small, white, dense blooms clothe the groundsel tree in cottony tufts late into October. Both the groundsel and the marsh elder are often wreathed by common dodder. This parasitic plant, with reduced leaves and minute flowers, grows in a tangle of yellow-orange vines over the shrubs.

The saltmarsh snail, *Melampus bidentatus,* a resident of the high marsh, is fascinating because of the behavioral mechanisms it has adapted to meet its demanding physiological needs. It is a small, herbivorous gastropod about half an inch long, shiny brown to amber in color, with three or four darker narrow bands. The saltmarsh snail is an

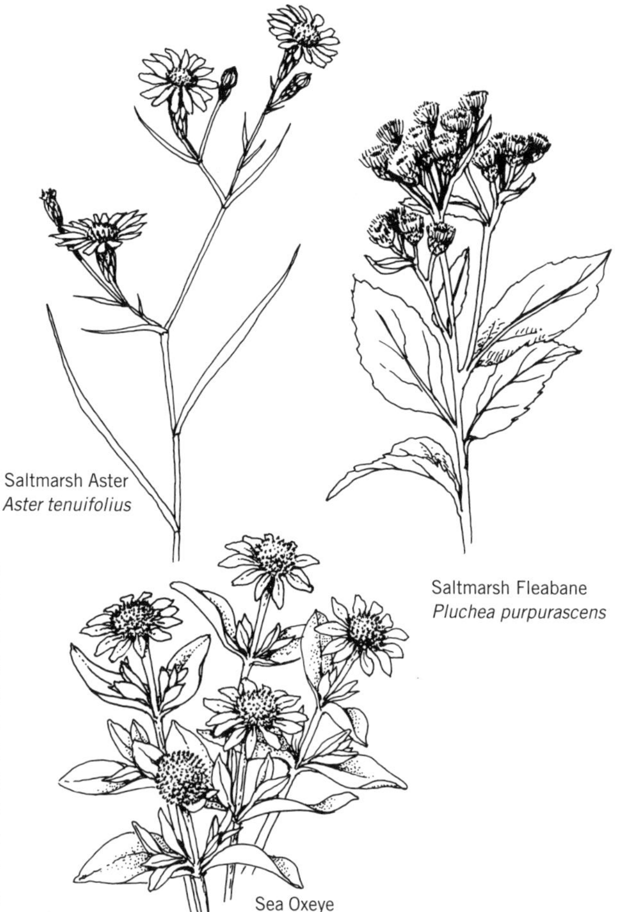

rigid stems. Saltmeadow hay grows in the higher parts of the marsh. Its stems are weak and subject to bending and to intertwining with adjacent plants when blown by winds and inundated by very high tides. This gives a distinctive whorled or, as one scientist wrote, "cowlick" appearance to the high marsh. Marsh elder, *Iva frutescens,* and the groundsel tree, *Baccharis halimifolia,* are shrubs or small trees that grow on high spots in the marsh and demarcate the transition zone from high marsh to upland. The two shrubs are similar in appearance, but a comparison of their leaves permits separation of the species. The marsh elder has fleshy leaves, which are arranged in pairs opposite each other on the stem, whereas the groundsel tree's leaves abundant species, which lives in the higher levels of irregularly flooded marshes under layers of litter and on the stems of grasses. It has an air-breathing lung and cannot tolerate immersion. However, it has retained its ancestral need to reproduce as a marine snail, and its eggs develop into free-swimming veliger larvae.

The most intriguing aspect of this species is the strategy it has evolved to deal with the dilemma of being an obligatory adult air-breather while maintaining a larval form that extracts oxygen from the water. The snail has developed a mechanism that brings into play the place where it lives as an adult and a precise synchrony with spring tides. The adult lives in the saltmeadow hay and salt grass meadow. It lays an average of eight hundred fifty eggs encased in a gelatinous covering. Spring tides (exceptionally high tides, which occur every two weeks, caused by the positioning of the new and full moons) wash organic debris over the egg masses, which are thus kept moist. The eggs are timed to hatch in about 13 days, with the arrival of the next spring tidal cycle. Enormous numbers of veliger larvae are released into the estuary and become part of the plankton. The veligers spend about 14 days as plankton and, again synchronized with the spring tides, are transported back into the marsh. They then develop into air-breathing adults and spend much of their time moving up grass stems to avoid the flooding water.

The saltmarsh snail, *Melampus bidentatus,* often found under flotsam on the floor of the marsh, is less than half an inch long.

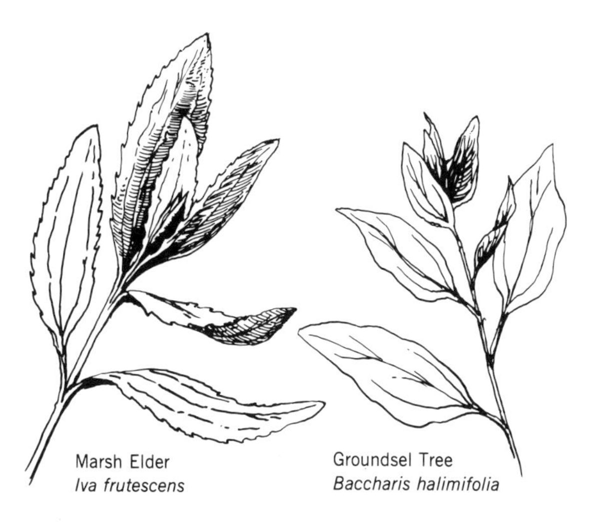

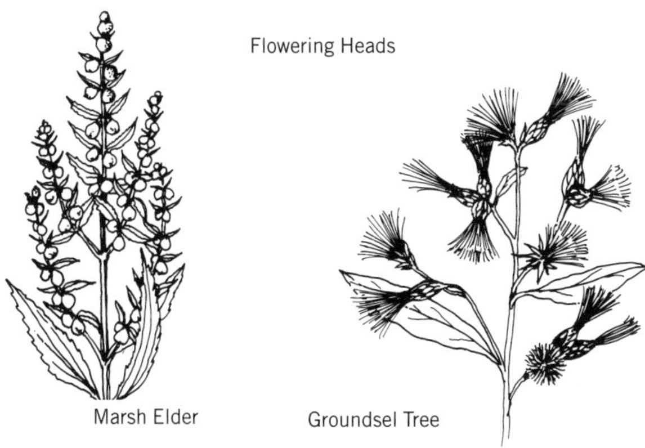

THE MARITIME FOREST

The maritime forest, or bay edge forest, is a transitional zone of shrubs, grading into trees as the land rises in elevation from the marsh. None of the species of shrubs and trees are unique to the maritime forest; they are simply species that can endure salt spray and moist soils. In areas, particularly along the ocean coast, where the winds blow constantly, carrying soil and salt particles, the trees are shaped and pruned into an angular canopy. In an estuary, such as the Chesapeake Bay, nature's forces are somewhat more yielding and kinder, so that we don't often see the sheared, contorted trees so evident along the shores of an ocean barrier island. As the high marsh rises and begins to connect to fast land, the marsh elders and groundsel trees rim the edge and grow in association with plants such as saltmeadow hay, salt grass, seaside goldenrod and saltmarsh asters.

The slope imperceptibly rises and the leathery leaved wax myrtle, *Myrica cerifera,* and the bayberry, *Myrica pensylvanica,* merge with the groundsels and marsh elders to form a shrub zone. Wax myrtle and bayberry are closely related species collectively known as candleberry. They are shrubs or small trees with gray bark and small berries tightly bound to the stem. The bayberry is a reluctantly deciduous tree that does not always shed all of its leaves at the same time. It is a stiff-branched shrub that reaches a height of eight feet or more; it has long, round-tipped leaves. The leaves may have smooth borders or they may be finely toothed; they are dark green above and paler with yellow resinous dots beneath. The wax myrtle is an evergreen and grows somewhat taller than the bayberry. The twigs are deep reddish brown and heavily covered with resinous dots. The leaves are wonderfully aromatic when crushed and are narrow and more pointed at the tip than the bayberry's. Both species have bluish white, round, hard, wax-covered, and persistent berries. The wax-bearing berries can be boiled to release the wax, which is then skimmed off the surface. When the wax cools, it is made into a greenish candle that, when lit, gives off the fragrant aroma of bayberry. Both species are abundant and can be found in sandy swamps, moist woods, and poor dry soils of the Coastal Plain.

Farther up the slope, loblolly pine, *Pinus taeda,* stands sentinel, forming a backdrop to the tidal salt marshes. Loblolly is a rapidly growing southern pine of the Coastal Plain; it is rare north of the Chesapeake Bay but common and abundant to the south. This large, straight-growing pine grows in soil types ranging from well-drained, poor upland soils to low, poorly drained areas. It thrives on abandoned, exhausted old fields. It is grown commercially and logged extensively in this region. Loblollys grow best where the summers are long, hot, and humid, the winters are mild, and the rainfall ranges from 39 to 59 inches; the

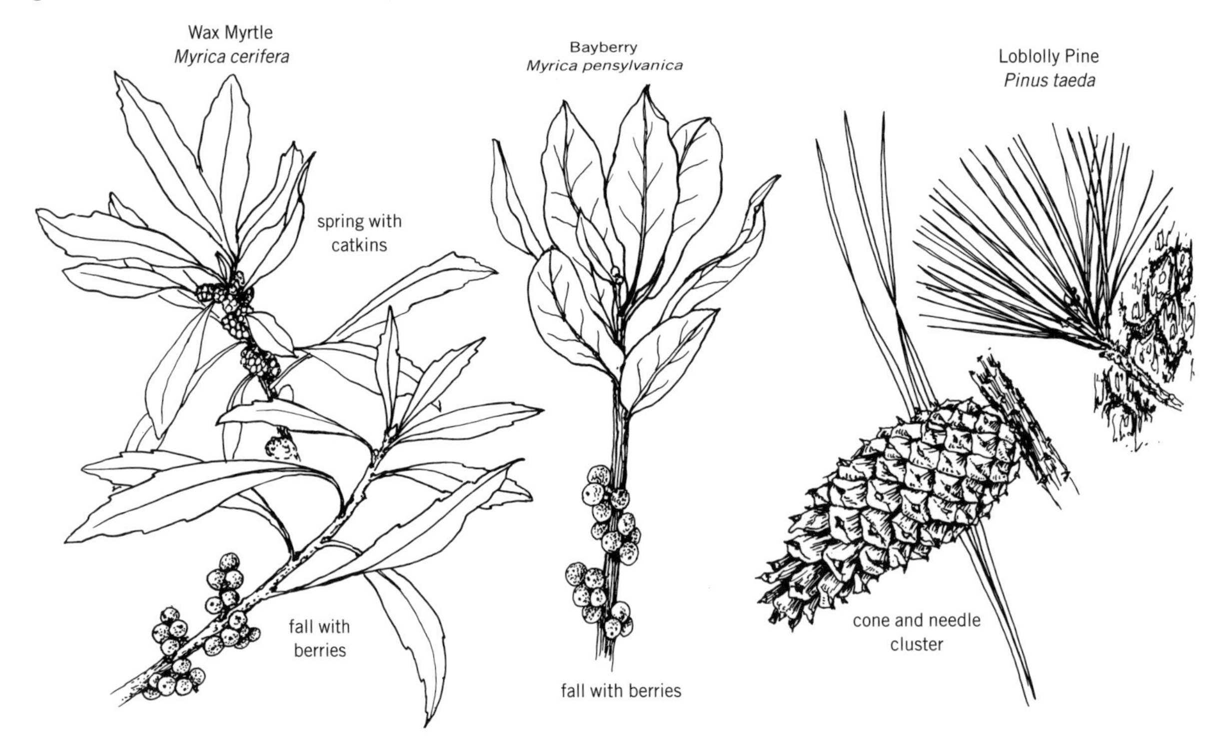

A forested wetland at the edge of a small stream.

Chesapeake Bay area fulfills those growing criteria quite well. Loblollys have a dark, scaly bark when young, which turns reddish brown to cinnamon as the tree matures. The bark, as the tree ages, splits into plates or irregular scaly blocks, which gives it the appearance of alligator hide. The needles are 5–10 inches long, usually occur in threes, and are sharp pointed and stiff. The cones are approximately 3–6 inches long, are sometimes conical but often oblong, and have a stout, short, sharp spine on each scale.

FORESTED WETLANDS

Forested wetlands are often overlooked as estuarine wetlands, let alone as areas of natural beauty, and are not usually even considered as vital habitats of the Bay. These somber and mysterious places are not particularly inviting to the casual visitor—wet during the spring, often muddy and strewn with fallen trees, nicely populated with mosquitoes and biting flies, and festooned with a confusion of poison ivy, trumpet-creeper and honeysuckle vines. Usually, wooded wetlands are avoided or are in someone's plans for future "improvement" and development.

Wet woods are not sharply separated from adjoining upland forests, and they frequently occupy the upper edge of fresh and brackish marshes as well. They are, in fact, a continuum of the uplands; as the land slopes to sea level, certain woody and herbaceous plants commonly thought of as species of drier forests are capable of taking hold in the wetter, poorly drained soils. Many shrubs and trees usually referred to as pioneer or invader species, such as willows, buttonbush, silky dogwood, river birch, red maple, and gums and oaks, establish themselves rather quickly in moist bottomlands. As in most hardwood forests, a typical canopy or overstory of ashes, sweet gums, oaks, and tupelo dominates the understory of young trees, shrubs, and a herbaceous layer of ferns and flowering plants. The species composition of forested woodlands is quite variable depending on the amount and duration of surface water saturating the soil. Generally, flood plains and contiguous woodlands are inundated in the spring and remain relatively dry the rest of the year; however, if the wooded area is frequently soaked by above-normal tides several times a year, a different forest species complex may well arise. Further complicating the occurrence and timing of surface water is the salt content of the water: some species can adapt to brackish water; others drop out.

The Chesapeake Bay lies astride that nebulous line that separates northern forested wetlands from southern forested wetlands. The Chesapeake Bay wooded wetlands contain species typical of northern as well as southern habitats. Consequently the species diversity is rich, and the forests range from a mix of northern and southern hardwoods to southern pines and bald cypress stands. The understory is also quite variable.

Forested wetlands have an abundance of breeding songbirds because the lush growth provides nesting cover, protection from predators, and an ample food supply. Owls and hawks dwell in these woods and feed on songbirds and on rodents, which are everywhere on the forest floor. Turtles, amphibians, and the northern water snake are also common inhabitants of these wooded wet areas.

Forested wetlands, similar to the more familiar salt and freshwater marshes, accumulate nutrients and minerals and build rich organic soils and flourishing plants. At the end of their annual growing cycle, these plants drop leaves and stems, which partially decay on the forest floor. This material is then flushed out with the high waters of spring flooding and tides into the estuary, where it also contributes to the productivity of the Bay.

An example of a representative forested wetland overstory in a tributary of the mid-Bay might include black ash, cherrybark oak, swamp white oak, sycamore, river birch, and water tupelo. The smaller trees and shrubs that comprise the understory quite often will include swamp rose, silky dogwood, red maple, buttonbush, smooth alder, shadbush, and winterberry.

Black ash, *Fraxinus nigra,* a member of the olive family, is a broad-leaved tree that grows to be 75–80 feet tall. The bark is corky and grooved. It has compound leaves divided into 5–9 stalkless leaflets, which turn brown in the autumn. The flowers are rather inconspicuous; its winged fruits are broad and flat and hang in drooping clusters. The wood is dark brown, coarse grained, and splits easily. Thin strips of wood split from logs of the black ash are used for weaving baskets and for chair seats. Black ash grows in moist soils and forested wetlands.

Cherrybark oak, *Quercus falcata* variety *pagodifolia,* is a form of southern red oak that grows in moister conditions than the southern red oak. The cherrybark oak is a large tree that can grow to 100 feet in height. The bark is relatively smooth with short ridges, more like a cherry tree's than the typical bark of the southern red oak, which in mature trees is dark gray and ridged or fissured. The leaves are pagoda-shaped, bristle-tipped, and have 5–11 lobes. Cherrybark oak leaves, like most oak leaves, turn brown in the fall and remain on the tree late into the winter. Another oak of wet soils, the swamp white oak, *Quercus bicolor,* is also a large tree which can reach 100 feet in height. The lower branches tend to droop, and the bark is light gray and ridged. White oaks, in contrast to red oaks such as the cherrybark oak, do not have bristle-tipped leaves. Leaves of the swamp white oak are round-toothed or lobed, blunt or rounded at the tip, and broader beyond the middle of the leaf. The flowers are borne on catkins, a scaly spike of flowers without petals, and produce an acorn which is three-fourths to one and one-fourth inch long. This is the only oak with acorn stalks longer than its leaf stalks. Acorns, particularly acorns produced by white oaks, constitute one of the most valuable foods for wildlife. Apparently the acorns of white oaks are tastier to wildlife than those of the red and black oaks, which are bitter because of the concentration of tannins. Mallards, pintails, and wood ducks, as well as doves and many species of songbirds, feed abundantly on acorns. Both the cherrybark and the swamp white oak are important indicators of low-lying areas that retain water, sometimes called hydric forests and commonly called swamps.

The sycamore, *Platanus occidentalis,* is one of the largest hardwood trees in eastern North America. The London plane tree of Europe is closely related. This rapidly growing tree can reach a height of 175 feet and a diameter of 12 feet, but is generally 6 feet or less in diameter. It is a dominant and beautiful tree of the floodplain forest, with characteristic mottled gray, green, brown, white, and yellow bark that easily peels and flakes. The maplelike leaves are lobed and coarsely toothed, turning brown in the fall. The drooping, globelike fruits bear hundreds of seeds which are scattered by the wind and germinate on river bars and mud flats. The sycamore's wood is hard and is used for furniture, boxes, pulpwood, butcher's blocks, and flooring.

River birch, *Betula nigra,* is sometimes called red or black birch for the color of its peeling bark, which is pinkish to reddish brown, becoming dark, fissured, and shaggy on larger, more mature trees. The leaves are arrowhead-shaped, coarsely saw-toothed, and alternately arranged. The flowers are inconspicuous and are borne on catkins which develop flattened, winged fruits. The river birch is a competitive pioneer species of low bottomlands, quickly taking hold on river bars and growing to a sizable tree of up to 100 feet in height. It is the most southerly distributed species of birch. The wood of the river birch is light brown and not considered very valuable; it is sometimes used for furniture, but more often for pulpwood or firewood.

Swamp tupelo, *Nyssa sylvatica* variety *biflora,* also known as black gum or sour gum, is a tree of the bottom-

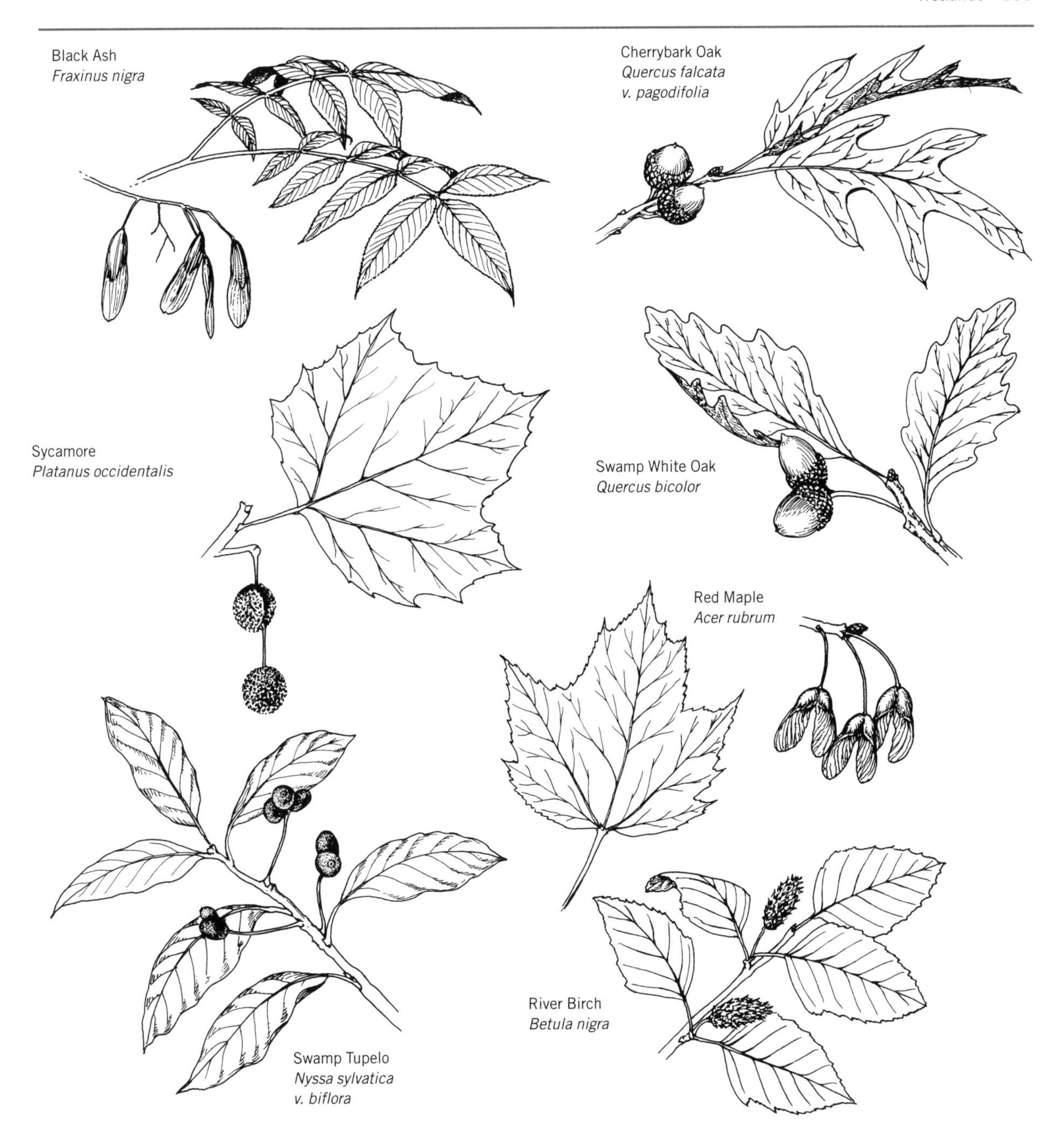
Black Ash
Fraxinus nigra
Cherrybark Oak
Quercus falcata
v. pagodifolia
Sycamore
Platanus occidentalis
Swamp White Oak
Quercus bicolor
Red Maple
Acer rubrum
River Birch
Betula nigra
Swamp Tupelo
Nyssa sylvatica
v. biflora

lands along the Coastal Plain. Tupelos are long, straight trees with narrow crowns and a dark brown or gray, deeply furrowed bark. In areas where they are submerged, the base of the trunk is often swollen. The alternately arranged leaves are shiny green, smooth bordered, and somewhat leathery. The leaves begin to turn dark red in late summer and early fall. Wood ducks and songbirds feed on the small, oval, bluish black, pulpy fruits, and beavers browse on the twigs.

The understory of forested wetlands is often a mix of shrubs, small trees, and vines. When large trees fall, creating gaps in the canopy, sunlight penetrates to the forest floor, encouraging the growth of vines and some of the smaller trees. Depending on the amount of light penetrating the canopy, the soil type, and the amount of surface water, the types of plants in the understory will vary.

Red maple, *Acer rubrum,* also known as swamp maple, is the most widely distributed maple in North America and is a common species in the Chesapeake Bay region. Where conditions allow, red maple trees often grow in pure stands. They grow both on uplands and in the moist soils of tidal freshwater marshes. They can reach heights of 100 feet or more, but more often they are smaller and quite spindly. The bark is gray and smooth in young trees, becoming fissured and ridged as the trees mature. The leaves are broad with three pointed lobes and are oppositely arranged. The reddish flowers grow in clusters and bloom in late winter and early spring before the leaves emerge. The fruit is a double winged seed, called a samara, about an inch long. The seeds, flowers, and buds are fed on by squirrels, raccoons, deer, and songbirds.

Silky dogwood, *Cornus amomum,* also called red willow or silky cornel, is a highly branched shrub which grows in clumps to a height of 10 feet. The leaves, oppositely arranged, are egg shaped with pointed tips, and the twigs are green to reddish brown or purple. Silky dogwoods bloom from May through June. Their broad clusters of flowers are variably colored from white to greenish white or creamy yellow. They have small, fleshy, bluish fruits, which ripen in late summer and are fed on by ducks, songbirds, quail, and beaver. Silky dogwoods grow in forested and shrub wetlands, and along stream banks.

Buttonbush, *Cepthalanthus occidentalis,* is a broad-leaved deciduous shrub that grows up to about 10 feet or more. This is a fast-growing, short-lived species that can quickly establish itself in forested wetlands and fresh tidal marshes. The bark is grayish brown and flaky; the leaves are egg shaped tapering to a short point. The leaves may be oppositely arranged or can be in whorls of threes and fours. The foliage is poisonous to some grazing animals. Buttonbush gets its name from the dense, ball-shaped blooms formed of numerous small, tubular flowers, which bloom May through August. They are sometimes called globe flowers or honey balls because they are an excellent source of nectar for butterflies and bees. The ball-like fruit clusters bear nutlets, which mature in the fall and are fed upon by mallards, teals, wood ducks, and shorebirds.

Smooth alder, *Alnus serrulata,* is a shrub or a small tree that grows to 25 feet. Alders are closely related to birches; like birches, most alder species grow in the north. The bark is dark gray and the leaves are egg shaped, fine toothed, and alternately arranged. The minute flowers are borne on catkins, March through May. Alders do not do well in shaded areas and tend to die out if shaded by taller trees. Sometimes one may encounter an alder with leaves rolled into tubes. These leaf tubes are created by small caterpillars that secrete a sticky silk that they weave back and forth on the edges of the leaf to form a cylindrical chamber where the caterpillar feeds and pupates. The fruit of the smooth alder looks like small cones attached in small clusters. Songbirds feed on the seeds, and beavers feed on the wood and foliage of the smooth alder. The seaside alder is a closely related species found only in Delaware and the Eastern Shore of Maryland. The catkins of the seaside alder are slightly longer than those of the smooth alder.

Shadbush, *Amelanchier canadensis,* is a member of the rose family, which characteristically has flowers with five sepals and petals. The rose family is a large group of related plants which includes apple, pear, plum, peach, apricot, cherry, hawthorn, mountain ash, and, of course, the garden rose. The shadbush or juneberry, as it is sometimes called, is a shrub or small tree that grows as tall as 25 feet. The alternately arranged leaves are roundly oblong and fine toothed. The shadbush blooms along the edge of forested wetlands and in thin woods from March through June, usually earlier rather than later. It begins blooming in the spring when the shad return from the sea to spawn in the freshwater tributaries of the Chesapeake—thus its name. The numerous white-petaled flowers grow in clusters before the leaves appear. The fruits are red to purple and mature in early summer. Songbirds feed on the fruit, and rabbits, deer, and beaver gnaw on the twigs.

Winterberry, *Ilex verticillata,* is a deciduous holly that grows as a shrub or small tree. The bark is smooth and dark gray; the leaves are deciduous, leathery, longer than wide or egg shaped, and coarse toothed. Like most hollies, the

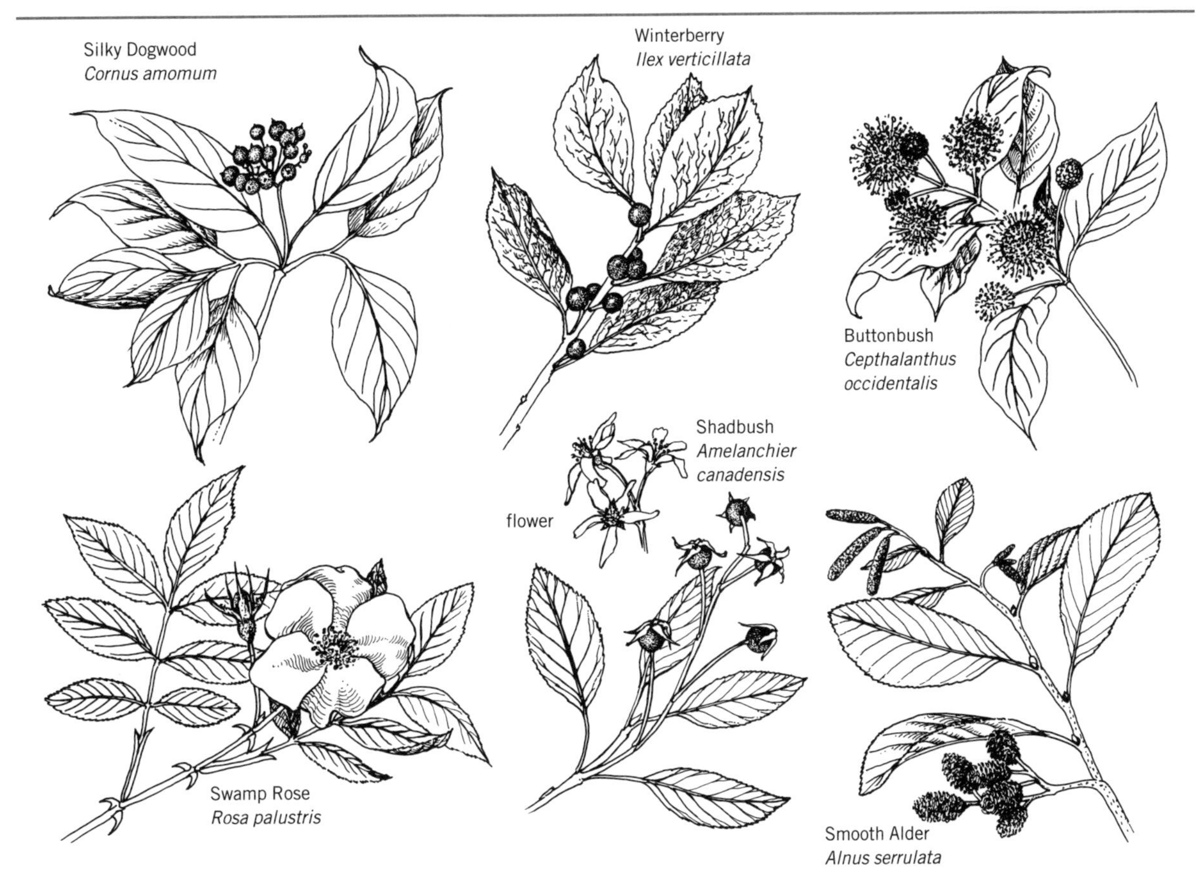

winterberry is a slow-growing species. Winterberries bear very small, white, clustered flowers in the leaf axils, the angle where the leaf joins the stem. The bright red fruit persists through most of the winter, since it is not a favored food of wildlife. Apparently the high-quality fruits of the dogwood, magnolia, and tupelo, which are rich in fat content, are eaten before those of the winterberry, which is a poorer quality food. The winterberry grows in forested wetlands and tidal swamps.

The lovely pink to pale crimson flowers of the swamp rose, *Rosa palustris,* are always a welcome sight in wooded wetlands, particularly if your path is not guarded by this thorny relative of raspberries and blackberries. This shrub has alternately arranged, compound leaves. The leaflets are dull green, fine toothed, and egg shaped. The flowers, undeniably roses, are pink with yellow centers; they are five petaled and up to two and one-half inches wide. The swamp rose blooms from June through August. The fruits, the red, bristly hips, and the flower buds are eaten by songbirds and rodents; deer feed on the twigs and foliage. The tangled thickets of the swamp rose provide excellent nesting and protective cover for birds and small mammals. Swamp roses often grow in forested wetlands and swamp shrub communities along with red maples and willows. They also grow along stream banks and in freshwater tidal marshes where cattails and purple loosestrife may be found.

In certain areas of the Bay the majestic bald cypress, *Taxodium distichum,* towers over tupelos, sweet gums, red maples, and oaks. These tall, straight-trunked, pyramidal crowned relatives of the redwood and sequoia trees seem out of place to those who think of the bald cypress as a tree of the Deep South. Yet stands of bald cypress grow in the Chickahominy River, off the James River, in Battle Creek in southern Maryland's Calvert County, and in the Pokomoke River, which flows along the Maryland-Virginia boundary on the Eastern Shore. The bald cypress grows naturally in wet, deep woods; its trunk is often swollen and

Bald Cypress
Taxodium distichum

buttressed when it grows in water. The characteristic cypress knees protruding above the water are extensions of the root system. The bald cypress also grows well, however, as an ornamental on dry, loamy soils—usually without the familiar swollen base. As a cypress slowly matures, it loses its conical shape and the crown becomes somewhat flattened. The bald cypress is needle-leafed and deciduous with gray-brown, fibrous bark. The soft, feathery needles turn brown in the fall and, with the twigs, drop in winter; hence the name bald cypress. The bald cypress produces gray, ball-shaped cones three-quarters to one inch in diameter, with one or two at the end of a twig. Bald cypress swamps are inhabited by turtles and snakes and are important feeding and shelter areas for waterfowl.

INSECTS OF THE WETLANDS

Slowly canoeing or sailing along a winding tributary, bordered by luxuriant wetland plants, on a hot summer day provides a feast for one's eyes, not to mention a feast for biting flies and mosquitoes. But there are other insects in the wetlands as well: metallic-colored dragonflies, fluttering butterflies and moths, and a variety of beetles feeding on the rich, organic debris along the shore. On a calm day, particularly in the shallows, furiously active insects, sometimes by the hundreds, skitter across the surface of the water, creating tiny ripples as they dart and circle almost endlessly.

Insects develop in various ways from egg to adult. A female may lay eggs in a specific plant stem and only in a specific location; other females of different species may deposit their eggs only under the water; still others cement their eggs to the back of a male. Insects are undoubtedly the most successful group in the animal kingdom. There are thousands of species of insects throughout the world, and they have wondrously adapted to their environment.

Flies and Mosquitoes—Wetland Pests

Some insect larvae begin their existence in standing pools of water in freshwater marshes and nearby ditches. Practically any container such as a discarded tire, a can, and a tree cavity can be a nursery area for fly and mosquito larvae. Flies, mosquitoes, gnats, and midges all belong to the insect order Diptera. Dipterans use only their forewings for flight; the hind pair have been reduced to flight-balancing knobs. Their mouth parts are adapted for sucking: many females of the group are able to pierce and suck blood. There are hundreds of species of mosquitoes and flies common to Bay wetlands, three of which are the American horse fly, *Tabanus americanus,* the deer fly, *Chrysops* spp., and the saltmarsh mosquito, *Aedes solicitans.* The females of these species are persistent, vicious biters and suck blood in order to produce viable eggs. The males feed on pollen and nectar of flowers.

The two species of flies, the horse fly and the deer fly, are quite different in appearance. The stout-bodied American horse fly is larger than the deer fly, about one inch long. It has a black body and large, bright green, iridescent eyes. The smaller deer fly is about one-half inch long and varies in color from black or brown to a yellowish green. The eyes may be green or yellow and the wings often have dark patches. Both flies lay their eggs on vegetation overhanging the water. As the larvae hatch they drop into the water, where they live, feeding on other aquatic invertebrates such as worms and other insects, until they pupate later in the season—or, in some cases, the next year. The adult fly emerges from the pupa, mates, and begins the cycle again.

The saltmarsh mosquito breeds in brackish and saltwater marshes, where the female deposits her eggs on floating vegetation or sticks or on the bare mud of the marsh floor during the lowest, or neap, tide. The eggs actually require a period of drying before they are ready to hatch. The eggs hatch two weeks later, in synchrony with the next lunar spring tide. The resultant larvae, called wrigglers, feed on algae and single-cell organisms. They are often found in dense aggregations in still waters. Adult saltmarsh mosquitoes eventually emerge from the pupae. Saltmarsh mosquito adults are golden brown. There are some color differences between the sexes: the female's abdomen is striped or banded with gold or silver. The wings

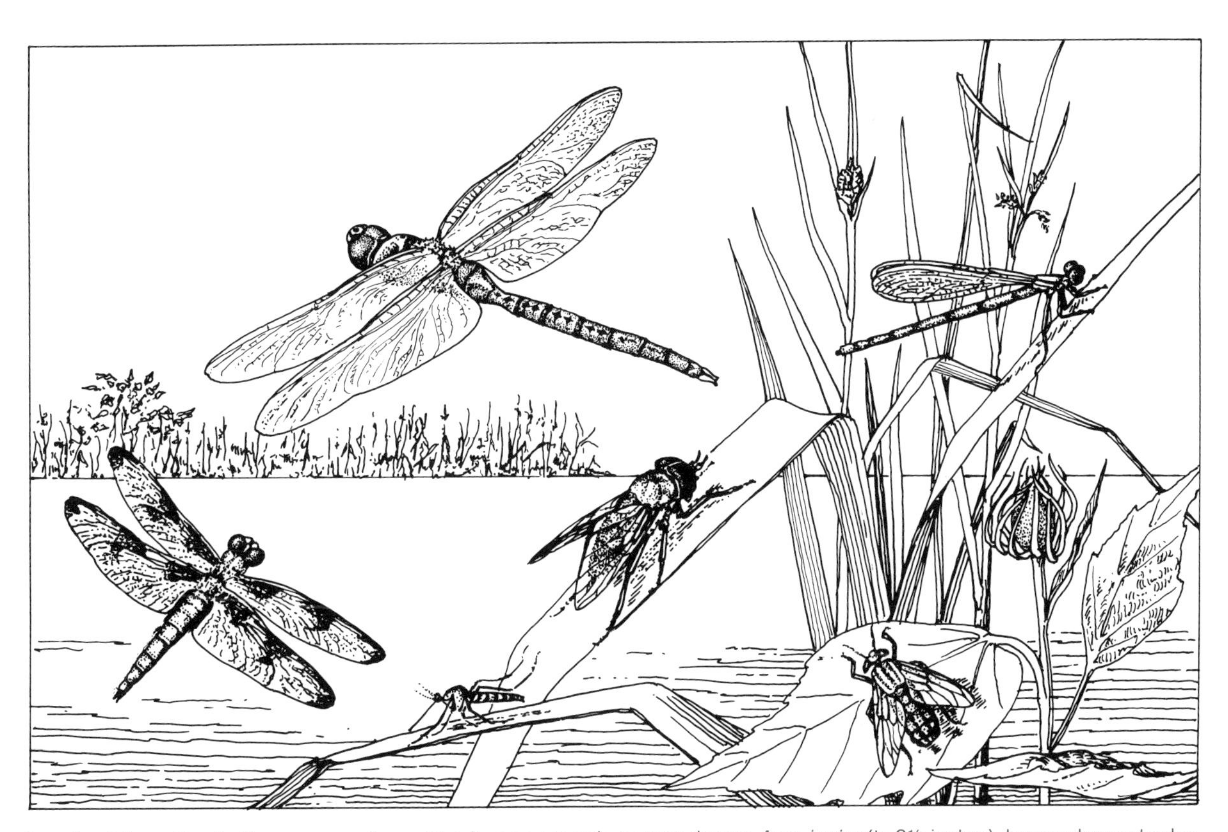

Insects of the wetlands. Two common dragonflies fly over a marsh; a green darner, *Anax junius* (to 3⅞ inches), hovers above a twelve-spot skimmer, *Libellula puchella* (to 2¼ inches), to the left. An American horse fly, *Tabanus americanus* (to 1⅛ inches), rests on a threesquare blade in the center above a saltmarsh mosquito, *Aedes solicitans* (to ¼ inch), and a deer fly, *Chrysops* sp. (to ⅝ inches), on a marsh hibiscus leaf. Above to the right is a Doubleday's bluet, *Enallagma doubledayii* (to 1¼ inches), characteristically holding its wings vertically above its body.

of the male are smoky gray; the female's wings are spotted. The American horse fly, which is a strong flier, the deer fly, and the saltmarsh mosquito are common throughout the bay.

Dragonflies and Damselflies

The brilliantly colored dragonflies and damselflies that hover over the marsh and then quickly dart away are predacious feeders on other insects. These are some of the most beautiful and obvious insects, other than butterflies and moths. They have large net-veined wings and enormous compound eyes. The eyes of the dragonfly cover the head, whereas the damselfly's eyes bulge to the side. Dragonflies and damselflies are closely related; when at rest, dragonflies hold their wings horizontally and damselflies hold them vertically above the body. Damselflies are feeble fliers compared to dragonflies. Both catch their prey on the wing by loosely shaping their legs into a basket to snare insects.

Depending on the species, the female drops her eggs directly into the water, attaches them to a stick or plant stem, or inserts them within soft plant tissues. The aquatic larvae, or naiads, which hatch from the eggs, are fascinating in appearance—in fact, one might say hideous or bizarre. These predacious larvae are equipped with a specialized lower labium, or underlip, that springs out with lightning speed and is capable of capturing small fish, insect larvae, and tadpoles. This protractile organ is unique in the animal kingdom. These grotesque creatures, depending on the species, may be stout or elongated, variously colored, and adorned with spines or covered with algae and organic debris. Damselfly naiads have three leaflike gills projecting from the tip of the abdomen, whereas dragonfly naiads have internal rectal gills.

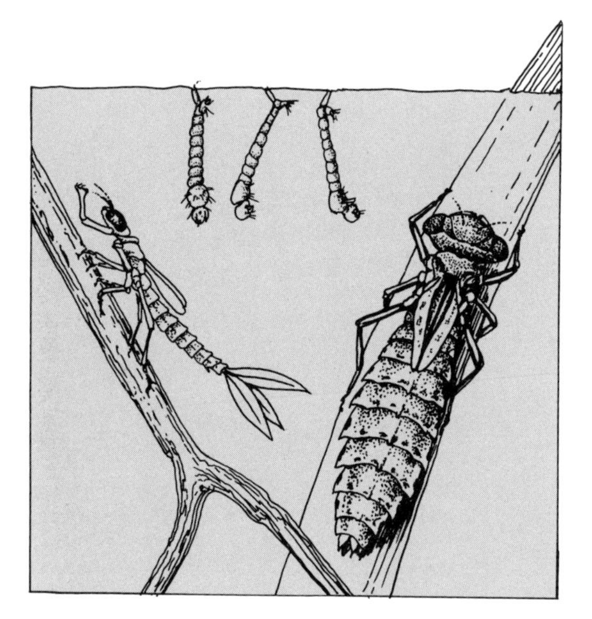

Aquatic larvae of wetland insects include a predacious damselfly naiad to the left with leaflike gills at the tip of its body and underlip extended ready to catch its prey; a larger dragonfly naiad to the right; and some small mosquito wrigglers at the surface.

Along the shores of the Chesapeake Bay, there are many species of dragonflies and damselflies: some are known as darners and skimmers, others as bluets. The green darner, *Anax junius,* is a large, fast-flying dragonfly that can exceed three inches in length. It has large transparent wings with yellowish tips, a green thorax, and a blue to purple abdomen. The twelve-spot skimmer, *Libellula puchella,* is smaller than the green darner, with a light or dark brown head and thorax and a lighter abdomen. Each wing is marked with three brown spots. Doubleday's bluet, *Enallagma doubledayii,* is just one example of the damselflies in this area. It measures less than one and one-half inches. It gets its name from its brilliant cerulean blue abdomen. Its head and thorax are also brightly colored with blue and yellow and marked with black.

BIRDS OF THE WETLANDS

There are many birds that roam the wetlands of the Bay. Some, like the seclusive rails and bitterns, seldom leave the marsh; the herons and ducks, on the other hand, dabble along the shallow, bordering shores, often in company with stalking herons. Bald eagles habitually perch on a favorite tree limb overlooking a wetland and the open water beyond. Sometimes snow geese will wander to the edge of a marsh to feed or loaf in the sun. The red-winged blackbird, the hallmark of a cattail or bulrush marsh, does not belong to the waterfowl or raptor groups at all but is in fact a songbird tied closely to the wetlands.

Waterfowl

Puddle ducks, dabblers, dippers, are all descriptive names for a group of ducks that tip over and extend their bills down to feed on aquatic vegetation and organisms.

BIRDS OF THE WETLANDS

Snow Goose (28″)
Chen caerulescens

American Black Duck (23″)
Anas rubripes

Northern Pintail (26″)
Anas acuta
male

Northern Shoveler (19″)
Anas clypeata
male

Blue-winged Teal (15½″)
Anas discors
male

Wood Duck (18½″)
Aix sponsa

American Coot (15½″)
Fulica americana

They are capable of diving, some deeper than others, but they typically graze from the surface rather than swimming underwater. Some waterfowl, such as wood ducks, nest in wooded wetlands and are more commonly found in those habitats, whereas other waterfowl are typically seen in fresh- and saltwater marshes.

The blue-winged teal, *Anas discors,* is a small, "half-sized" marsh dabbler. It is a rather drab, brownish duck, with blue shoulder patches. The male has a gray head with a large white crescent in front of the eye. The female is a mottled brown with a lighter brown or tan head and a brown stripe through the eye. The male calls in a soft, peeping voice; the female quacks softly. Blue-winged teals are agile fliers; they usually fly in small, tight flocks, wheeling and twisting, which reveals their blue shoulder patches. They are quite common in the Bay and nest in cattails, sedges, and other freshwater and brackish marsh vegetation. Blue-winged teals are fond of quiet water where there is little or no current. They feed on seeds, vegetation, grain, snails, and insects. These little ducks arrive early in the fall and may be seen in the upper Patuxent and Potomac rivers, in Blackwater Marsh in Dorchester County, Maryland, and along Virginia's Eastern Shore.

The northern pintail, *Anas acuta,* is a long-necked, slender-bodied duck with a long, filamentous tail extending from its wedge-shaped rump. It is one of the most graceful of all the ducks, whether flying or on the water. The male has a deep chocolate brown head and a white neck and breast which extends as a narrowing finger of white to the head. The female is a mottled brown with a lighter head and neck and a shorter tail than the male. Both sexes have a metallic brown speculum with a white trailing edge. The male's call is a soft or mellow whistle; the female's call is a harsh "quack." Swift and suspicious, they are very wary birds and will extend their necks to improve their range of vision; like other puddle ducks, they can literally hurtle into the air when alarmed. They normally frequent the freshwater areas of Chesapeake Bay tributaries, but they will also feed in flooded salt marshes on sedges, pondweeds, seeds, and a goodly portion of insects such as flies, beetles, dragonflies, and mosquitoes. Pintails have the greatest breeding range of any North American duck; they nest throughout the center of the country from east to west, and in all of Canada and Alaska to the Arctic Circle. Like the blue-winged teal, they are early ducks: they arrive in the Bay in early fall and leave early in the spring on their migration north.

The duck with the large shovel- or spatula-shaped bill longer than its head can only be the northern shoveler, *Anas clypeata.* Northern shovelers, or spoonbills, are also one of the earliest ducks to arrive in the Bay area, and they leave the Bay later than most other ducks. The male has a green head, a white breast, and reddish brown flanks; the bill is dark gray or black. Females are brown; their bill is grayish yellow marked with orange on the edges. The bill of both sexes is distinctive: besides being oversized, it is equipped with a row of bristles or comblike structures—imagine a mustache attached to the inside of the upper bill. Shovelers use this bill to good advantage as they swim and feed in shallow water over mud bottoms. They filter out the mud with their bill fringes and feed on the remaining crustaceans, mollusks, seeds, and aquatic plants. They are strong and direct fliers. When alarmed, like the closely related blue-winged teals, they twist and turn, showing their blue forewing, or shoulder patch. Shovelers may be seen along the Eastern Shore and from time to time in the upper Potomac River near Washington, D.C.

Ducks and Canada geese often feed along the edges of forested and shrubby wetlands, where they graze on small seeds, fleshy fruits, and acorns. American black ducks, *Anas rubripes,* are closely related to mallards—in fact, male black ducks look very much like female mallards, described in Chapter 5. Both the male and the female black duck are a uniformly sooty, dark brown with slightly lighter heads and white underwings. Their bills are yellow, and their feet and legs are red to deep orange. The female's feet and legs are not as brightly colored, and the bill is mottled. Typical of puddle ducks, they have bluish purple speculums bordered by black and sometimes white on the trailing edge of the wing.

Once one of the most abundant puddle ducks in the Chesapeake Bay, the American black duck has been in serious decline since the 1950s. Waterfowl experts point to the loss of nesting habitat and submerged aquatic vegetation, heavy hunting pressure, and hybridization with mallards as factors in their decline. Black ducks are extremely wary of humans and do not often nest near areas of human activity, which is much different than the nesting behavior of their close relatives, the mallards. Black ducks begin pairing early in fall and begin nesting in mid-March. They nest in forested wetlands with thick overstories and dense understories of shrubs, poison ivy, and other vines, and in secluded marshes. Their nesting areas generally are concentrated from the Chester River along the Eastern Shore to Crisfield, Maryland, and along the Potomac, Rappahannock, York, and James rivers in Virginia.

They lay an average of 6–12 creamy white to greenish buff eggs in nests built in trees or on the ground. The eggs hatch in 28 days, and the young are ready to fly in about 60 days. Black ducks have a diverse diet of the seeds of threesquare sedges, wild rice, and redhead grass; acorns, nuts, and corn; the stems and leaves of eelgrass, widgeon grass, and other plants; and clams, snails, mussels, and fish.

The startlingly beautiful wood duck, *Aix sponsa,* is also an inhabitant of forested wetlands, wooded swamps, and the adjoining marshes. The male is vividly marked with a white-striped face, a head dressed in iridescent greens and purples with a smooth sweptback crest, a white throat, chestnut breast, dark glossy back, and buff flanks. The sharp, gooselike bill is marked with red or orange, and the feet are equipped with sharp claws enabling the wood duck to perch in trees. The female is mottled brown with a white teardrop-shaped eyepatch and a short dark crest. Both sexes have long squared tails, which allow them to maneuver through the trees. The translation of the scientific name of the wood duck is a meaningful "waterfowl in a bridal dress."

The wood duck was on the verge of extinction owing to heavy market hunting for its plumage, meat, and eggs. In addition, heavy logging in the early 1900s laid waste to its essential habitat. Hunting restrictions were placed on waterfowl, including the wood duck, followed by an active, widespread program of placing artificial nesting boxes in the bottomland nesting areas that wood ducks require. Natural nests are usually in a tree cavity; a two-day-old duckling can climb several feet from the bottom of the nest to the opening by using its needle-sharp claws and the hook at the end of its bill. The ducklings jump from the cavity when the mother calls and are soon swimming in woodland ponds. The wood duck has now recovered and is once more abundant in the Chesapeake bottomlands, voicing its finchlike rising and buzzing "jeee" or "woo-eek."

Estuarine marshes vegetated with cordgrass, cattails, and sedges were favorite wintering grounds for the snow goose, *Chen caerulescens,* in the past. Now snow geese seem to be deserting the marshes for agricultural uplands, where they feed extensively on grains and grasses. These very social geese, often numbering in many hundreds or thousands, are quite noticeable in open fields, where they use their serrated bills to good advantage by grubbing out the rootstocks, rather than grazing like Canada geese. The blue goose, long thought to be a distinct species, is classified as the dark phase of the snow goose and not a separate species. Snow geese are readily distinguished from any other bird in the region by the partial or extensive white plumage contrasted with dark flight feathers at the wing tips and a pinkish bill with a dark "grinning patch." Snow geese constantly call with a shrill "la-uk," which has been likened to a dog's bark, and they continually gabble in resonant waves throughout the flock. Snow geese migrate from the Arctic to the Chesapeake Bay in long diagonal or V-shaped formations; they arrive in peak numbers by late November and are gone from the Bay by early March. They are scattered throughout the Bay area, but are most common and abundant on the Eastern Shore.

Herons, Bitterns, and Rails

Many birds search the shallow waters along the marshes of the Bay, slowly and with measured and precise steps stalking their prey, or stirring the water and quickly stabbing a startled fish. These are the familiar and often-seen herons (some of which we discussed in Chap. 5). There are other marsh-dwelling birds, more often heard than seen, such as the bitterns and rails, that live and feed in the marsh and are extremely secretive and wary.

The black-crowned night heron, *Nycticorax nycticorax,* is a stocky gray-and-white heron with a black crown and back. It has two to three white plumes trailing from the back of the neck. The bill is relatively short, for a heron, and has a slight downcurve. The legs and neck are also short as compared to those of the blue heron and other "day herons." The stocky appearance of this bird is enhanced further by its habit of hunching its neck into its body. The black-crowned night heron roosts in trees, usually in groups, and generally tracks its prey at night. They begin breeding as early as February, but the most active egg laying occurs between mid-March and late April. Their fragile nests are placed on shrubs, in dense undergrowth, and in cattail marshes. There are several breeding colonies in the Bay. One of the largest along the Atlantic Coast is on Fishermans Island at the mouth of the Bay. There are also breeding colonies near Baltimore Harbor, on islands off Pocomoke Sound and Tangier Sound, and on Virginia's western shore near Mobjack Bay. The yellow-crowned night heron, which has a creamy white crown and a black face with a large white cheek spot, is also present in the Bay, but in fewer and widely scattered numbers.

The green heron, *Butorides striatus,* is a small, crow-like heron with brownish red sides, a bluish green back, and a dark green crown that is sometimes raised into a scraggly crest. Their legs are a dull yellow, but during

breeding season the male's legs are a bright orange. Immature green herons are brown above with brown and white-streaked underparts. These are very vocal birds and will often make their presence known by uttering a "kyow" or "skow." The alarm call is a piercing "skeow." Green herons do not generally wade when hunting; rather, they stand motionless at the edge of the water, often almost horizontally or in a crouched posture, then quickly lunge for their prey. They also rake the bottom in shallow water to stir up fish. But perhaps one of the most unusual feeding methods of the green heron, or any other bird, is the use of lures to attract their prey. These wily little birds have been seen dropping twigs and leaves into the water where small minnows customarily patrol. The minnows, also always on the lookout for food, quickly come to the surface to investigate their next meal and meet, all too late, their dinner partner. Green herons nest in wooded or bushy habitats between mid-April and late June. They are commonly seen throughout the Bay.

The little blue heron, *Egretta caerulea,* is a slender heron that resembles the great blue heron in contour but is only about half the size. The adults, both male and female, are a dark, slate blue with a brownish maroon or purple neck. Immature blue herons are not blue at all; they are almost entirely white except for dark primary wing tips. They can easily be confused with the immature snowy egret and the cattle egret. Immature snowy egrets differ by having a slimmer, darker bill and plumage that is entirely white. Cattle egrets are smaller birds with a shorter, thicker bill that turns yellow in late summer.

Little blue herons hunt by slowly walking, stopping to peer, and then methodically resuming their deliberate pace. They feed on fish, amphibians, and insects in shallowly flooded marshes, freshwater ditches, and flooded grasslands. Little blue herons build flimsy nests of sticks on cypress, willows, shrubs, and other small trees along the water. Their peak egg-laying period is mid-April through mid-May. Two to five pale blue eggs are laid in the nest, and both sexes incubate the eggs and feed the chicks. Little blue herons nest in several places in Maryland and Vir-

A juvenile and a breeding black-crowned night heron, *Nycticorax nycticorax* (25 inches), sit on a branch overlooking a marsh scene. A tricolored heron, *Egretta tricolor* (26 inches), leans forward as it wades. A little blue heron, *Egretta caerulea* (24 inches), is below. A green heron, *Butorides striatus* (18 inches), peers into the shallow water from its perch while a least bittern, *Ixobrychus exilis* (13 inches), hides among the grasses.

ginia, including Poplar Island, several locations in the Potomac River, and on Watts and Fishermans islands.

The tricolored heron, formerly called the Louisiana heron, *Egretta tricolor,* is only slightly larger than the little blue heron. Both sexes have similar plumage: the upper parts of the head, neck, body, and tail are slate gray-blue; the underparts are white. The neck may range from a pale khaki to dark chestnut or maroon. The bill and neck are long and slender. The tricolored heron is not a major breeder in the Chesapeake Bay; however, a few birds do nest in patches of scrubby or second growth trees close to salt marshes. They are scattered throughout the Bay but are most abundant on the barrier island marshes along the Maryland and Virginia ocean coast. Primarily fish eaters, tricolored herons often wade deeply in shallow water, typically wading slowly like most herons. They also run, making headlong dashes, than abruptly stopping and spreading their wings before striking a fish. They inhabit salt and freshwater marshes and often stalk along mud flats and tidal shallows.

The smallest heron, the secretive and wary least bittern, *Ixobrychus exilis,* is a bird of dense cattail, reed grass, needlerush, threesquare, and saltmarsh cordgrass marshes. The least bittern is about the size of a blue jay or smaller. The male least bittern has a greenish black crown and back and tail. The sides of the head and the tips of the flight feathers are chestnut, the underparts are buff marked with white, and the neck and breast are lightly streaked. Female least bitterns are about the same size as the males, but the greenish black coloration is replaced by purplish brown, and the throat and breast are streaked with dark brown. Immature birds are paler brown and streaked, somewhat resembling the young of black-crowned night herons and green herons. The least bittern haunts dense marshes and hunts for food by walking over tangled reeds and branches in a crouched position, neck extended and bill nearly touching the water. They are loath to fly and frequently will furtively slink away or freeze, pointing their bills skyward and becoming as one with the straight marsh plant stems, literally blending into the landscape. When they do fly, their flight is ungainly and weak. With rapid wingbeats and dangling legs, they travel only a short distance before dropping to the cover of the marsh. Often the only signs of the least bittern are the dovelike coos of the male, the answering ticking call of the female, or the "tut-tut-tut" alarm call.

"As thin as a rail"—is it the narrow steel ribbon of a railroad track or the slim boards that make up a fence? Just where did that old saw come from, anyway? It pertains to certain members of the Rallidae family, the rails, which also includes coots and gallinules. The rails have thin, compressed bodies that allow them to thread their way through seemingly impenetrable thickets and literally to disappear into the marsh. Coots are more like ducks; they are divers and excellent swimmers that use their lobed toes to propel themselves through the water. Rails are usually brown and patterned or mottled with white, while coots are slate or soot colored. Rails are found in the Bay wetlands year-round; coots are mostly winter migrants.

The clapper rail, *Rallus longirostris,* a chicken-sized bird (in fact, it is often called a marsh hen), is a creature of the salt marshes. The clapper rail has a long, slightly decurved bill and is grayish brown with lighter underparts. The flanks are barred with grayish brown and white stripes. They are active at night and agilely walk on marsh reeds, often grasping the vertical shafts with their oversized feet. Clapper rails are secretive birds and are usually not seen unless forced off the reed floor by high tides. Then they are frequently seen along the edge of the marsh and even along nearby roads. Even though they are in the open and quite

A clapper rail, *Rallus longirostris* (14½ inches), feeds along the edge of a marsh.

visible, clapper rails apparently think they are still in the marsh, unseen and safe. Like the least bittern, they are reluctant fliers and when flushed will make brief sorties, legs dangling, then drop and disappear into the marsh vegetation. Curiously enough, rails are quite capable of making long migratory flights. The best way to "see" a rail is with your ears: listen for the clattering "kek-kek-kek," especially in the early evening and at dawn. The clapper rail is widely distributed throughout the Bay. The Virginia rail and the king rail are closely related to the clapper rail and also occur in the Chesapeake Bay. The Virginia rail is smaller than the king rail; both are rust colored and similarly marked. The Virginia rail is primarily an inhabitant of fresh and brackish marshes, but may also be found in saltier marshes in the winter. The king rail, in the main, lives in freshwater marshes and sometimes in brackish marshes.

A slate-colored, ducklike bird with a bill shaped like a chicken's, a bird that bobs its head back and forth as it swims, can only be the American coot, *Fulica americana.* It is blackish or slate gray with a whitish gray bill, greenish or yellowish orange legs, and large lobed feet. Coots are sometimes called spatterers because of their habit of skittering and scrambling across the surface of the water before they gain altitude. They are noisy birds whose typical call is a "kuk-kuk-kuk." They also communicate by arching their wings, angling their white tails, and swelling their forehead shields (a bulbous patch at the base of the bill). They are good divers but often pirate choice morsels of aquatic plants from canvasbacks and other diving ducks. American coots are found here and there throughout the Chesapeake Bay, particularly in low salinity areas where they feed on wild celery, redhead grass, sago pondweed, and some small fish and invertebrates.

Red-winged Blackbird (8¾")
Agelaius phoeniceus

Other Birds of the Wetlands

Red-winged blackbirds, *Agelaius phoeniceus,* are songbirds very much at home in the marshes, and they seldom nest far from the water. Male red-winged blackbirds arrive in early spring and claim their breeding territories along the upper marsh. The males have distinctive red epaulets with buff or yellow borders, which are often concealed, on their shoulders. They sing their liquid call, "Oka-leeee," in the plume-topped reed grass meadows, flashing their bright crimson shoulder patches both as a display of territorial behavior and as a courtship signal to the female. The female and immature red-wings are brownish with bold stripes below. The female builds a cup-shaped nest from just above the ground to about eight feet high in the freshwater and brackish marsh reeds. There the pale bluish green eggs are incubated for 10–12 days, and about two weeks later the young begin to fly. Red-winged blackbirds may produce as many as two to three broods in a season. Red-winged blackbirds are relatively long-lived and may reach an age of about 15 years. The generic name, *Agelaius,* quite appropriately means "gregarious" or "living in large groups." As fall approaches they throng in huge mixed flocks with grackles and cowbirds. They gather in great clouds and spiral in long, seemingly never-ending smokelike plumes—twisting and turning, they suddenly descend into a grove of trees. Here they will stay for several days or more, noisily twittering and whitening the branches with droppings before an unknown signal spurs the flock to move on.

Bald eagles, *Haliaeetus leucocephalus,* seldom fail to

evoke a thrill, even in the most experienced observer of nature. Bald eagles are not only a symbol of this country, they are also a symbol of what man did to the environment in the first half of this century and what man has done to right that wrong. When Rachel Carson first sounded the alarm about DDT, bald eagles were already in decline. Perhaps it was necessary that this majestic bird become the emblem for a tragic environmental problem, since it is unlikely that any other living creature could have provided the rallying point around which society could take a stand. The stand was a success, and now bald eagles are again flourishing and are commonly seen throughout the Chesapeake Bay.

The bald eagle is one of ten species that are known as sea eagles or fish eagles. The adult is instantly recognizable by its white head and tail, large, hooked yellow bill, and yellow feet and eyes. Immature birds are more difficult to identify, since they are brown all over with varying markings of white. There are approximately two hundred or more pairs of breeding bald eagles in the Chesapeake, and they are increasing yearly. The largest breeding concentrations are along the Potomac and Rappahannock rivers and in Dorchester County, Maryland. They are early nesters and begin laying eggs in a huge nest made of sticks in January. An average of three eggs are laid; the eaglets leave the nest in late spring to midsummer and by winter are independent. In addition to the permanent population of bald eagles in the Chesapeake Bay, eagles from the north, the Canadian Maritimes and Maine, winter on the Bay from late fall to midspring, and bald eagles from the south summer in the Bay. The bald eagle's primary food is fish, usually taken as the eagle swoops down from a perch, skims the surface of the water, and grasps its prey in its talons. They will also feed on carrion and waterfowl when fish are unavailable. Occasionally, a golden eagle will be spotted in this area, far from its usual haunts in mountainous and hilly country. They are brown with a lighter tinge on the head and neck. Golden eagles are feathered to their feet and are known as booted eagles; the bald eagle's feathers reach only halfway down their legs.

Bald Eagle (37″)
Haliaeetus leucocephalus

MAMMALS OF THE MARSHES

Beavers—Wetland Engineers

A sharp thwack reporting across a still pond on an early evening, chilly with impending autumn, is a sure sign of beaver. *Castor canadensis,* the beaver, is truly a busy creature of the forested wetlands. A strange mammal, largest of the rodents in this area, with a broad, scaly, flattened paddle for a tail; webbed feet; and a magnificent deep brown, thick coat—this unlikely creature was a major factor in opening the unknown north country of the Great Lakes and Canada. *Voyageurs,* French traders, paddling huge canoes into this vast territory traded goods with the Hurons and other Indians for great loads of beaver pelts. There were many valuable fur-bearing animals in the North Country—marten, otter, mink, bear, and rabbit—but the beaver was prized by Europeans, who had depleted its population on their own continent. The beaver was special because the soft underfur could be processed and brushed to a flat, feltlike material and fashioned into stylish hats that were impervious to wind and rain.

By the early 1900s, another species, much like the egrets and wood ducks, had been so unrelentingly hunted, and its habitat so devastated by logging, that there were only a few isolated pockets of beavers in North America. Hunting regulations were put in place to protect the beaver; now it has made a successful recovery in many parts of the continent, including the Chesapeake Bay region.

Beavers are engineers that rearrange habitat to suit their needs; a still pond where a lodge can be built to raise their young is essential. Deep, quiet ponds offer protection from predators and enable the beaver to float logs and food from the nearby woods to the lodge. An impoundment is created when the beaver constructs a dam across a stream

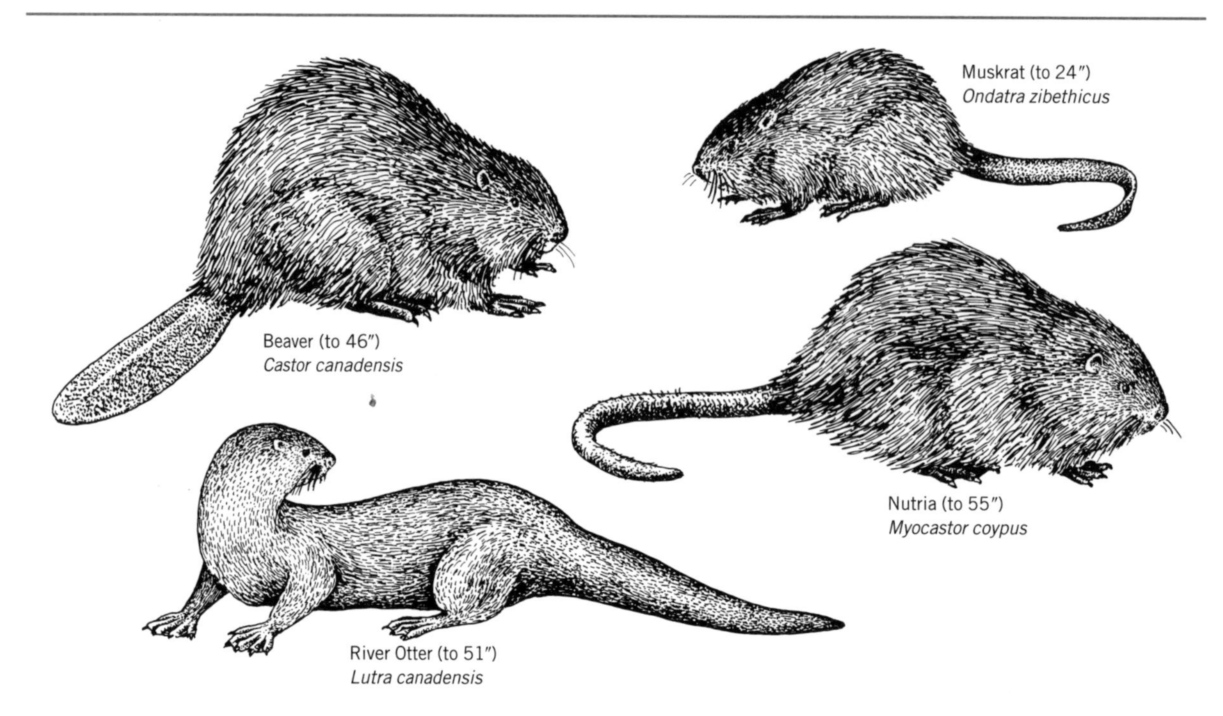

which flows into the many tidal freshwater tributaries of the Chesapeake. The upstream surface of the dam is smoothly plastered with mud and sticks, and a few rocks when available; the downstream side is mostly a projection of saplings which form the framework of the dam. Dams and lodges, which are houses for the beaver colony, are built by a family of beavers consisting of the parents, yearlings, and the kits, or cubs. The construction of a dam floods the surrounding forested area and the meadows beyond. Maples, willows, alders, river birches, and poplars are some of their important construction materials, and their favorite food. Beavers can fell a small tree in a matter of minutes. They are constantly cutting down trees for maintenance of their dams and lodges and for food. Beavers can propel themselves through the water with the aid of their webbed feet and rudderlike tail at six miles per hour and can remain submerged for a quarter of an hour. They are primarily nocturnal, but are often active late in the afternoon. An average of two to four kits in a litter are born between April and July. They are born well furred and weigh about one pound, and they are able to swim quite well in a week.

Where beavers live, in that tension zone between human settlements and the wild bottomlands, there is no question, as far as most humans are concerned, that a flooded backyard and a felled prized, flowering tree are matters of serious contention. Beavers have indeed returned to their native haunts and are not always welcomed by newly arrived suburbanites. However, beaver ponds also collect sediment and silt from runoff largely emanating from human development and stem the flow of eroded materials into the Chesapeake Bay. In addition, beavers create important waterfowl and fish habitat wherever they are present.

Muskrats—Marsh Lodge Builders

Just as the beaver is generally considered a denizen of woodland ponds, the muskrat, *Ondatra zibethicus,* is at home in fresh, brackish, or saltwater marshes, as well as in ponds, lakes, and rivers. Muskrats are about the size of a small house cat or a large squirrel; their dense, glossy fur ranges in color from a light silvery brown, to dark brown, or sometimes black. *Ondatra* is the North American Indian name for the muskrat, and *zibethicus* stems from Greek and means "musky odored." Muskrats do indeed emit a pronounced musky secretion, which is deposited as a scent trail within its territorial boundaries. The long scaly tail is almost furless and is flattened from side to side, just the opposite of the beaver's. Their eyes and ears are small, and their hind feet are partially webbed to aid in swimming. Muskrats, as well as beavers, have fur mouth flaps that seal the mouth opening behind their protruding incisor teeth, which allows them to chew underwater without getting water in their mouths. Muskrats are excellent swimmers and can travel long distances underwater only to emerge into the underside of their beaverlike lodge. The muskrat's house is often dome shaped and constructed of aquatic vegetation, particularly American threesquare and cattails. The female can produce several litters a year with an average of about five young in a brood. They are fully furred in approximately two weeks. A month later they are thrust from the lodge by the mother and are on their own. The population densities of muskrats often fluctuate: some studies show that there may be 2–4 per acre in open ponds and as many as 30–35 per acre in cattail marshes. Muskrats feed on sedges, cattails, and other aquatic vegetation, and they sometimes feed on small fish, crayfish, and frogs. They are trapped for their fur pelts, and, in some areas of the Bay, they are called "marsh rabbits" and seasonally sold as food for human consumption. Raccoons, mink, and otter prey on the young.

Otters—Sinuous Swimmers

River otters, *Lutra canadensis,* are remarkably graceful and powerful swimmers. Their elongate bodies, webbed feet, and muscular furred tails allow them almost to pour through the water—twisting, diving, and turning, seeming to frolic like few other animals in the wild. River otters have dark brown, sleek fur, lighter underparts, and long whiskers. Otters are as comfortable on land as in the water. They often have "rolling places"—areas of flattened vegetation or bare spots where they playfully roll and tumble, often with other individuals. Otters are energetic and busy animals; they habitually slide down to the water, creating paths in the snow or soft dirt of the riverbank. Mating burrows are dug into riverbanks, and an average of two to three young are born fully furred and gone from the nest to new areas in less than a year. Otters inhabit rivers, marshy ponds, and wooded areas; they commonly feed on fish, crayfish, crabs, and frogs. They also prey on small mammals, and where beavers and otters coexist, otters can become serious predators of beaver kits. They are quite vocal and make whistling and chattering noises, grunts, snorts, and growls. The otter population is slowly increasing on Maryland's Eastern Shore. In fact, some individuals have been captured from the Eastern Shore and released in other parts of the state from which they have all but vanished.

Nutria—South American Immigrants

Nutria, *Myocastor coypus,* look like large, bulky rats; they can weigh up to 20 pounds. Their eyes and ears are relatively small and their legs are short; the webbed hindfoot is much larger than the forefoot. The incisor teeth are large and are colored a brilliant orange. Nutria are semiaquatic and usually inhabit freshwater and brackish water marshes. They are quite comfortable on land, but are at home in the water and are more active at night than during the day. Nutria live in pairs, and although there may be several nutria in a marsh, the pairs do not live in a colony. The female may produce two or three litters a year with an average of five young per litter. They were introduced into Louisiana in the 1930s from South America in hopes that their soft, thick fur, particularly the plushlike underfur, would be a boon to the fur trade. However, instead of becoming an important fur producer, they have become a destructive pest of the marshes, undermining streambanks, overgrazing marsh vegetation, and competing with native furbearers. Most nutria live in the wetlands of Louisiana and are descendants of 20 individuals originally released there. There are other populations throughout the South as well as small pockets in the Chesapeake Bay area, namely in the Dorchester marshes of Maryland's Eastern Shore and just below Norfolk in the Back Bay area. Recently a Louisiana entrepreneur has been shipping nutria to Japan, where they are apparently enjoyed as food. Oh, well—one person's fur coat is another's sukiyaki.

● The wetlands, the shoreward extension of the estuary, tie water and forested wetlands to the uplands. Wetlands supply an abundance of food and habitats which enable uncountable plants and animals to live there. They bring beauty and peace and a sense of enduring constancy, reflecting the changing seasons and daily moods of the Bay, from the smallest tributary to the main stem of the estuary.

3

OYSTER BARS

Oysters cluster together, spreading in dense colonies over the surface of the bottom and creating a special habitat, the oyster bar community. The hard surface of the shells on the bottom increases manifoldly the area available to the innumerable sessile organisms in the Bay that must find a firm attachment to survive. It has been estimated that the surface area over an oyster bed, across the dips and folds and crevices, may be fifty times greater than that over an equally extensive flat mud bottom. The oyster bar, then, acts very much like a pier piling or a seagrass plant: it provides a habitat where sessile organisms thrive; where mobile epifaunal animals, worms, snails and the like come to feed; and where small crabs and fishes find food and protection among the crevices. Many of the same species found on pilings and in seagrass meadows are occupants of the oyster bar community. Some other species are found more often here than in other habitats and can be considered typical oyster bar organisms.

The oyster of the Chesapeake Bay is the American oyster, *Crassostrea virginica,* which, in the Bay, lives subtidally, mostly in water depths between 8 and 25 feet. In more southerly regions and in isolated areas along Maryland's and Virginia's seaside, this same species of oyster is intertidal, forming dense oyster reefs along the edges of marshes and streams in the estuaries of the southern coastal states. Oysters do not generally tolerate the freezing temperatures found in Maryland and Virginia and so must form their "beds" in deeper waters. This means that this special habitat cannot ordinarily be viewed directly by the shoreline visitor; however, small isolated clusters or single oysters may often be plucked from the shallows, giving ample indication of the community life that exists on an oyster bar. The shell may be covered with bryozoans and sea squirts, and a clamworm may be crawling over the surface, a small mud crab hiding among the shells, or a small

8. OYSTER BARS

oyster skilletfish protecting a patch of amber eggs inside a gaping dead shell.

Oyster bars are located throughout most of the Bay and its tributaries, upstream to low-brackish waters. Oyster populations in upstream bars are prone to marked fluctuations, as they are vulnerable to freshwater from heavy spring rains or floods. Oyster bars can survive over a variety of bottoms, from hard sand or shelly areas to firm bottom muds. However, they cannot survive on soft, silty sediments or in areas with shifting sands. Thus, they are not evenly distributed but are concentrated in restricted areas called bars, rocks, or beds. In the Chesapeake, the locations of major oyster bars are well known to commercial oystermen. They have been charted on maps and each has a name. Not every oyster concentration is charted, however, and small clusters of oysters are scattered in areas where environmental conditions are suitable. Not only must the bottom be firm enough that the oyster does not sink below the surface, but water movement and exchange must be adequate to flush the area of wastes and bring new supplies of planktonic food and larval oysters. In addition, sediments settling out from the water onto the oysters may smother the entire bar if the sediment load is too heavy.

Oysters formerly were one of the most widespread and

abundant organisms in the Bay. Scientists have estimated that in the not too distant past, oysters actually processed the waters of the Bay through their efficient filtering system every seven days or so. They were, in effect, miniature treatment plants that were capable of continuously removing excess nutrients that took the form of algae. Small wonder that the Bay waters were clearer than they are now. Sadly, the famous oyster resource of the Chesapeake Bay is in precipitous decline. No longer are commercial harvests numbered in the millions of bushels; the harvest is now tallied in thousands of bushels. The reasons for this catastrophic decline are longstanding and additive. For over a century oysters were overharvested and literally wasted, sometimes taken only for their shells, which were used for fill and roadbuilding material. Heavy fishing over the years remolded the oyster bars from large, irregularly shaped mounds to thin layers of shells lying on the muddy bottom of the Bay. The reduction in the size of the bars resulted in fewer young oysters every year, with the bars slowly diminishing in size or completely disappearing in some areas of the Bay. Then, in the 1950s, insidious and debilitating oyster diseases appeared on the scene.

At first these diseases, which do not threaten humans, seemed to be limited to the saltier waters of the Bay. Virginia suffered the first early losses, only to be followed by serious desolation on Maryland oyster bars as well. The burgeoning human population around the shores of the Bay has also contributed to the sad plight of the oyster. The construction of housing developments, shopping centers, and roads to service society's insatiable needs has accelerated the deposition of silt and chemicals in the Bay, which undeniably harm oysters and create havoc in their habitat. Scientists and resource managers are working hard to solve the many problems assailing the oyster and its unique habitat. Surprisingly, there are still productive areas of oysters in the Bay; with continuing research and wise management, the oyster may yet rebound in the future.

THE FAMOUS OYSTER

If there is one Chesapeake Bay animal that everyone recognizes immediately, it is the oyster. Without grace, beauty, or charm, it has nevertheless been praised for centuries for its delectable flavor. An oyster fresh from the water and covered with muck is not an appetizing sight, and one wonders who was the first brave soul to taste one.

Oysters come in various shapes and sizes. The form of the shell reflects the type of bottom on which it grew; oysters growing on softer bottoms tend to be elongated and narrow, whereas those growing on hard bottoms and in calm water tend to be rounded. The American oyster has two valves, a deep and cup-shaped left, or bottom, valve and a flat right, or top, valve. The lower valve is heavier, so if an oyster is dropped into the water it usually settles into its normal functional position, with the cup-shaped side to the bottom. Examine an oyster on the half shell, and you will be able to see its smooth outer mantle and gills. An oyster's life style is entirely different from that of the burrowing clam, so it has no need of a foot. It orients itself on the bottom with the outer, flared edge of its shells tilted upward. The shells open periodically to feed on plankton and, perhaps, detrital matter, which is brought into the mantle cavity and drawn by ciliary currents along the gills to the mouth. The mantle cavity is always filled with seawater so that the oyster can survive when clamped shut for long periods of time. In cooler weather, oysters can live for extended periods out of the water.

A question commonly asked is why oysters are traditionally eaten only during months whose name contains an *r*. This tradition is based on two factors: first, lack of refrigeration in earlier times—oysters spoil quickly in warm-weather months (those without an *r*) unless chilled—and second, the quality of oysters is poor during this season for biological reasons. Oysters spawn in early summer, during and after which they become thin and watery as stored food reserves are used up to make spawn. They grow more robust toward fall, as the weather cools. With proper refrigeration, oysters can be eaten year-round, but they are still decidedly at their best from September through April.

Oysters mature at a very early age. At two to three months old, they are bisexual, but by their first winter most of them will have become strictly male; in another year, most will become females. The ability to change gender occurs in all species of this genus. In the Chesapeake Bay region oysters are ready to spawn by the end of June. Rising water temperatures trigger male oysters to release their sperm and female oysters to release their eggs into the water. This mutual stimulation soon results in a chain reaction of spawning which sweeps across the oyster beds, turning the water milky white with millions of eggs and with clouds of sperm. The eggs are fertilized in the water and soon develop into typical molluscan trochophore larvae, then into veliger larvae, which eventually settle, to search for a suitable spot to attach. Oyster larvae are gregarious and apparently are attracted to chemicals released by larger oysters on the bottom, thus building up beds, one layer of oysters on top of the other. Young, newly attached oysters are called spat, and in summers in which there has been a good spat fall, a single oyster shell may have dozens of tiny spat peppered over its surface.

LIFE IN, ON, AND AMONG THE OYSTERS

Many kinds of animals live on the shells of live oysters, some firmly attached, others crawling in and about the fouling organisms. A number actually live, either within the shell material or inside the mantle cavity, as commensals, in harmony with the oyster but not harming it.

Most of the attached colonial animals of the piling community are equally at home on the hard substrate of an oyster shell. Hydroids are there; the soft snail fur may cover portions of the shell with its pink fuzz, and long, thin stalks of tube hydroids and bushy clumps of white hair may cling to the edges of the shells. The common cushion moss and hair bryozoans attach to the oyster shells, while their cousins, the lacy crust bryozoans, may totally encrust entire shells. The small, creeping bryozoan also competes for space on the shells. White anemones and sea squirts are common on oysters. Their jellylike globs can be picked off the shells from almost any bushel of fresh oysters. Sea squirts may cover the oysters in great clumped masses during the early summer, but these masses are easily wrenched off by heavy currents and the oysters suffer no ill effects.

Boring sponges, *Cliona* spp., are one of the most persistent pests to oysters. They riddle the shells with a bright sulphur-yellow sponge material. As the sponge grows through the shell it emerges in buttonlike mounds at numerous spots along the surface. An empty oyster shell covered with pock marks is evidence of a formerly heavy infestation of boring sponge. A shell that breaks easily when touched indicates the weakening that occurred in the infested live oyster. A boring sponge can eventually cover a shell with a thick, irregular mass of sponge. Boring sponges at times break holes through to the interior of the shell. To combat this intrusion, a healthy oyster produces a thin coating of shell over each hole. At times, however, the sponge wins out over the oyster, and small dark spots will be seen on the mantle where the sponge has bored through the shell. Boring sponges can be found in oysters throughout the Chesapeake Bay; in some areas they may be heavily infested, in other areas only slightly or not at all. Other encrusting sponges, such as the sun sponge and the volcano sponge, grow on oysters in higher-salinity regions of the lower Bay. The redbeard sponge is often abundant, attached to bits of oyster shells; huge clumps up to two feet in diameter and a foot and one-half tall are not unusual on oyster beds.

Worms are plentiful in the oyster bar community as they are everywhere in the Chesapeake Bay. The oyster flatworm, which preys on barnacles and bryozoans on pilings

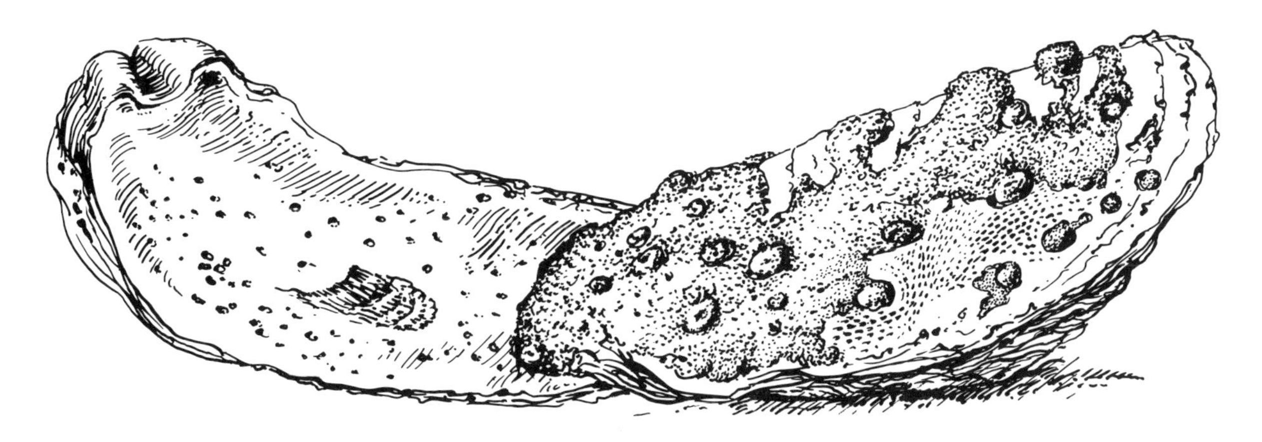

A live boring sponge, *Cliona* sp., protrudes from the shell of a living oyster, while a dead shell shows the characteristic pock marks of formerly infested shells.

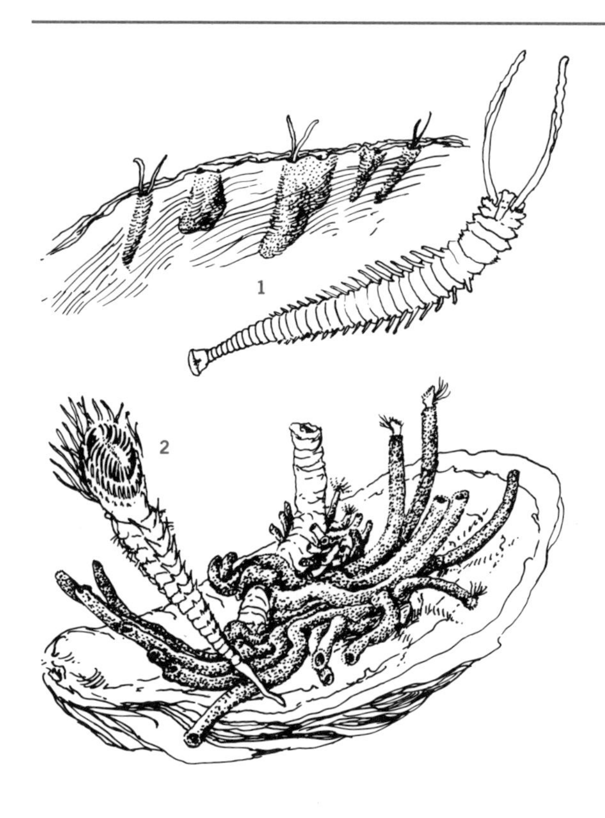

1. Oyster Mud Worm, *Polydora websteri* (to 1″)
2. Sandbuilder Worm, *Sabellaria vulgaris* (to 1″), growing over white tube of limy tube worm

Common worms of the oyster bars include the oyster mud worm, which builds its tubes within the shell's edge, forming characteristic black blisters, and the sandbuilder worm. Fragile sand grain tubes of the sandbuilder worm form thick mats on the exterior of an oyster, often in association with limy tube worms. The drawing shows a sandbuilder worm removed from its case.

or in eelgrass meadows, is frequently found moving over the oysters, feeding on spat and other prey. Errant worms of the intertidal flats, such as clamworms, thread worms, and paddle worms, wander over the oyster shells. Clamworms are such an integral part of the oyster bar community that it is almost impossible to search through a half-dozen oysters and not find at least one clamworm. They are hardy creatures, and live specimens can frequently be retrieved from oysters bought in seafood markets. The ubiquitous whip mud worm also abounds in the oyster beds. It builds its soft mud tubes by the hundred and covers some areas to the point where the oysters below the mat of mud tubes are smothered by the fine silt particles that are continuously drawn to the area by the worms. Mud worms build double-ended tubes and can quickly change direction within their tubes by folding their bodies in half and flipping themselves over. The oyster mud worm, *Polydora websteri,* is a close relative of the whip mud worm. Rather than building its tube on top of the oyster shell, this worm builds on the inside edge of the shell. The oyster responds to this intrusion by laying a thin shell coating over the tubes. Dark shell "blisters" along the edge of a freshly shucked oyster shell are signs of the worms that lived within. As the worm grows inward, it maintains contact with the outside and with its planktonic food source through the original entrance.

A number of other tube-building worms construct intricate, entangled tubes over the oyster shells. The structure of the tube and the head appendages of the worm emerging from the end of the tube are helpful in identifying the more common worms of the oyster beds. An oyster dredged up from lower Bay waters may have four or five different types of encrusting tube worms.

In addition to the small, soft mud tubes of the whip mud worm, with its two sinuous palps waving from each tube opening, there might be a network of well-cemented narrow sand grain tubes of the sandbuilder worm, *Sabellaria vulgaris;* small leathery tubes of the fan worm, *Sabella microphthalma;* and the large, white, calcareous tubes of the limy tube worm, *Hydroides dianthus.* Sandbuilder worms are widespread throughout oyster bars, where they build thick mats of fragile sand tubes over the shells. The tubes are built directly on hard bottoms in some areas, and populations are at times so thick that they form raised reefs of considerable size. They form in smaller patches on oysters. Sandbuilder worms within the tubes are small, conical worms only an inch long. Their body is pale, but their head is adorned with two semicircular pads fringed with bright, golden setae. A cluster of long filamentous tentacles lies below the pads. The tubes crumble easily when touched. The leathery tubes of fan worms are not so fragile. They are frequently covered with silt and sand and can easily be mistaken for mud worm tubes. Place the shell in a jar of seawater; as the head of the worm emerges it will display a crown of distinctive, graceful, feathery white tentacles. Compared to other tube worms on the oysters, the limy tube worm is large, up to three inches long. It too has ornate fringed tentacles, one of which is modified into an operculum cap, which closes the entrance of the tube when the worm withdraws.

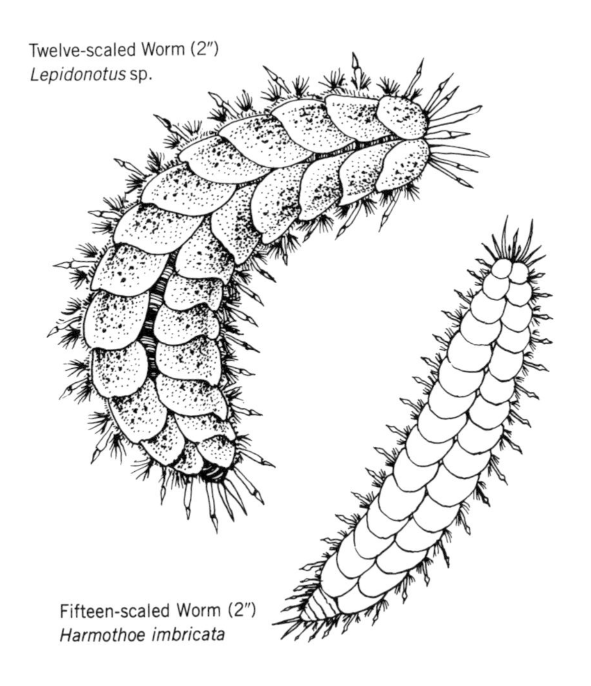

Twelve-scaled Worm (2″)
Lepidonotus sp.

Fifteen-scaled Worm (2″)
Harmothoe imbricata

Oysters that are dredged from high-salinity waters may have an unusual looking armor-plated creature crawling among the barnacles, worm tubes, and other attached life on the oyster shell. It is one of the scale worms; perhaps a twelve-scaled worm, *Lepidonotus* sp., or a fifteen-scaled worm, *Harmothoe imbricata.* There are several species of each genus recorded in the Chesapeake Bay, but none is common. They are unique among polychaete worms, bearing thin plates across the back, twelve or fifteen pairs according to the species. Scale worms are about two inches long with variable mottled dark gray to brownish plates. The worm often dislodges the plates when disturbed, and its pinkish body can be seen beneath. The margins of the body are fringed with bulbous pointed cirri.

CRABS AND OTHER CRUSTACEANS

Oyster Crabs

In past years it was considered quite a treat to discover a little round oyster crab tucked within a raw oyster on the half shell or perhaps floating in an oyster stew. Oyster crabs, *Pinnotheres ostreum,* are small crabs that live within oysters as adults. The young are pelagic until late summer, when they enter the interior of oysters and take up residence attached to the oyster's gills. Females become permanent residents, whereas males soon leave to search for mates in other oysters. The quarter-inch males are rarely seen, and die after mating. Females grow to half an inch or more, about the size of a large pea (they are commonly referred to as pea crabs) and live for two to three years. They cause some harm to the host oyster by eroding its gills and depriving it of food. In the Chesapeake Bay, oyster crabs are more abundant in higher-salinity regions, although they may be found in lower salinities. Oyster crabs live protected lives within the oyster shells and have lost many of the characteristics of other crabs. Their shell is soft and pale, their legs are poorly developed, and they are limited in their ability to move about.

Mud Crabs

Small, brown mud crabs can often be picked out from the crevices of oyster shells, where they are well hidden and protected. Mud crabs are common predators of oyster bed communities. They crush small oysters, barnacles,

MUD CRABS OF THE CHESAPEAKE BAY

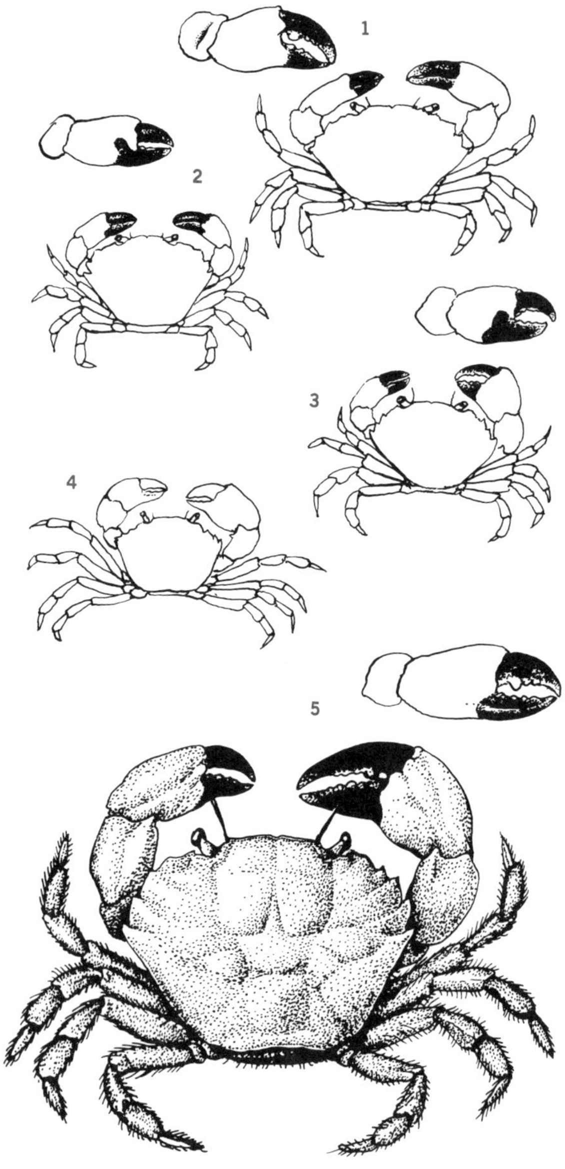

1. Grooved-wristed Mud Crab, *Hexapanopeus angustifrons* (to 1⅛″)
2. Equal-clawed Mud Crab, *Dyspanopeus sayi* (to ⅞″)
3. Flat Mud Crab, *Eurypanopeus depressus* (to ¾″)
4. White-fingered Mud Crab, *Rithropanopeus harrisii* (to ¾″)
5. Common Black-fingered Mud Crab, *Panopeus herbstii* (to 1½″)

and mussels with their heavy, stout claws. They can also be found among sponges, bushy hydroids or bryozoans, in the midst of fouling organisms on pilings, or secreted within the roots of sea grasses; but in the Chesapeake Bay, mud crabs are most typically inhabitants of oyster bars.

Five species of mud crabs occur in the Bay. All are similar in appearance, so it takes some study to identify them. The two most typical estuarine species found in tidewaters from very low to higher salinities (Zone 2 to upper Zone 3) are the white-fingered mud crab, *Rhithropanopeus harrisii,* and the flat mud crab, *Eurypanopeus depressus.* These are the smallest of the mud crabs, only three-fourths of an inch wide or less. The white-fingered mud crab is the only species with pale-tipped claws; all the others have black tips. The flat mud crab has claws of unequal size and a flat but rough shell. It can be distinguished from other black-fingered mud crabs by the spooned-out inside edge of the fingers of its smaller claw. The largest mud crab is the common black-fingered mud crab, *Panopeus herbstii,* although it, too, is still a small crab at one and one-half inches wide. It has black-tipped claws of unequal size and can be identified by a large tooth on one finger of its larger claw which is visible when the claw is clamped shut. This common mud crab is found upstream into moderately brackish waters. The equal-clawed mud crab, *Dyspanopeus sayi,* is more widely distributed in the lower Bay. It is a small mud crab about an inch wide with two black-tipped claws of similar size. It has a smooth back, speckled with pigmentation. The inner margins of the fingers of the somewhat smaller claw are chisel-edged and meet like a scissors. The grooved-wristed mud crab, *Hexapanopeus angustifrons,* is a large, black-fingered mud crab of the higher-salinity waters of Zone 3. It can be distinguished by a groove on the inside of its "wrist," the segment to which the claw is joined. It also has a large tooth on the finger of the major claw, as does the common mud crab, but this tooth in this species is hidden when the fingers are clamped.

Other Crustaceans

Bay barnacles and ivory barnacles settle and grow over oyster shells as profusely as they do on just about any other hard surface. Live barnacles and empty shells, from tiny young to full-grown adults, may cover a single shell. Amphipods, such as the spine-backed scud, other scuds, and

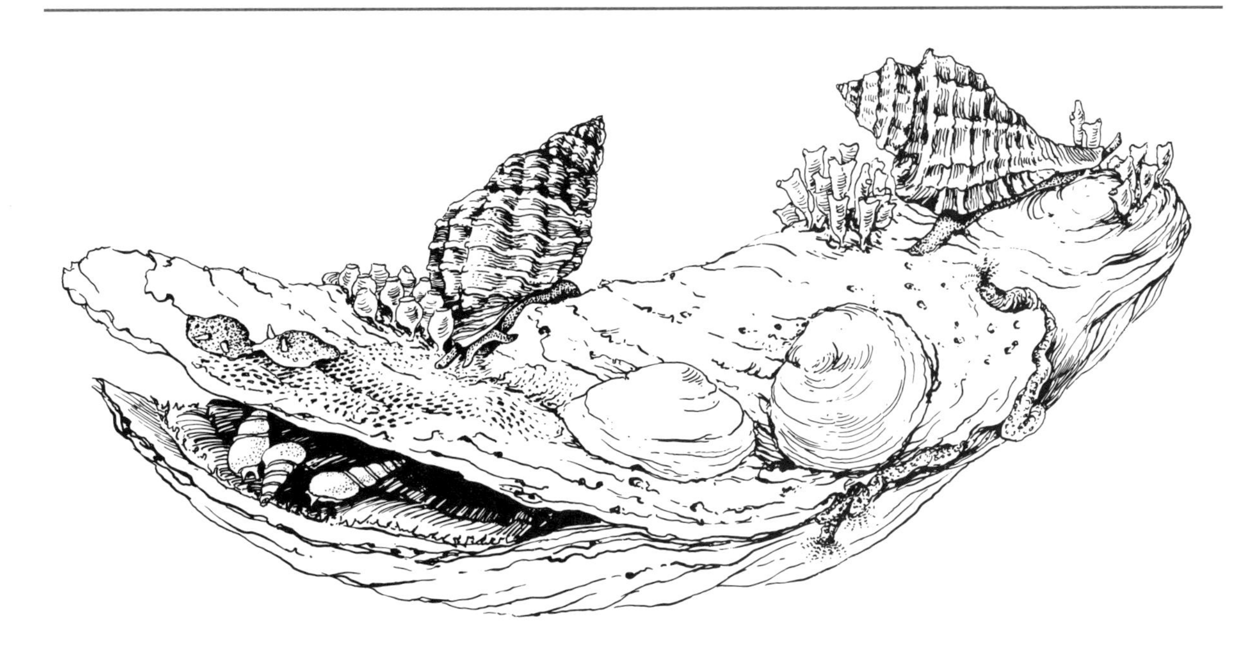

skeleton shrimp, are common grazers of oyster bars. Blue crabs search in and among the oysters, and spider crabs and hermit crabs may also be found here.

MOLLUSKS OF THE OYSTER COMMUNITY

Familiar mollusks common to other habitats of the Bay are also members of the oyster community. Hooked mussels and slipper shells are often attached to oysters. The tiny, parasitic odostome snails congregate just within the edge of the shell, sucking on the oyster mantle with their extended proboscis. In addition to the impressed odostome, so frequently found ravaging the small cerith snails on the seagrass meadows, is the two-sutured odostome, *Boonea bisuturalis.* Tiny limpet nudibranchs are there, too, feeding on encrusting bryozoans, while striped nudibranchs move over the beds preying on sea nettle polyps and other small hydroids and sponges. The flat, circular valves of jingle shells, *Anomia simplex,* which adhere tightly to the surface of oyster shells, are often mistaken for oyster spat. Jingle shells are thin and translucent, much like spat but without their characteristic dark radial stripes. Jingle shells are a common beach shell. They are marked by a hole in the lower flat valve through which a bundle of byssal threads pass to anchor them to the shell. Jingle shells occur only on oysters in the lower half of the Bay and in lower Bay tributaries.

Oyster drills are snails that prey directly on the oysters by drilling pinholes through the shells and then feeding on the soft oyster meat within. Drilling is accomplished by two protuberances, one of which secretes an enzyme that softens the shell material, after which the other rasps away at the softened layer. Oyster drill holes can be identified by their shape, which tapers from a larger surface opening to a smaller interior hole. These snails are particularly harmful to smaller oysters. They attack barnacles on the oysters as well. Two species of oyster drills are found in the Chesapeake Bay, in the higher-salinity waters of Zone 3 and lower Zone 2 only. The Atlantic oyster drill, *Urosalpinx cinerea,* is the more abundant and widely distributed species. It is a small, grayish snail about an inch high with a pointed spire and knobby shell. The thick-lipped or rough oyster drill, *Eupleura caudata,* is similar in size and appearance but has a shorter spire, a larger siphonal canal, and a thick, flaring outer lip. Oyster drill predation has seriously affected the oyster beds of the lower Bay. Oyster drills also attack barnacles, mussels, and other small bivalves attached to pilings and sea grasses. They lay distinctive, urn-shaped leathery egg cases, which are firmly anchored by a pedestal. The capsules are large enough to be easily noticed. Those of the Atlantic oyster drill are spherical, those of the thick-lipped drill more spatulate.

FISHES

The Ugly Toad

Toadfish may lay claim to being the ugliest fish in the Chesapeake Bay, a vision for nightmares, slimy and ragged, with fleshy flaps hanging from their lips and over their eyes, covered with warts, and with threatening, wide-gaping jaws armed with sharp teeth. Oyster toadfish, *Opsanus tau,* or dowdies, as they are also called, are known to anyone who has dropped a fishing line into Bay waters. Toadfish are omnivorous feeders and quickly take to bait. The unhappy angler must be wary of this pugnacious fish. When caught, it erects sharp spines on its dorsal fin and gills and snaps viciously with its powerful jaws. Fortunately, it is not very large, attaining a maximum length of a foot or so.

Along with its formidable appearance and habits, the toadfish is also vocal. These loquacious creatures are capable of making two distinct sounds. On a warm summer night, when below deck on a moored boat, it is not unusual to hear distinct but plaintive foghorn calls through the sides of the hull. The foghorn or boatwhistle call is made by spawning male toadfish calling for a mate. Females and nonspawning males apparently do not make this call, but they are all capable of making loud grunting noises. The grunts arise out of annoyance or fear, and express warning and alarm. When taken out of water, toadfish often respond with loud grunts. Oyster toadfish are residents of oyster beds, but they are also abundant, during spring and summer, in both vegetated and unvegetated shallows. During this season, males are found inside old tin cans, under broken bottles and boards, or within any dark, secretive place guarding their young. Females are attracted into the nests by the foghorn calls. They slip in upside down to attach eggs to the top side of the nest and then leave, many moving to seagrass beds for the rest of the summer. Meanwhile, the males keep watch over the eggs, cleaning the nest by puffing out debris and fanning the eggs with their fins.

Their task is prolonged over a period of a month or more, but they do leave occasionally to feed. Toadfish lay the largest eggs of any Chesapeake Bay fish, almost one-fourth inch in diameter. Eggs are attached to the nest by an adhesive disc. After hatching, the young tadpolelike larvae remain attached, sitting atop a huge bulbous yolk until the yolk has been absorbed and they become little toadlets. Even after the young are free-swimming, the male guards his flock, harboring them within the nest as they swim in and out among his fins. This parental care sometimes lasts for three or four weeks after the toadlets have hatched.

Gobies, Blennies, and Skilletfish

Gobies, blennies, and skilletfish are small fishes that live together in and among the shells of the oyster bar community. They are abundant but reclusive and solitary in habit and are not seen as often as the more visible schooling shore fishes. With luck, you can pick up an empty, gaping oyster shell from the shallows and discover one of these tiny fishes hiding within. Gobies, blennies, and skilletfish are all similar in size and behavior. They are generally only an inch or two long and are closely associated with oyster bars, although they are found at times in shallow flats and seagrass meadows. All species lay their eggs on the inside of dead oyster shells, usually shells that are still hinged together and gape only wide enough for the fish to enter. Males guard the eggs, which are as frequently attached to the upper valve as to the bottom. Eggs are small and amber-colored and are often laid in large clutches of a few hundred or more. Gobies lay narrow, elliptical eggs attached by fibrous bundles; blennies lay round eggs, which adhere to a flattened disc; and skilletfish lay ovate eggs closely packed and attached by a sticky mucus.

Gobies have prominent eyes set close together on the top of their head, two separate dorsal fins, and pelvic fins shaped into suctionlike discs for adhering to shells. Blennies have deep, broad heads, a single long dorsal fin, and small pelvic fins, which are not formed into suction discs. They grow somewhat larger than the others, up to four inches. Skilletfish are shaped like a skillet with a broad, flat head narrowing abruptly into the body. The eyes are small and widely spaced, and the pelvic fins are formed into a large, broad suction disc.

Three gobies are commonly found in Chesapeake waters: the naked goby, *Gobiosoma bosci,* the seaboard goby, *Gobiosoma ginsburgi,* and the green goby, *Microgobius thalassinus.* a few other goby species are less frequently seen in the lower Bay. Naked gobies are the most abundant and widely distributed, occurring even into tidal freshwaters. They are devoid of scales—thus their name. Naked gobies are dark greenish brown on top, pale below, with eight or nine light, vertical bars along their sides. Seaboard gobies are similar in appearance, with only subtle differences. They are without scales, except for two on each side just at the base of the tail fin, and are more irregularly marked, with less distinct bars and generally with dark spots along the lateral line. Their ventral disc is distinctly longer than that of naked gobies. Green gobies are quite different in appearance, with longer second dorsal and anal fins and with scales on the back part of the body. Their head and belly do not have scales. Green gobies are colorful fish. Males are greenish blue with an intense reddish hue on a spotted dorsal fin, vivid orange to yellow pelvic fins, and a row of dark spots on the border of the white-edged anal fin. Females have no spots along the anal fin but have a large, black spot on the back of the dorsal fin, iridescent blue-green sides, gold-blue bands under the eyes, and golden eyes and head. Green gobies often find refuge deep among the entwining fingers of redbeard sponges.

Two blennies occur in the Chesapeake, the striped blenny, *Chasmodes bosquianus,* and the feather blenny, *Hypsoblennius hentz.* Blennies belong to a large family of typically tropical fish and, like many tropicals, are brightly colored. Male striped blennies are marked with bright blue longitudinal lines that converge toward the tail. The head is spotted and an orange stripe runs through the dorsal fin, which also has a brilliant blue spot near the front edge. Females are a darker olive-green with a network of paler green lines. Feather blennies are so named because of two branching tentacles that arise over the top of their eyes. They are not as distinctly marked as striped blennies, but have indiscriminate brown spots along their body, with those toward the back tending to form vertical bars. Males, however, have a brilliant blue spot at the front of the dorsal fin.

The skilletfish, *Gobiesox strumosus,* or oyster clingfish, as it is also called, is not likely to be confused with either gobies or blennies because of its peculiar shape. Skilletfish are evenly speckled with brown and are well camouflaged against the oysters or bottom sands and

SMALL FISHES OF OYSTER BARS

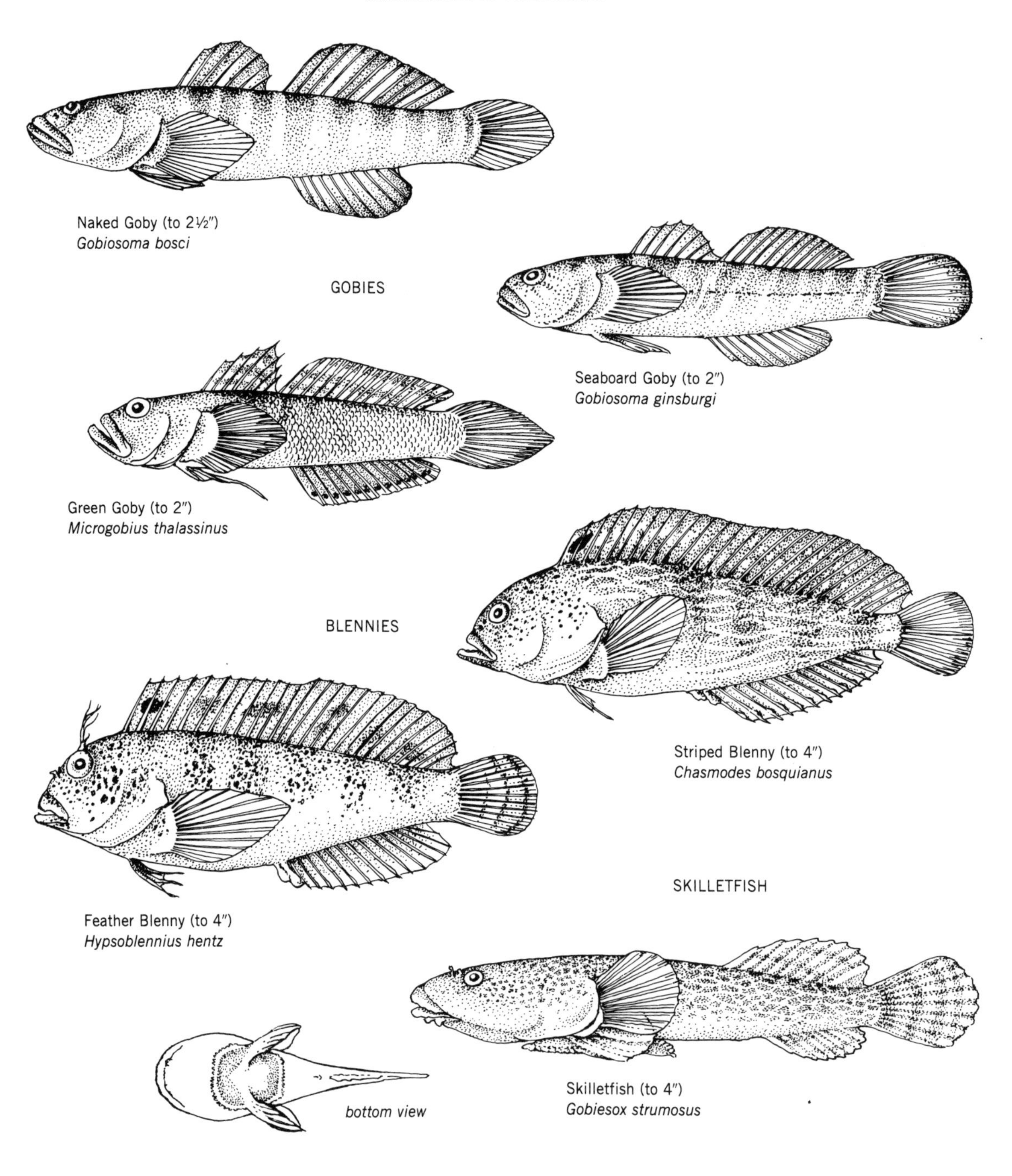

Naked Goby (to 2½″)
Gobiosoma bosci

Seaboard Goby (to 2″)
Gobiosoma ginsburgi

Green Goby (to 2″)
Microgobius thalassinus

Striped Blenny (to 4″)
Chasmodes bosquianus

Feather Blenny (to 4″)
Hypsoblennius hentz

Skilletfish (to 4″)
Gobiesox strumosus

muds. They spend most of their time clinging to the outside of shells, often on the underside, patiently awaiting any stray amphipod or isopod that passes by. They are nearly as widely distributed as naked gobies from the lower Bay to upstream low-brackish waters.

● The oyster bar community is an island of intense biological activity, with all the organisms that make up the community living in close association with one another. The intermingling and interdependence of the many residents reflects complex relationships and requirements that have developed over eons. If you chance to visit an oyster bar in shallow water at low tide, reflect on the thousands of gallons of plankton-enriched water pumped by the oysters every day. Wonder at the vast number of hard, limy tubes secreted by tube worms, and contemplate the pulsating activity within the oyster bar community, which at first glance and many glances thereafter looks like a pile of inert oyster shells imbedded in nothing but muddy ooze.

DEEPER, OPEN WATERS

Beneath the surface of Chesapeake Bay waters is an unseen world. You may glimpse an occasional sign of life as the surface is rippled by a school of fish, or catch the shadowy form of a blue crab as it ghosts by. Some people come in close contact with this world; veteran anglers pride themselves on their knowledge of fishes of the Bay, and commercial watermen who dredge, seine, trawl, pot, or trotline are well aware of the variety of life in deeper, open waters. For most other people, however, the shoreline habitats and the organisms that inhabit them are far more accessible and familiar. You may wander over a beach or mud flat, peer closely at a piling or marsh plant, or wade in the shallows; but the deeper-water habitat cannot be observed as closely, yet it is important to know that it includes the greatest amount of bottom and water area in the Bay.

In this book, this habitat is distinguished from others, but in reality it should not be separated from them because together the habitats form the Bay. As we have stated a number of times, each habitat is continuous with others, and many of the same species we have already talked about also thrive in deep waters. This chapter covers the vast areas of the estuary beyond the beaches and shallow depths of six feet or so and beyond the regions where aquatic plants grow. Oyster beds occur in deeper water and should be considered specialized areas within the broader reaches of the open Bay.

Just as in the shallows, benthic animals live in or on the bottom sediments, and pelagic forms swim or float in the water column. As the water deepens, the physical environment changes considerably; sediments become finer, low-oxygen conditions prevail during certain seasons, and light does not penetrate to the bottom. The benthic community responds with a decrease in the number and variety of organisms. Beyond 30 or 40 feet, many mollusks, such as soft-shelled clams, hard clams, and oysters, are sparse or absent. Most of the sessile colonial forms—the

9. DEEPER, OPEN WATERS

hydroids, sea anemones, and bryozoans, so abundant in shallower regions—are fewer in number. The dominant organisms are polychaete worms, small crustaceans, and certain small mollusks. In winter, blue crabs and other crustaceans that spend warmer months inshore seek the deeper channel regions to hibernate in bottom muds.

Bottom sediments of the deeper waters are not uniformly consistent along the length of the Bay. Sediments are sandier at the head of the Bay, near the mouth, and along the lower Eastern Shore; most of the mid-Bay region tends to be silty. Salinity approaches ocean strength concentration in the deepest part of the lower Bay, and a number of typically marine animals, rare elsewhere in the Bay, are common here.

Pelagic organisms in open waters include the whole spectrum of phytoplankton and zooplankton described for waters of the shallows. The water is rich in microscopic diatoms and dinoflagellates, copepods and cladocerans, as well as the larvae of worms and mollusks, barnacles and shrimps, and crabs and fishes. Hydromedusae, jellyfishes, and comb jellies float freely across the Bay and throughout the tributaries. Large schools of fishes are concentrated here. Some are small forage species such as anchovies and menhaden, others are large predators—bluefish, striped bass, seatrout, and other favorite species of sport and commercial fishermen. Where the anchovies, silversides, and menhaden school is often where terns, gulls, and sea ducks wheel from above to feed on the clouds of fish and bottom-dwelling invertebrates. Beneath the surface of the water, dolphins and huge loggerhead turtles

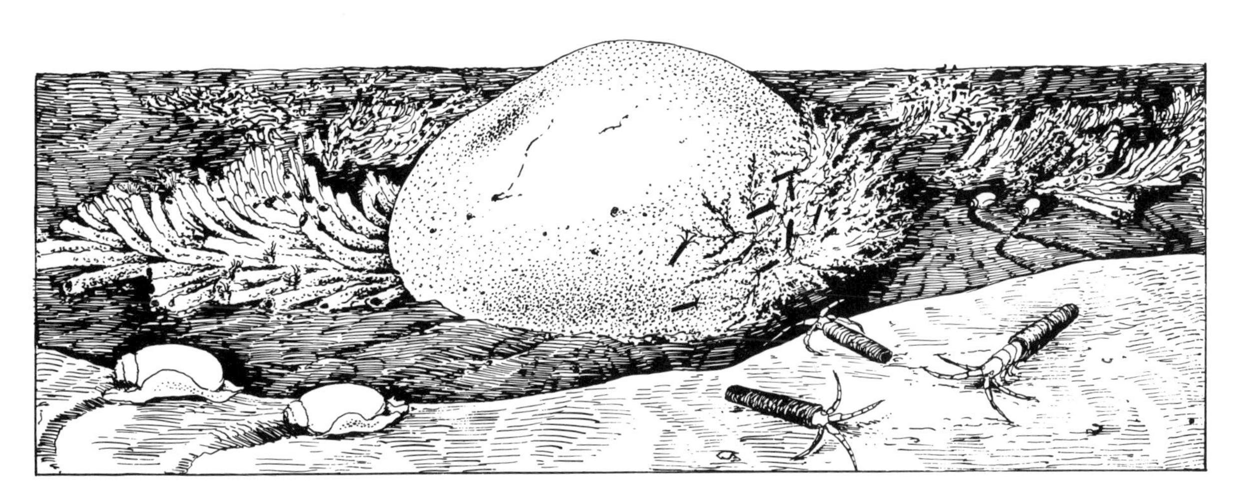

A sand bottom community of deeper, open waters showing the potato sponge, *Craniella* sp. (to 4 inches or more), in the center with hydroids growing on it, and house-carrier amphipods, *Cerapus tubularis* (to 1/6 inch), grazing on the hydroids. These small amphipods are shown in greater detail in the right foreground. Clumps of sandbuilder tube worms surround the potato sponge. The barrel bubble, *Acteocina canaliculata* (to 1/4 inch), plows through the bottom in the left foreground.

graze on the rich offerings of the Bay from the shallows to the deeper, open waters.

DWELLERS ON THE BOTTOM

With the exception of oyster bars or old shell beds, hard substrates are at a premium over the broad stretches of bottom in deeper waters. Consequently, epifaunal communities are sparse. Isolated clumps of sessile colonial animals scattered on the bottom may attract a host of other motile animals. Redbeard sponge and sea squirts can develop large masses and cover a considerable area. In the lower Bay, large clumps of the rubbery dead man's fingers bryozoan may attach directly to the floor of the deeper Bay. Expansive growths of the white hair hydroid flourish anew in winter. Sandbuilder tube worms often grow in such profusion over the bottom surface that they form large reefs. Limpet nudibranchs are common on these colonies, as are scuds (for example, the spined-back scud), skeleton shrimp, and tube-building worms. Potato sponges, *Craniella* spp., live on coarser sandy bottoms of the lower Bay. They look and feel like a hard, wet potato and may be quite abundant locally. Although they are smooth and dense, without the inviting nooks and crannies found in other colonial animals, they too serve as a home for other creatures. Amphipods live deep inside, buried in tiny openings within the sponge. Creeping bryozoans and hydroids may cover the surface, and skeleton shrimp manage to find a grip. Occasionally, potato sponges are washed up on the beach.

Typical roamers over open bottoms of deeper waters include minuscule snails, such as the tiny, spire-shelled interrupted turbonille, *Turbonilla interrupta,* and the pitted baby-bubble snail. The barrel bubble snail, *Acteocina canaliculata,* is also common to deeper waters. In the lower Bay, large whelks plow through the surface leaving a furrowed trail behind as they probe with their long proboscis for small buried clams. Spider crabs and hermit crabs move into deeper waters, and many amphipods and isopods, such as the mounded-back isopod, wander over bottom sediments. A strange little amphipod, the house-carrier amphipod, *Cerapus tubularis,* carries its tough tube case along with it wherever it goes. This curious animal is easily passed over. When disturbed, it retires completely within its brown, twiglike home, which looks like bits of vegetation. House-carriers may also be found among hydroids, where they cling to the branches with their antennae.

Large numbers of bay opossum shrimp congregate

The star coral, *Astrangia astreiformis* (colony to 5 inches), forms small stony patches attached to shells and rocks. Individual polyps extend their tentacles and capture prey with stinging nematocysts. Star-shaped polyp cups of recently dead coral animals can be seen on the right. The common eastern chiton, *Chaetopleura apiculata* (to 1 inch), is found on hard bottoms and is often found in association with star corals. Two chitons, tightly clinging to a rock, are grazing on encrusting bryozoans.

over the bottom or are partially buried in the loose, unconsolidated sediments during daylight hours. At night, they emerge to rise up through the water and feed on zooplankton close to the surface; with dawn they move again to the bottom.

Star corals, chitons, starfish, and cancer crabs are oceanic animals that may be found close to the mouth of the Chesapeake. They are rarely encountered elsewhere in the Bay.

People are usually surprised to learn that corals grow in the Chesapeake Bay. One immediately thinks of tropical reef corals, but the star coral, *Astrangia astreiformis,* is a common cold-water coral distributed all along the Atlantic coast as far north as Cape Cod. It is found in the Chesapeake Bay firmly attached to shells or rocks in small stony patches of perhaps 5 to 30 individual polyps. The living coral is composed of delicate little polyps that expand out approximately one-fourth inch from their cuplike homes. Each has 18 to 24 tentacles, tipped with a whitish knot, which hold the stinging nematocysts. Other nematocysts speckle the tentacles with white spots. Feeding is accomplished in the same manner as in sea anemones, close relatives of corals. Passing prey are stunned or killed by the poisonous stinging cells and then brought toward the central mouth by the tentacles. Bits of star coral skeleton attached to a broken shell or rock may wash up onto the beaches of the lowermost Bay. If the star coral has only recently died, the starlike polyp cups will be sharply etched. If the coral has been tossed about on the beach for a while the cup will be smoothly rounded, but the structure of the interior septa can still be seen.

The common eastern chiton, *Chaetopleura apiculata,* is familiar to beachcombers along the Atlantic Ocean coastline, where it lives in shallow tidal pools and among the crevices of rock jetties. In the Chesapeake Bay, it occurs only in deep water, where it slowly slides over rocks and shells grazing on encrusting bryozoans or other minute organisms. The eastern chiton belongs to one of the most ancient groups of mollusks. It is elliptical-shaped, with eight shells, or valve plates, in a row across its back, and surrounded by a fleshy girdle. The girdle is equivalent to the mantle of other mollusks. When disturbed, chitons cling tenaciously to the substrate with a broad, sole-shaped foot, clamping down with the entire edge of the girdle to create strong suction. They are difficult to pry loose and, once freed, curl up into a ball, much like a sea pill bug. Common eastern chitons are rather small, one- to three-fourths inch long and indiscriminately colored grayish brown, which camouflages them well against a rock or shell.

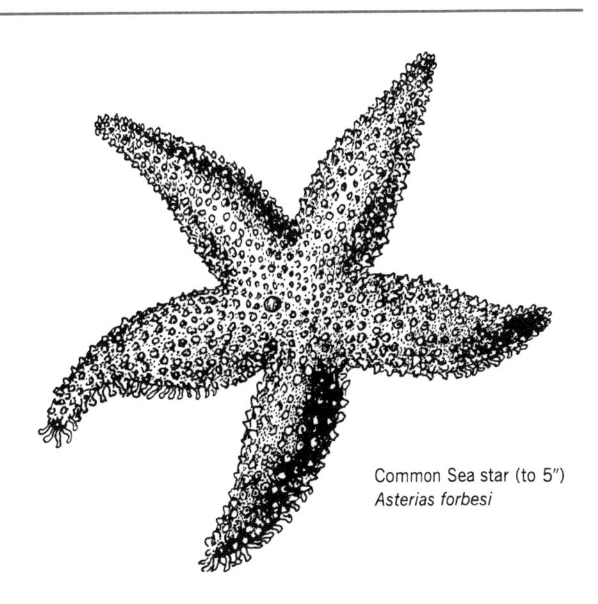
Common Sea star (to 5″)
Asterias forbesi

Starfish, a familiar symbol of the sea, may occur in the Bay in large numbers for a time and then diminish for some period. A single species, the common sea star, *Asterias forbesi,* is often caught by crab dredgers working in deeper waters over sandy-bottom areas close to the mouth of the Bay.

Sea stars typically have five somewhat pointed arms, although specimens with six, seven, or eight arms may be encountered. They have a strong regenerative ability, and if one arm is lost it is quickly replaced, sometimes with the animal overcompensating and growing an extra one or two. The top surface of the common sea star is rough-skinned, with tiny spines and bumps. A single bright orange "eye," or madreporite, is located toward the center of the back. The madreporite, is actually a sieve plate through which water enters into the internal water-vascular system of the sea star. Sea stars are variably colored, from yellowish orange to deep purple. The bottom, or underside, is pale, with a tiny mouth in the middle. A deep groove bordered by four rows of tube feet radiate down the center of each arm. If you turn over a live sea star, fresh from the water, you will see the hundreds of little tube feet waving in synchronized pulses. Sea stars are carnivorous predators on snails, clams, mussels, and barnacles. In northern waters, where they are vastly more abundant, they have caused great havoc on oyster beds. They attack the oysters by attaching two arms on one valve and three arms on the other by means of their tube feet. Their strength and hold are so great that they force open the valves by pitting the suction-strength of the tube legs against the strong adductor muscles of the oyster. The sea star is usually the victor. The shells are pried open just wide enough for the sea star to evert its stomach into the oyster and devour it.

Cancer crabs are round-backed crabs common to tidal pools of New England, but in the Chesapeake Bay they are only found in deeper waters close to the mouth. They have a smooth, pentagonal-shaped carapace (back shell) with nine teeth on the front margin on each side of the eyes. Two species occur in the Bay: the rock crab, *Cancer irroratus,* and the jonah crab, *Cancer borealis.* They are similar in appearance and can be identified only upon close inspection. The rock crab is yellowish, with small, purplish brown spots sprinkled over its back. The edges of the carapace teeth are smooth. The jonah crab is yellowish on the underside, but reddish on the top, with lines of yellow spots across its back. Its carapace teeth are roughly edged with small denticulations. Both crabs attain the size of a moderate-sized blue crab, about five inches across the back shell.

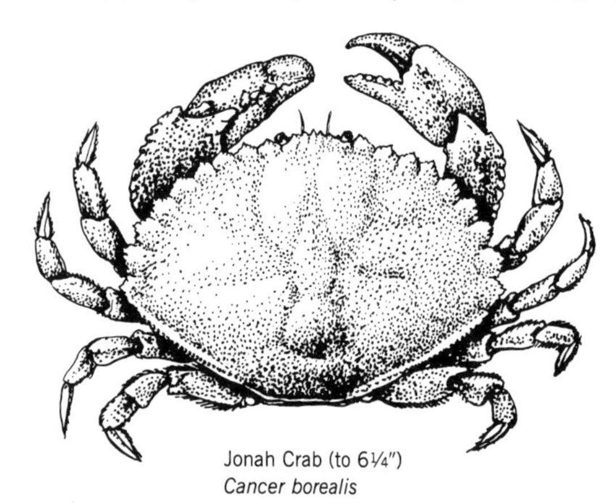

Jonah Crab (to 6¼″)
Cancer borealis

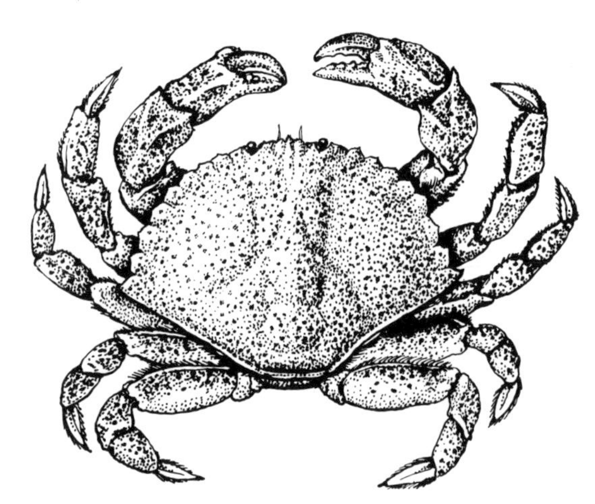

Rock Crab (to 5¼″)
Cancer irroratus

DWELLERS IN THE BOTTOM

Worms of all types are the most abundant creatures living below the surface muds and sands of deeper, open waters of the Chesapeake Bay. The most diverse group, as elsewhere in the Bay, are bristle worms, but ribbon worms and oligochaete worms are also there; oligochaetes are particularly abundant in upper Bay regions, as is the small red ribbon worm.

The unusual phoronid worm is common in the deeper waters of the higher-salinity regions of the lower half of the Bay. Phoronids have long been an enigma to scientists. Wormlike in appearance, they are most closely related to bryozoans because they possess a horseshoe-shaped ring of tentacles (a lophophore) similar to that of bryozoans. They might be considered merely stretched-out bryozoans. Phoronid worms, *Phoronis* spp., build thin, straight, sand-covered tubes that extend down vertically six inches or so into the bottom. The soft, stringy, unsegmented worm remains within the tube as the lophophore extends out of the top. Minute cilia on the tentacles create water currents that draw microscopic planktonic food particles to the tentacles, where the particles are entrapped by a mucus film and eventually passed to the mouth. When startled or disturbed the phoronid immediately withdraws completely into its tube.

In the lower half of the Bay, the spectacular large milky ribbon worm digs deeply through the substrate foraging for food and leaving behind an oval tunnel, shaped by its body. The tunnels are gradually filled with sediments as the ribbon worm ventures into new territories.

Among the most ubiquitous and numerous bristle worms in deeper waters, throughout all regions of the Bay, are the various mud worms that build straight mucus-lined tubes down from the surface, out of which they extend long, probing palps that comb the flocculent surface particles for food. The common species of the intertidal flats are often the dominant deep-water species. The large red-gilled mud worm and the fringed-gilled mud worm construct long tubes, penetrating 8 to 10 inches into the substrate. The minuscule barred-gilled mud worm builds small, fragile tubes only an inch or so deep.

The clamworm is as abundant in deep-water habitats as elsewhere. However, it spends more time burrowing through its network of mucus-lined burrows below the bottom surface than wandering over surface muds, since suitable epifaunal prey is scarce. Bloodworms, chevron worms, and capitellid thread worms all move through deep galleries of mucus-lined burrows, while paddle worms move freely through the substrates without benefit of burrows.

The red-spotted worm, *Loimia medusa,* is one of the most impressive tube-builders in the Chesapeake Bay. A tiny worm, only an inch or so long, it builds tremendously large, soft, U-shaped tunnels complete with a thick mud coating. The tunnels may be a foot or more long. The red-spotted worm is green with a bright red blotch just below its gills and long, fringed tentacles. It is a close relative of the large ornate worm more common to inshore habitats.

In the lower half of the Bay, thick colonies of the common bamboo worm are distributed in deeper waters as well as in the shallows. Colonies of its larger cousin, the elongated bamboo worm, *Asychis elongata,* are also found here. They live in long, narrow, sand-encrusted tubes that penetrate down some 20 inches; the worm within looks very like the red-jointed, common bamboo worm but is twice the size, up to a foot long, and has a scoop-shaped tip to its tail.

Other common bristle worms in deeper-water substrates include the trumpet worm, with its trumpet-shaped sand tubes, and the parchment worm, which lives inside its long, U-shaped parchment burrow.

BURROWING WORMS OF DEEPER WATERS

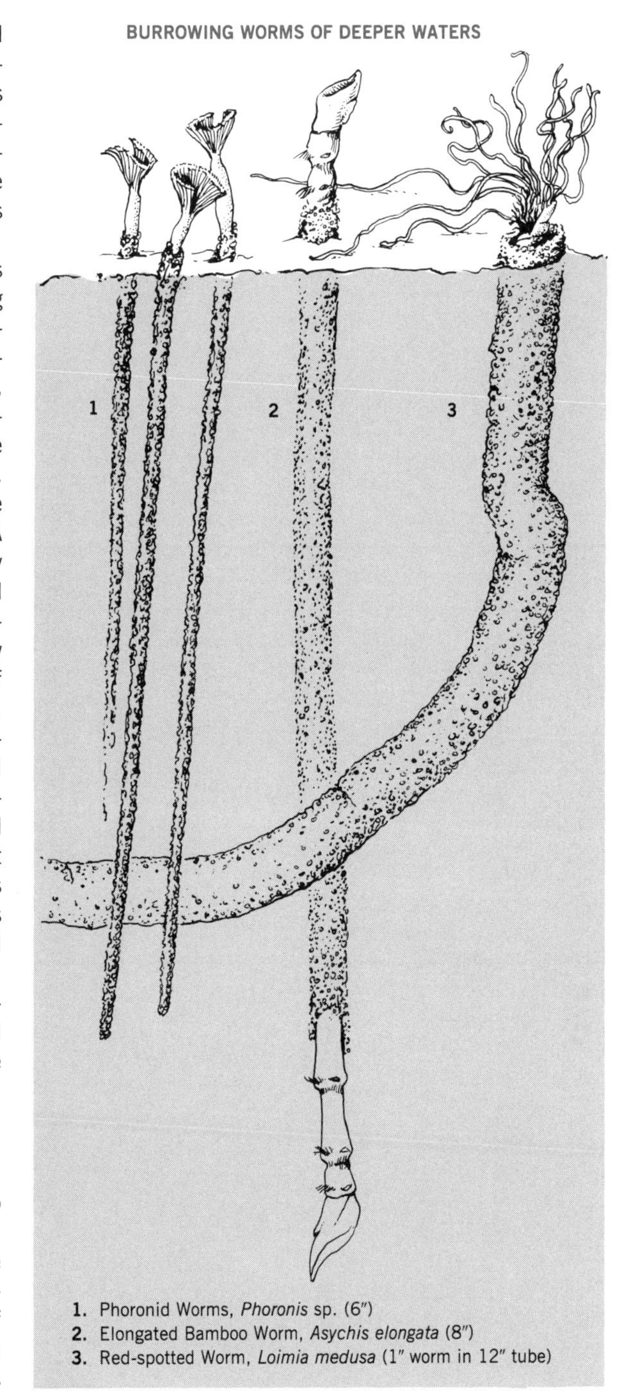

1. Phoronid Worms, *Phoronis* sp. (6″)
2. Elongated Bamboo Worm, *Asychis elongata* (8″)
3. Red-spotted Worm, *Loimia medusa* (1″ worm in 12″ tube)

Clams and Such in Deeper Waters

Many of the bivalves buried in inshore areas are also found in deeper waters. Soft-shelled clams and tiny gem clams are found to depths of about 20 feet but are sparse or absent in deeper waters. The most abundant and widespread clams are the Baltic macoma and the little surf clam. The Baltic macoma moves up and down within a long, open tunnel, whereas the little surf clam remains

BURROWING CLAMS OF DEEPER WATERS

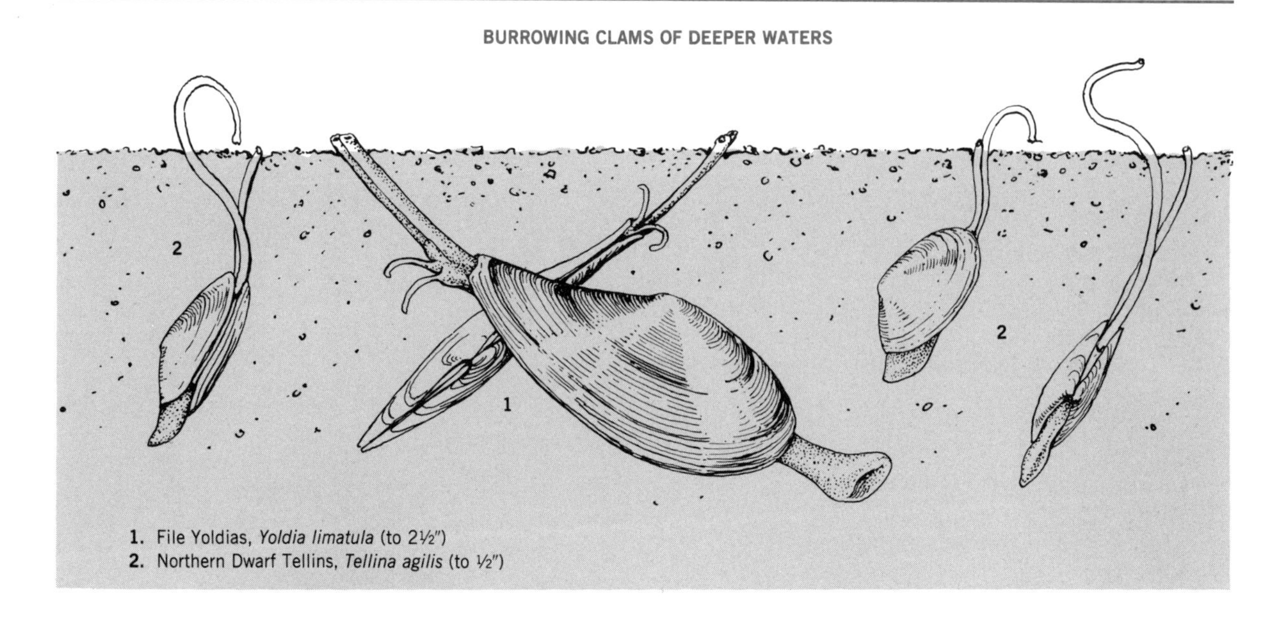

1. File Yoldias, *Yoldia limatula* (to 2½")
2. Northern Dwarf Tellins, *Tellina agilis* (to ½")

buried just under the surface. Other species occur in deeper waters: the northern dwarf tellin, *Tellina agilis,* is a common resident in the lower half of the Bay, as are the stout razor clam, the blood ark, and the transverse ark. Here, too, is the strange little bivalve, the file yoldia, *Yoldia limatula,* which can be quite abundant. These are thin, fragile clams, distinctive in color and shape. The shell is a highly glazed brownish green color, drawn out to a narrow, flared edge on one end.

Other Burrowers

Many crustaceans that live below the surface on intertidal flats in the shallows live in deeper waters as well. The slender isopod and the common burrower amphipod are often dominant species in these habitats in the upper and mid-Bay regions. They live in unlined burrows that extend into the bottom for two or more inches. In higher salinities, all three species of four-eyed tube-building amphipods may cover the surface of the bottom with thick mats of parchment tubes. Both the flat-browed and the short-browed mud shrimp construct their elaborate burrows here. The large, bizarre mantis shrimp are generally commoner in deeper Chesapeake Bay waters than they are inshore.

The small burrowing sea anemone and the large, sloppy-gut burrowing sea anemone extend into the deepest parts of the Bay; the latter are far commoner in waters deeper than 20 feet.

Brittle stars are an unusual inhabitant on the bottom muds of the lower Bay. They are five-armed and prickly, like their close relatives the sea stars, but they differ in many ways. The arms of a sea star are not really appendages, but extensions of the body. The arms of brittle stars are separate appendages attached to a small, flat disc of a body. Brittle stars are so named because their arms break off so easily; however, they have remarkable regenerative powers. They can lose all appendages, plus the top half of their body, and still grow anew all lost parts. The burrowing brittle star, *Micropholis atra,* is a common Chesapeake Bay species living in soft muds in waters usually deeper than 15 feet. It lies below the surface in a burrow with two of its arms stretched upward. Water is circulated through the burrow bringing with it microscopic food particles. The disc is a mere half-inch in circumference, but the long, thin arms may be over three inches, so the stretched-out brittle star may extend six or seven inches into the bottom. The arms are jointed internally by interlocking plates that allow only a sideways movement. A series of external plates extends along the top, bottom, and sides of each arm. The mouth lies in the center of the underside of the disc body and leads directly into a saclike stomach.

BURROWING ECHINODERMS

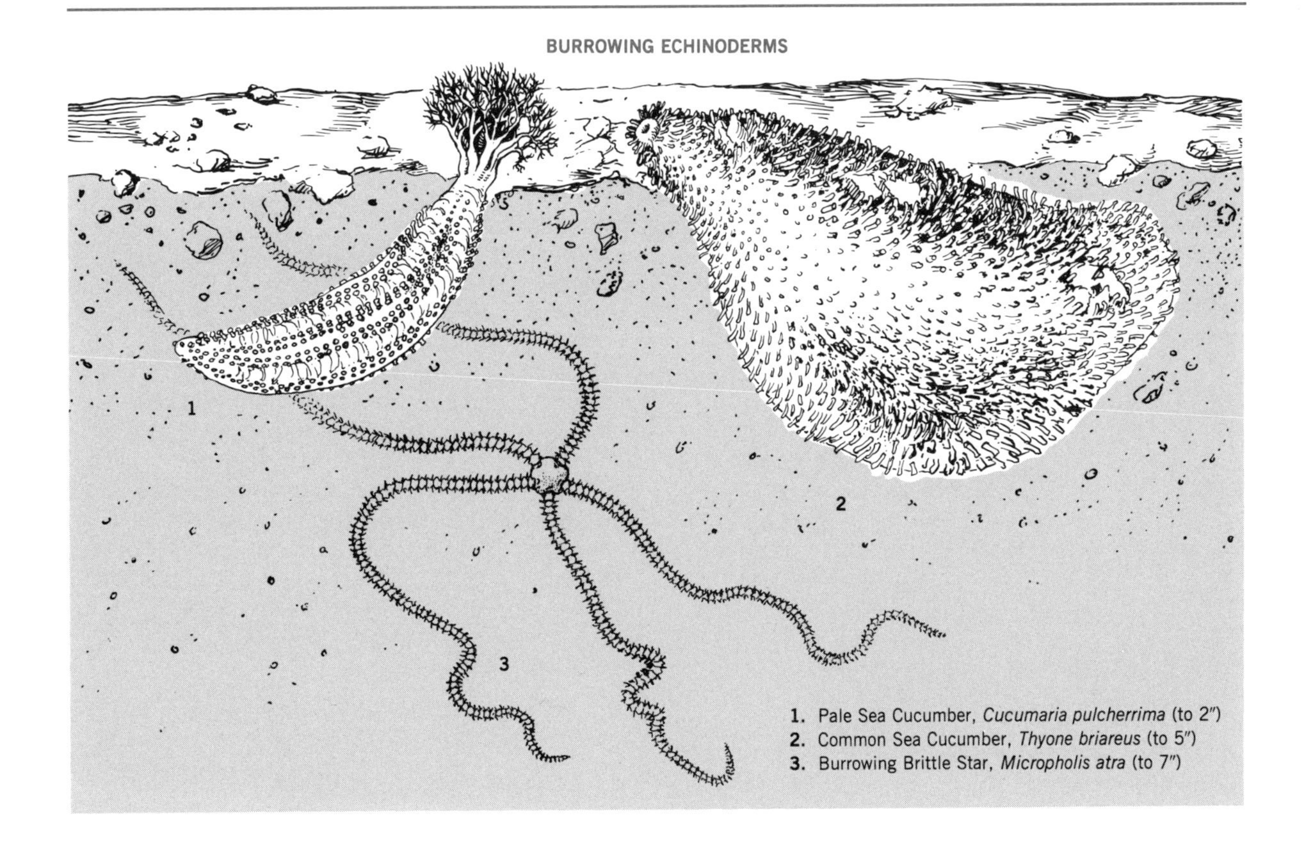

Sea cucumbers are another unusual resident of deeper waters. In appearance they look nothing like brittle stars, yet they too are spiny-skinned animals and are characterized by a five-part radial symmetry. They are found only in higher salinities. Shaped like a squat cucumber, they are highly contractible and have a definite head and rear end. A series of 10 treelike tentacles, which can be retracted within the body, surround the mouth at one end. Sea cucumbers, like sea stars, have tube feet, which they use to creep over the surface or to bury themselves. They burrow by lying flat on the surface and removing sediment particles below their body with their tube feet. They eventually sink into the bottom with only their head and rear openings exposed. They sweep their tentacles over the surface to capture food. Sea cucumbers contract when picked up and eject a jet of water from their rear. If greatly disturbed, they may eject some of their innards—startling to anyone holding the animal at the time! In some instances, the sea cucumber can recover from this violent protest; in others, it dies. There are two species of sea cucumbers in the Chesapeake Bay: the common sea cucumber, *Thyone briareus,* and the pale sea cucumber, *Cucumaria pulcherrima.* The common sea cucumber has a fat, rounded body, which is studded with thickly scattered tube feet. It is quite large, up to four or five inches long, and is a dull brown in color. The pale sea cucumber is one to two inches long and white to pale yellow in color. Its tube feet are arranged in five distinct tracts along its body.

CREATURES OF THE OPEN WATERS

Schools of the brief squid, *Lolliguncula brevis,* often move into the Bay from the ocean, commonly moving upstream as far as Tangier Sound. This is a small squid, but it is similar to the squids sold as bait to ocean-shore an-

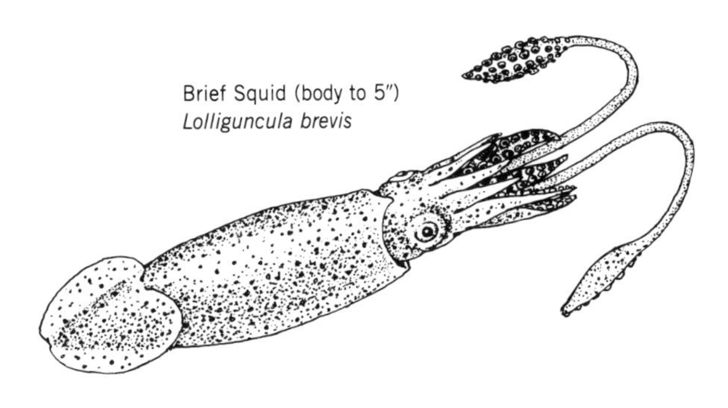
Brief Squid (body to 5″)
Lolliguncula brevis

glers or in the marketplace and known as calamari to Italians, Greeks, and Spaniards. Squid belong to an advanced group of mollusks with little resemblance to their progenitors. The only remnant of a shell is a transparent, horny pen, which lies internally and gives support to their soft bodies. Squid are agile and quick in the water, swimming in jet-propelled spurts. Water is drawn in through openings around the head and then forcefully squirted out through a siphon. They swim more rapidly backward but may move forward by directing the siphon to the rear. Squid have streamlined bodies, large, human-looking eyes, and long tentacles extending forward from their head. The tentacles are used for feeding and mating. They have a fantastic ability to change color quickly, and if one is fortunate enough to capture a live specimen, it is fascinating to see how this feat is accomplished. Dark spots just under the skin will appear like blinking stars. These stars are chromatophore cells, filled with pigment and equipped with a muscle system that can expand or contract to limit the amount of color exposed.

Arrow worms, *Sagitta* spp., are unusual pelagic ocean visitors to Bay waters. They are small, bizarre creatures of an isolated group, kin to no other kind of animal. Arrow worms have changed little from ancestors known to have lived 500 million years ago. The scientific name of the group is Chaetognatha, meaning bristle-jawed (recall that polychaete worms are many-bristled worms), referring to the fringe of stiff bristles on each jaw used as grasping weapons. Arrow worms are entirely transparent except for two black dots of eyes. They are less than an inch long, yet they are fierce predators willing to attack larval fishes almost their own size. They hover motionless in the water waiting for some luckless prey and then quickly lurch forward by rapidly vibrating their tails.

FISHES OF THE OPEN WATERS

Fishes in deeper waters include pelagic schooling predator fishes, bottom-feeding fishes, and reef-type fishes, as well as small forage species. Pelagic fishes cruise through the Bay swimming in more-or-less concentrated schools. Some swim just under the surface, some swim near the bottom, some deep, and others move freely throughout the water column. Pelagic schooling fishes usually have streamlined bodies adapted for rapid swimming. Most are predators on smaller fishes and invertebrates, such as shrimps and squid. They are equipped with large, strong mouths with jutting lower jaws and sharp teeth. A fish's mouth often indicates its feeding habit. In contrast to the large mouths of pelagic predators, bottom-feeding fishes usually have either small mouths positioned under the snout to probe for food on or within the bottom or large mouths at the top of the head to snap upward at passing prey. Reef-type fishes are nibbler species, usually with thick-lipped, nipping-type mouths. Sunken wrecks, floating buoys, deep pilings, and the like are the "reefs" of the Chesapeake Bay.

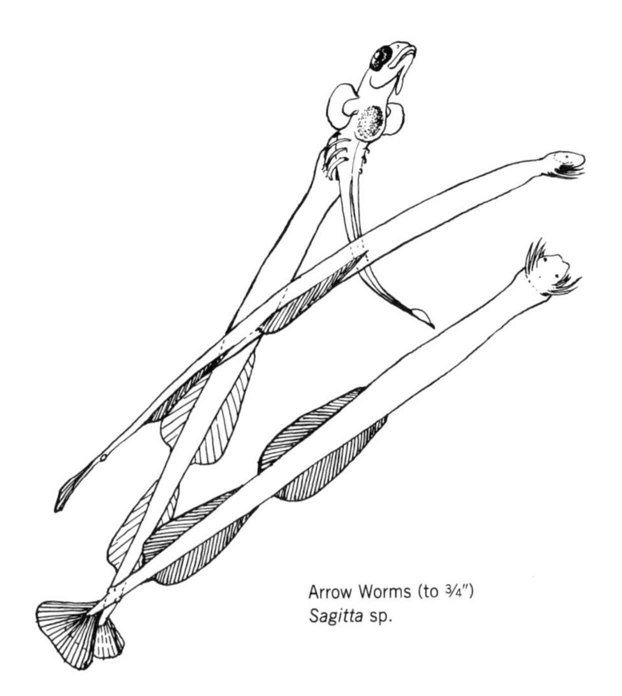
Arrow Worms (to ¾″)
Sagitta sp.

Marauding bluefish, *Pomatomous saltatrix* (to 3 feet), feeding on menhaden.

Pelagic Fishes

Most of the fishes seen along the shallow shorelines as juveniles are found as adults schooling in deeper, open waters. The striped bass, or rockfish, is considered the king of Chesapeake Bay fishes by many. It is a noble game fish, a delicious food, and the prized target for commercial and sport fishermen. Its close relative, the white perch, is also found in the deeper waters, particularly during winter.

Schools of adult shads and herrings move upstream through deeper waters on their annual spawning runs. Adults are present in the Bay for a relatively short period of time, returning again to sea soon after spawning. The juveniles, as we have discussed, remain in the shallows throughout the summer.

At times, large eels are so thickly congregated in deeper waters that a single trawl haul by research vessels has covered the deck with hundreds of slithering creatures. Great numbers are often encountered in the autumn, when sexually mature eels migrate down the Bay to the ocean to spawn. Adult spot, croaker, silver perch, needlefishes, and other species of the shallows are also typical residents of deeper waters. Schools of anchovies and menhaden are abundant here. These two forage species provide a major food source for most of the large predator fishes, such as striped bass, bluefish, and seatrouts. Most of the large Bay fishes are marine species that enter the Bay in spring and summer and return to the ocean in autumn. Some marine fishes are regular visitors, others rare stragglers; we have included only the commoner species.

Bluefish. Bluefish, *Pomatomus saltatrix,* are perhaps the most voracious predators in the Chesapeake Bay, killing just for the sake of killing even when they have had their fill. A good look at the large, strong jaws and sharp teeth of a bluefish tells you that it is well equipped for its predatory life. You might wonder, however, where the name comes from, since these fish are more sea-green than blue in color. Schools of large adult bluefish and smaller juveniles called snappers or tailer blues enter the Bay from the ocean in spring. In years when populations are large, they move well upstream into low-salinity waters; when populations are smaller, schools do not move so far into the Bay. Historically, bluefish have fluctuated widely. Some years they are extremely abundant in all size classes, ranging from the small snapper up to 20 pounds or so; at other times they are much less abundant and widely scattered throughout the Bay. By late fall, bluefish leave the Bay, moving offshore and southward for the winter.

Seatrouts and Red Drum. Seatrouts and red drum are members of the drum family, more adapted to a pelagic foraging existence than to the bottom-feeding habits of their fellow drums, such as spot, croakers, and black drum. Seatrouts vie with bluefish and striped bass as the major predators on schools of menhaden, anchovies, and other small fishes. They are, in turn, as popular with sport fishermen as the other two species. Two species of seatrouts are common in the Bay: the weakfish, *Cynoscion regalis,* also commonly called gray seatrout or squeteague; and the spotted seatrout, *Cynoscion nebulosus.* Weakfish are so named because their mouths are fragile and easily torn when hooked. Seatrouts are streamlined, fast-swimming fish with large mouths. Since they are not bottom-feeders, they have no sensory chin barbels, as do others of their family. Both species are dark olive green above with an iridescent play of colors on their sides. The weakfish is marked with dusky, irregular lines; the spotted seatrout has dark spots scattered over its back and dorsal fin. Seatrouts, like all drums, enter the Bay in spring, moving into mid-Bay waters, sometimes as far upstream as Annapolis. They spawn near the Bay's mouth, and the larvae and juveniles are carried by deep-water currents upstream to brackish-water nursery areas.

Red drum, *Sciaenops ocellatus,* or channel bass, as they are popularly named, are one of the largest of the drums. They may reach 50 pounds or more. Red drum are coppery colored, with one or more black spots at the base of the tail. They travel in surface schools and can often be seen breaking the water. Young red drum are found in the shallows and move up into mid-Bay regions, whereas adults are usually confined to the lower Bay. They are a popular sport fish.

Black Drum. The granddaddy of all Chesapeake Bay drums is the huge, lumbering black drum, *Pogonias cromis.* It is a bottom-feeding fish, like most of its family, with a large, underslung mouth lined with crushing teeth adapted for feeding on oysters, clams, and other mollusks. Its chin is lined with a fringed beard of barbels. The scientific name for this species is particularly appropriate. *Pogonias* means "bearded" and *cromis* means "to grunt" or "to croak." Black drums are not really black, but silvery with a brassy sheen; their fins, however, are usually blackish. Black drums are one of the largest fishes in the Bay. Fish of 100 pounds are caught, and 50-pounders are not uncommon. Schools enter the Bay in spring and summer

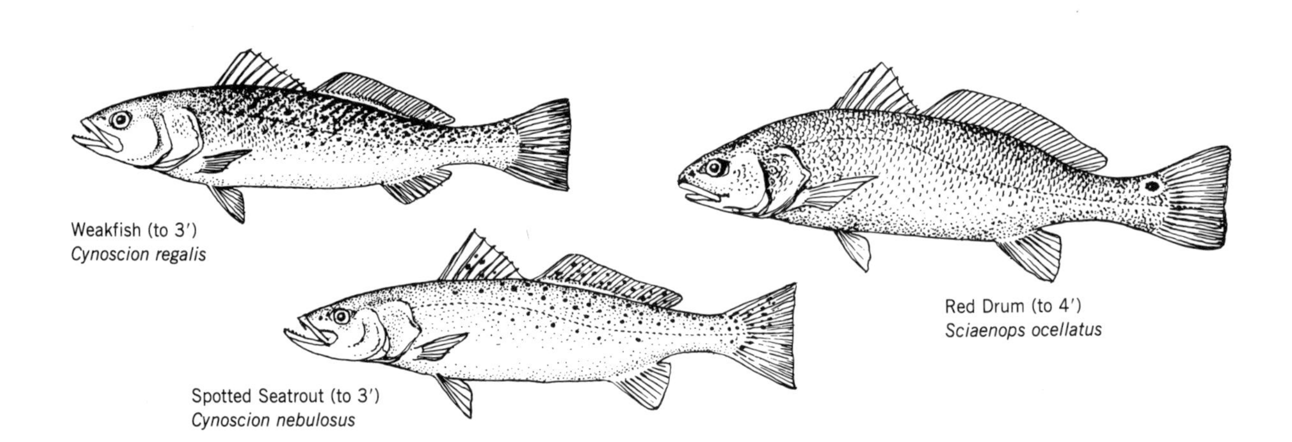

Weakfish (to 3′)
Cynoscion regalis

Spotted Seatrout (to 3′)
Cynoscion nebulosus

Red Drum (to 4′)
Sciaenops ocellatus

A young black drum, or puppy drum, *Pogonias cromis,* feeding on an oyster bar.

and move up into lower Maryland waters. The larger black drums are concentrated in deeper channel waters, often on clam or oyster beds or around wrecks or rock piles. Black drums of less than eight pounds, still good-sized fish, are called puppy drums. They move closer to shore and are marked with black vertical bars, which fade as the fish mature.

Jacks and Pompanos. Jacks and pompanos belong to the same family and are characterized by deeply forked tails on a very narrow tail base. They are plentiful sport and food fishes in more southern waters, but they occur only sporadically in the Chesapeake. Eighteen different species of this family have been recorded in the Bay, but most of these are rare stragglers. A few species, however, are quite numerous at certain times, particularly in the lower third of the Bay. These include the blue runner, *Caranx crysos,* crevalle jack, *Caranx hippos,* lookdown, *Selene vomer,* and the Florida pompano, *Trachinotus carolinus.* All are laterally compressed, silvery blue-green fishes. They are schooling, pelagic predators; the older and larger of each species are found in deeper waters, while the younger tend to move shoreward.

The blue runner and the crevalle jack look so much alike that they are often mistaken for the same species. They have large mouths with jutting underjaws, scimitar-shaped pectoral fins, and a row of scutes (keels with knife-sharp edges) running forward on each side of the tail. Fishermen use their names interchangeably and also call them Jenny Linds or rudderfish, which increases the confusion. They can be distinguished by certain key characteristics. Blue runners are slenderer than crevalle jacks, with more pointed heads and somewhat smaller mouths; the breast (the area just under their gill flaps) is fully scaled and the row of scutes extends well toward the head, with 38 to 45 scutes. Crevalle jacks are more truncate in shape with a blunt head and proportionately larger mouth; the breast is unscaled and the row of scutes is short, numbering only 26 to 30 scutes. Although both of these jacks occasionally move into the upper Bay, populations are commoner in the lower Bay.

The lookdown is a somewhat odd-looking fish with a deep, thin, sharp-edged body and a very low mouth positioned against a very high-placed eye. The fin ray at the front of both the second dorsal and anal fins is quite elongated. Young lookdowns also sport long filaments from the first dorsal and pelvic fins which are lost as the fish grows older.

The Florida pompano differs from the jack by having a more ovate, short body, a thicker tail base, small pectoral fins, a small blunt head, and a smaller, more inferior mouth. This species is at times plentiful in the Bay and moves well up into it. Schools of small Florida pompanos are not uncommon along the shores as far upstream as the Patuxent River, in Maryland.

Harvestfish and Butterfish. Harvestfish and butterfish are small, flat, round fish with deeply forked tails on a nar-

JACKS AND POMPANOS OF THE CHESAPEAKE BAY

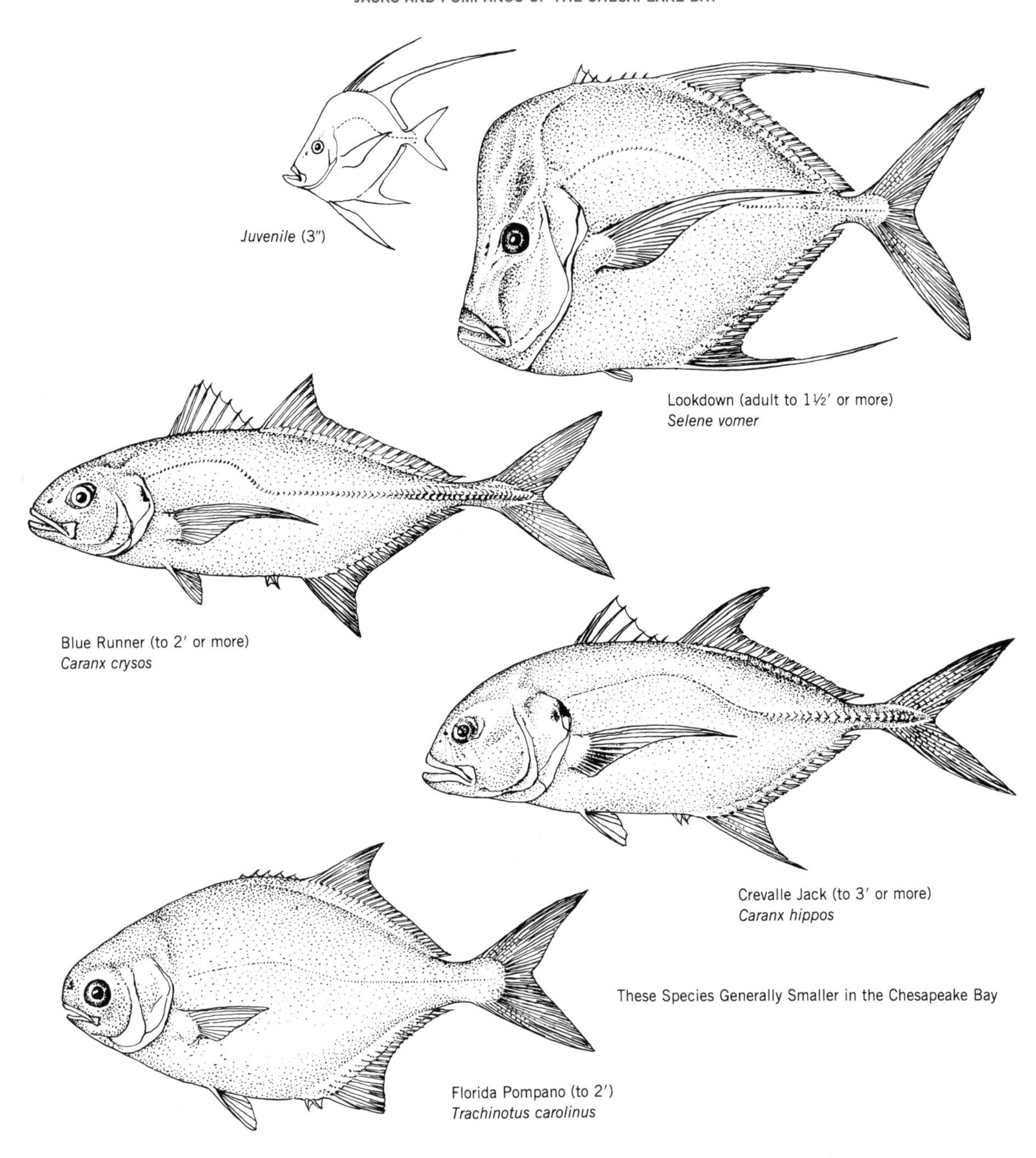

Lookdown (adult to 1½′ or more)
Selene vomer

Blue Runner (to 2′ or more)
Caranx crysos

Crevalle Jack (to 3′ or more)
Caranx hippos

Florida Pompano (to 2′)
Trachinotus carolinus

row base, like the jacks, but with small rounded heads, blunt noses, and without keels on their tail. They prey upon small pelagic fishes, shrimps, and squid in spite of having small, weakly toothed, mouths—somewhat out of character for a carnivorous predator species. Their scales are small and slough off easily when the fish is handled. Young harvestfish and butterfish are the fishes that are so commonly seen flashing in and out among the tentacles of sea nettles and other jellyfishes. They are known to eat both jellyfishes and comb jellies, but it is uncertain how much nourishment they get from this watery feast. Both are common commercial fish species of the Atlantic coast which move into the Bay in warmer weather. Harvestfish, *Peprilus alepidotus,* is the more widespread species in the Chesapeake, moving upstream into low-brackish waters. They are deeper-bodied than butterfish and are often referred to as silver dollars. Their dorsal and anal fins are high and curved in front, distinguishing them from butterfish, which have lower fins. Butterfish, *Peprilus triacanthus,* not as silvery as harvestfish, are leaden blue above and pale on the sides with dark splotches. Butterfish are more abundant toward the mouth of the Bay, where, in some years, the commercial harvest is substantial.

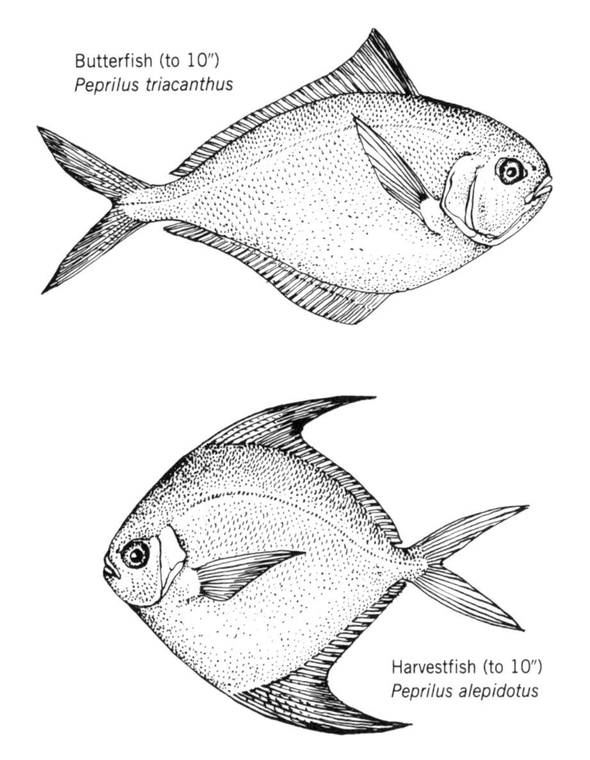

Butterfish (to 10″)
Peprilus triacanthus

Harvestfish (to 10″)
Peprilus alepidotus

Tunas. Tunas, albacore, bonitos, and mackerels all belong to the same family; all are characterized by sleek, streamlined bodies, deeply forked tails on a narrow tail base, and a series of finlets located behind the dorsal and anal fins. They are fast-swimming oceanic wanderers. A number of mackerels and tuna species forage into the Bay, but most are mere accidentals. At times, however, the Bay

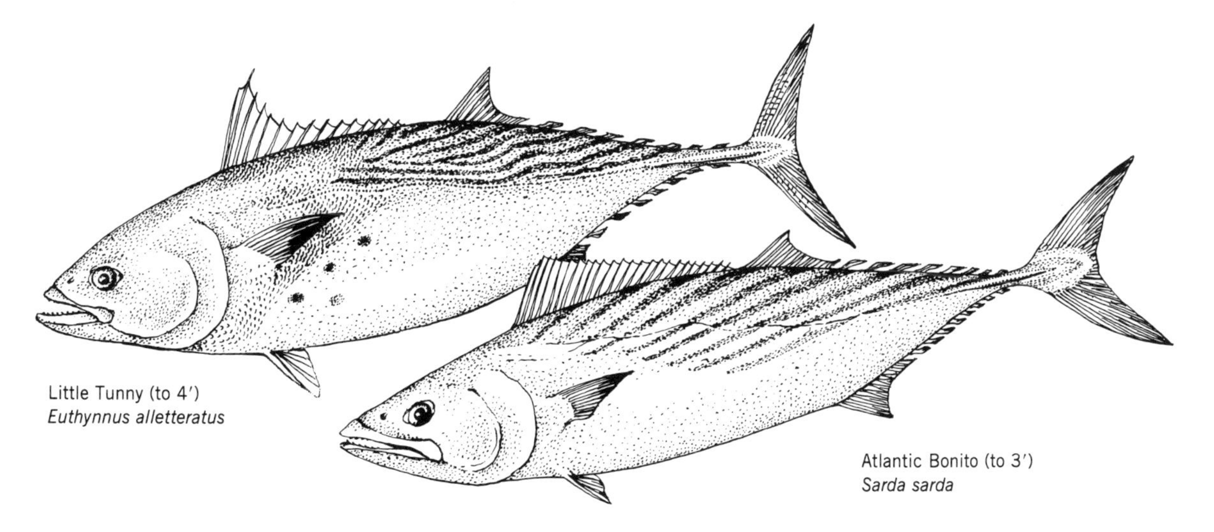

Little Tunny (to 4′)
Euthynnus alletteratus

Atlantic Bonito (to 3′)
Sarda sarda

angler is rewarded with a catch of one of these hard-fighting sport fish. The chances are it will be one of the small tunalike species, the little tunny, *Euthynnus alletteratus,* or the Atlantic bonito, *Sarda sarda.* Both of these species are typically tuna-shaped—streamlined and deeper bodied than mackerels. They are both steely blue on top, white below, and marked with darker blue bands. The little tunny has oblique wavy stripes, which do not extend down over the lateral line, while those of the Atlantic bonito are straight bands, which extend farther forward and across the lateral line.

Bottom Fishes

Bottom-oriented fishes tend to be solitary, although some are schooling species. They move slowly over the bottom searching for errant worms and small crustaceans or agitating the bottom sediment to uncover clams. Some bury themselves in the bottom, where they lay hidden, waiting for food to come to them. All of the flatfishes, including winter flounder, summer flounder, hogchoker, blackcheek tonguefish, and windowpane, are found in deeper waters as well as in the shallows.

Kingfishes. Kingfish, or whitings, as they are also called, are highly prized sport fish of the lower Bay. They are members of the drum family and are typical bottom-feeders concentrating over sandy bottoms along the edge of channels or over sand bars. There are two species in the Chesapeake Bay: southern kingfish, *Menticirrhus americanus,* and northern kingfish, *Menticirrhus saxatilis,* so alike that most anglers consider them to be the same species. Northern kingfish can be identified by an elongated filamentous spine on the first dorsal fin. Both species have a series of oblique bars along the side; the foremost bar is V-shaped only in northern kingfish. Both have a single chin barbel and an overhanging mouth.

Hakes. Two other bottom-feeding species with a single chin barbel frequent the same areas as the kingfishes. These are the red hake, *Urophycis chuss,* also called ling or squirrel hake, and the spotted hake, *Urophycis regia.* Hakes are related to the cods of the North Atlantic seas; the large cod and silver hake or whiting that are important commercial fishing species in northern waters are rare stragglers into the Bay. The two small hakes or lings of the Bay are generally only a foot long, with elongated bodies, soft fins, short first dorsal fins, and very long second dorsal and anal fins. Each pelvic fin is elongated into two filaments attached just below the gill flap. Both are brownish above and whitish below. The red hake has an extended filament on the first dorsal fin. The spotted hake has an identifying row of white spots connected with a black line along each side and a white border on the back side of the first dorsal fin.

Lizardfish. Inshore lizardfish, *Synodus foetens,* are unusual fish that usually mystify those who occasionally capture them. They are cigar-shaped, with broad, flattened heads and huge mouths, which open well beyond the eye. Their jaws are plated with numerous sharp teeth. They are relatively small (to one foot long) and are olive brown to gray. Young lizardfish have a pattern of dark crossbars over the back. A lizardfish "sits up" on sandy bottoms by propping itself up on its pelvic fins to feed on small fish and crustaceans. It remains still until it spies a passing prey

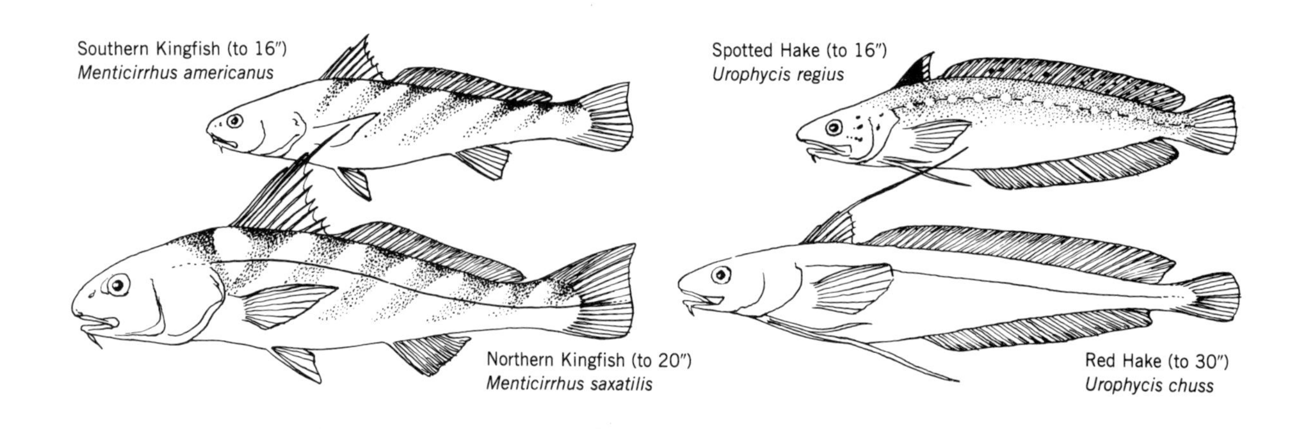

and then quickly darts up and engulfs it in one bite. Lizardfish also bury themselves, leaving only their eyes exposed; their eyes are positioned close together on the top of their head, allowing them to do this with ease. They may be found in the shallows as well as in deeper waters, and they range up into the mid-Bay.

Puffers. Puffers or swellfish fascinate anyone who has hauled one onto a boat and watched it quickly swell into an inflated ball. There are two kinds of puffers in the Chesapeake Bay, the commoner northern puffer, *Sphoeroides maculatus,* and the striped burrfish, *Chilomycterus schoepfi.* Northern puffers, when in their usual deflated state, are small, club-shaped fish covered with prickles. They have tiny beaked mouths, small slits for gill openings, and small dorsal and anal fins set far back near the tail. Their body is not well adapted for efficient swimming, which they accomplish by rapid wagging of the rear fins. They inflate themselves with air or water as protection. An inflated fish thrown back into the water will float upside down at the surface for a while, but it can deflate rapidly and soon be away. Puffers have the reputation of being poisonous. Some members of this large family are indeed poisonous, but not the northern puffer, which, in spite of its peculiar appearance, is a delicious and delicately flavored food. Puffers are colorful fish, yellowish with deep greenish blue bars on the side, dark above and pure white on the belly. Burrfish look like little thorny toadfish with short, round bodies. They too are colorfully marked with dark brown or black wavy stripes and spots on a yellowish green body. Burrfish are covered with long, sharp spines, which stick out menacingly when the fish swells. Burrfish are inefficient swimmers, but they help themselves along by jet propulsion, squirting water out of their small, round gill openings.

Stargazers. Northern stargazers, *Astroscopus guttatus,* are not common in the Bay—or anywhere within their range—but they are one of the most unusual species in the Bay. Picking up a live stargazer by the head will provide a memorable experience. Stargazers have electric organs on the top of their heads, and they can give a perceptible shock. Stargazers are very sedentary and solitary and are probably more numerous than is believed because of their secretive behavior. They are prehistoric-looking creatures, with a heavy, trunklike body and a large flat-topped head with an upward opening fringed mouth and large pectoral fins, which they use to dig themselves into the bottom, leaving only their eyes and mouth exposed. The electric organs are smooth, flat, squarish plates located just behind the eyes. The stargazer's scientific name is easy to remember; *Astro* means "star" and *scopus* means "to

Unusual fishes of deeper waters include, on the left, striped burrfish, *Chilomycterus schoepfi* (to 10 inches), one uninflated, the other fully inflated; the northern puffer, *Sphoeroides maculatus* (to 14 inches), swimming partially inflated above the inshore lizardfish, *Synodus foetens* (to 1 foot), one propped up on its pelvic fins, the other buried in the background, both waiting for a meal.

Strange-looking bottom fish of deeper waters include northern stargazers, *Astroscopus guttatus* (to 22 inches), reposing on the bottom and buried in the sediments, and a northern searobin, *Prionotus carolinus* (to 12 inches), creeping along on its fingerlike pelvic fin rays.

watch." *Guttatus* means "speckled," referring to the many small white spots on the upper body.

Searobin. Searobins are strictly bottom-dwellers that walk about with the fingerlike rays of their pelvic fins helped along by huge pectoral fins, which fan out from the side. Searobins have large, sloping, spine-covered heads and a body that tapers back to the tail. They root out crabs, worms, and mollusks with their broad-nosed snouts.

The northern searobin, *Prionotus carolinus,* is the common Chesapeake Bay species. It is a rather small fish, usually less than one foot long in the Bay. Perhaps its most unmistakable feature is its peacock-blue eyes. Northern searobins are usually found over deeper flats and channels and have been collected upstream at least to the Potomac River. Other species of searobins occur in the Bay, but they are rare.

Sturgeons

Huge, lumbering Atlantic sturgeon, *Acipenser oxyrhynchus,* used to be plentiful throughout the Chesapeake Bay and all its tributaries, but now they are seriously threatened here as they are all along the Atlantic coast. The sturgeon fishery was an important industry from colonial times until about the turn of the century, when the numbers of sturgeons declined precipitously. Before then, hundreds of thousands of pounds had been harvested annually in the Bay. Sturgeon meat was smoked and the black roe made into caviar, much of which was shipped to Europe. These large, sluggish fish were easily exploited. Their decline has been attributed to overfishing, to deteriorating water quality, and to the damming of rivers and streams, which prevents them from migrating upstream to spawning areas. Periodically, a huge sturgeon will lumber into a fisherman's net, an event that in these days usually warrants a picture in the local newspaper.

Sturgeons are prehistoric fishes. Their family has been in existence since the Cretaceous period. The sturgeon's body is covered with five rows of hard bony plates or shields: one row along the center of the back, one row along each side, and two rows along the belly. These plates are sharp-edged in young sturgeon, but they become blunt with age. Bony plates also cover the head. A soft, protractile, toothless mouth is located on the underside of an extended snout, and four sensory barbels project in front of the mouth. Sturgeons are bottom fishes that feed on mollusks, worms, and other bottom organisms. They root in the

Atlantic Sturgeon
Acipenser oxyrhynchus

mud or sand like a pig, nosing up worms and clams and sucking them into their mouth. The Atlantic sturgeon grows to a formidable size; the largest recorded catch was 14 feet long and weighed 811 pounds, sturgeon six feet or more are not uncommon even today. Atlantic sturgeon are anadromous fishes and must ascend the rivers to spawn in fresh or low-brackish waters. Mature adults migrate from the ocean into the Chesapeake Bay in April. These are huge fish, since they do not become sexually mature until they are at least 10 years old and weigh well over 150 pounds. Juvenile sturgeon, after hatching, remain within the estuary for up to 5 years, by which time they may be three feet long; then they gradually descend to sea.

A smaller sturgeon, the shortnose sturgeon, *Acipenser brevirostrum,* is so rare in the Chesapeake Bay and all along the Atlantic coast that it has officially been declared an endangered species in the United States. Shortnose sturgeon differ little in appearance from Atlantic sturgeon except for a somewhat shorter and broader snout and for dorsal plates that are separated from each other rather than meeting or overlapping.

Fishes Frequenting Buoys, Rock Piles, and Wrecks

Tautog. Tautog, *Tautoga onitis,* is a northern relative of the wrasses of tropical seas. They are regular summer visitors to the lower Chesapeake Bay, but are far more abundant in more northern bays, where they are often called blackfish. Tautogs are heavy, stout fish with thick tails, blunt heads, and a long, spiny dorsal fin. They are speckled dark gray to brown, the older fish being almost all black. Tautogs have small mouths with stout teeth, which they use to crush mussels, barnacles, small crabs, and all other crustaceans they find on wrecks, rock piles, or pilings. They apparently have a habit, when not feeding, of seeking out a hole and lying inert on their side. Tautogs in the Chesapeake are not large—about a foot long—but are a favored sport fish for those who know their habits.

Porgies. Two representatives of this family of deep-bodied, slab-sided fish are fairly common to the lower Bay. Porgies are a large family of primarily tropical species with a few northern strays represented in the Chesapeake by scup, *Stenotomus chrysops,* and sheepshead, *Archosargus probatocephalus.* They have small mouths with strong jaws and teeth that are used to crush the shells of crustaceans and mollusks. Scup congregate in small schools and nibble at barnacles, mussels, and small crabs near the bottom and around pilings and wrecks. Scup also move over shallow, inshore bottoms in search of food. Sheepshead have similar habits, but are more solitary. Scup, in the Bay also called porgy, maiden, and fairmaid, are rather plain-looking fish—dull silver with 12 to 15 indistinct vertical stripes, flecked with light blue on their sides. Their head is proportionately small, and the tail is markedly crescent-shaped with sharply pointed tips. Scup are small fish, generally under a foot in length. Sheepshead are far more distinctive, with black or dark brown "prison-stripes" on a grayish to yellowish background. The bands tend to

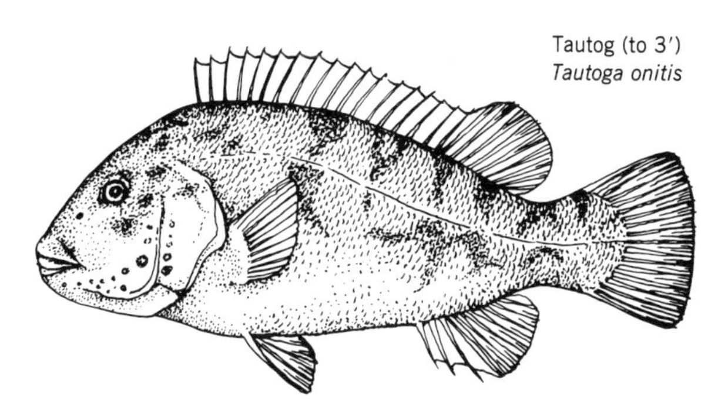
Tautog (to 3′)
Tautoga onitis

Planehead Filefish (to 10″)
Monacanthus hispidus

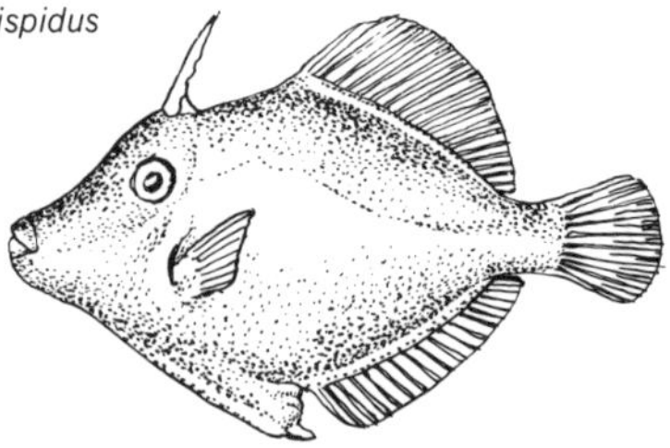

Orange Filefish (to 2′)
Aluterus schoepfi

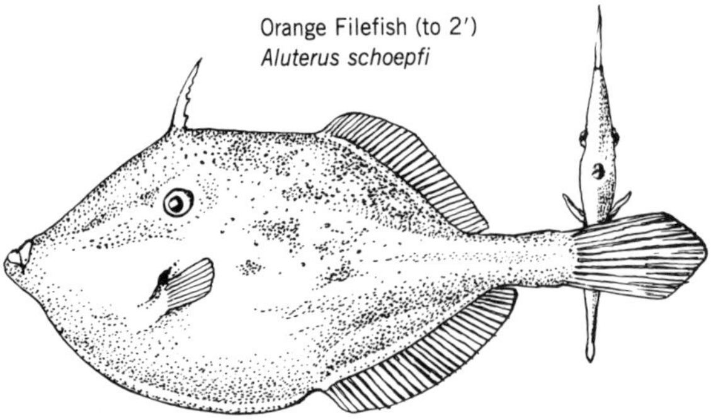

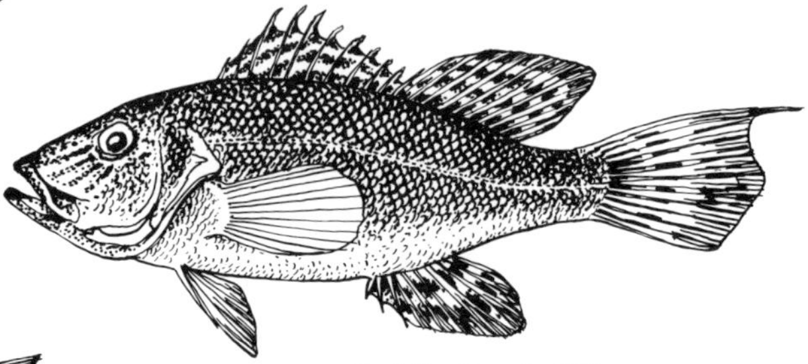

Black Sea Bass (to 12″)
Centropristis striata

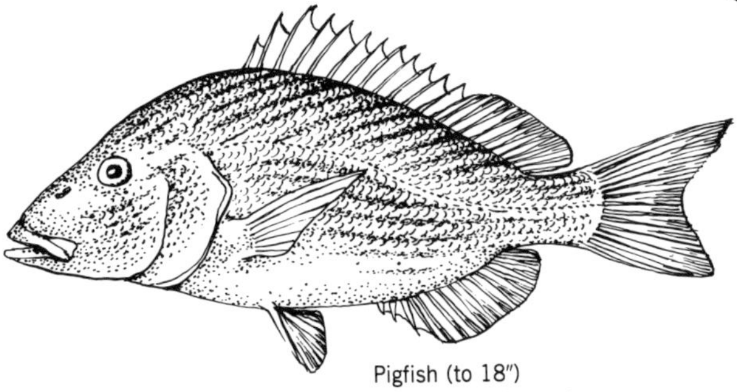

Pigfish (to 18″)
Orthopristis chrysoptera

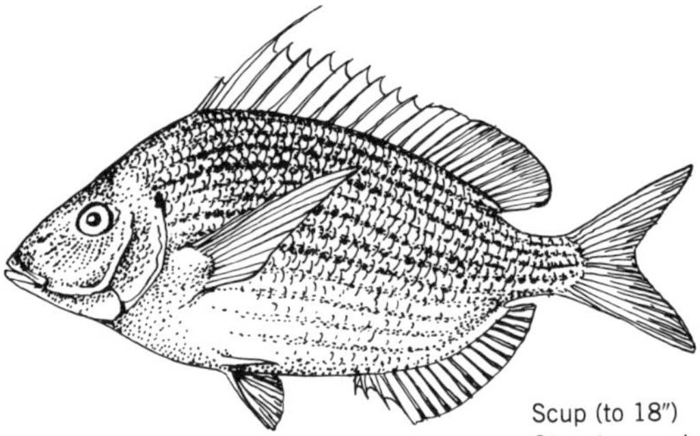

Scup (to 18″)
Stenotomus chrysops

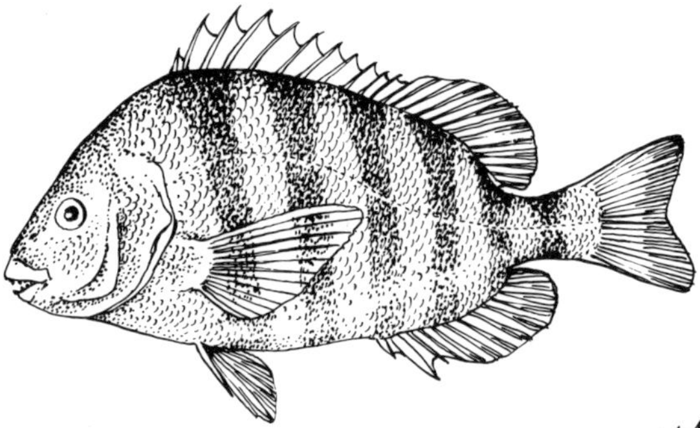

Sheepshead (to 20″)
Archosargus probatocephalus

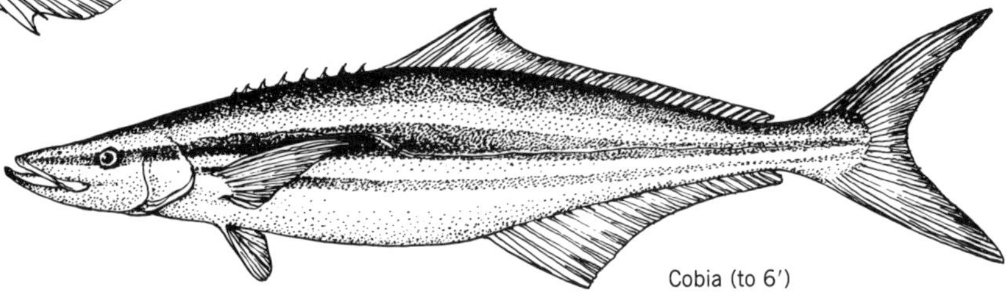

Cobia (to 6′)
Rachycentron canadum

become less conspicuous with age. Sheepshead are larger than scup and may exceed 20 pounds, although those in the Bay are usually considerably smaller.

Pigfish. Pigfish, *Orthopristis chrysoptera,* are similar in shape and size to scup and are often found in the same places. In contrast to the drab little scup, pigfish are splendidly colored, with dark blue to purplish backs marked with brilliant golden bars and spots over the head and sides. They have more pointed snouts and a forked tail with rounded tips. Pigfish belong to the tropical family of grunts, fishes characterized by small mouths with fewer and weaker teeth than porgies. They make grunting sounds, in and out of the water, by rubbing their teeth together. The sound is amplified by the adjacent air bladder, a mechanism different from that of the drums, which produce their sound by directly vibrating the air bladder.

Black Sea Bass. Black sea bass, *Centropristis striata,* is another solitary fish that hovers around pilings and wrecks of the lower Bay. Some few may stray upstream into Maryland waters. They are relatively small (usually to 12 inches) but robust fish with large mouths well armed with teeth. The black sea bass has a long dorsal fin with fleshy tips at the end of the spines and a round caudal fin with the top ray extended into a filament in larger fish. Their striking coloration immediately identifies them. They are deep blue-black with an intense lighter blue color in the center of each black-bordered scale, which gives the appearance of bluish horizontal stripes. This same blue color is streaked below the eyes. The dorsal fin is also striped. The brilliant blue color fades quickly after a black sea bass has been taken from the water, and the bass appears more or less as a white and black fish. Females, too, tend to be overall gray-blue rather than blue-black in tone. Black sea bass take bait quickly and are a favored sport fish.

Filefishes. Filefishes are odd fish—flat as pancakes, with a keel-edged belly, high-set eyes, and tiny mouths with projecting teeth, which they use to nibble at algae and small invertebrates. The skin is tough, with minute but very hard scales. It is said that their skin was sometimes used as sandpaper—hence their name. Actually, the scales are so small that the skin feels rather soft to the touch rather than rough. The keeled belly is formed by a long, thin pelvic bone which can easily be felt with the fingers. Two filefishes move into the Chesapeake Bay in summer, the orange filefish, *Aluterus schoepfi,* and the planehead filefish, *Monacanthus hispidus.* Orange filefish are by far the more common species and travel well up into Maryland waters. They are frequent visitors to Tangier Sound. Orange filefish have a propensity for drifting below jellyfishes and occasionally nipping at their tentacles. They are somewhat dim-witted, and if released into the water after capture, they will bob around on their sides for some time before it occurs to them that they are free. Many people call them foolfish. They may be found over deeper bars and spits as well as around pilings and such. Orange filefish are the larger of the two species and will grow to two feet or more in length, but smaller individuals are more common. They are grayish with dusky blotches and most of them are spotted with orange and yellow.

Planehead filefish are smaller, reaching a maximum of 10 inches, and drabber than orange filefish. The dorsal spine is sharply barbed in contrast to a smooth dorsal spine in the orange filefish. The body is deeper, and the bony keel projects slightly out of the angle of the belly with a soft dewlap of flesh above it. Planehead filefish do not generally move as far into the Bay as orange filefish.

Cobia. The cobia, or crab-eater, *Rachycentron canadum,* is a large, fierce-fighting fish prized by sportsmen. Cobia are usually found in deeper waters and seem to prefer the shade around wrecks and floating buoys. They are caught as far up in the Bay as the mouth of the Potomac River and into Tangier Sound. There is only one species of this family, which is distributed throughout the world in warmer waters. Cobia of 50 to 70 pounds and three to four feet are not uncommon in the Chesapeake Bay. It is a streamlined, torpedo-shaped fish with a broad, flattened head and a large mouth with a projecting lower jaw. A row of eight to nine short, stiff, separated spines is located in front of a long dorsal fin. Cobia are dark black to gray on top, changing rather abruptly to grayish white on the belly. Young cobia have black and white longitudinal stripes along the body which gradually fade as they grow.

Sharks, Rays, and Skates

Small sandbar sharks, cownose rays, and bluntnose rays, so common over eelgrass beds, move into deeper waters as well. Other sharks, rays, and skates are regular visitors to the Bay. Sharks here are rarely the large ocean behemoths that are so often the objects of fearful curiosity. Some of the larger ocean species do appear in the Bay, but they are generally the younger and smaller of the species.

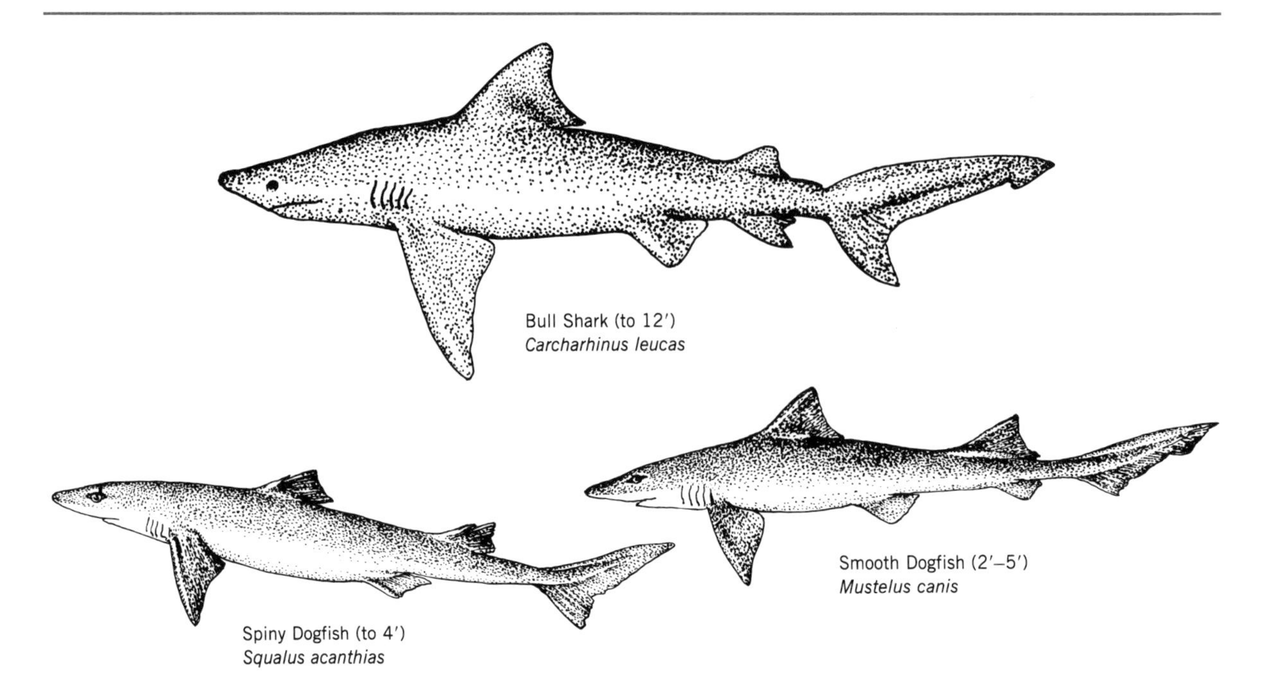

Bull Shark (to 12′)
Carcharhinus leucas

Smooth Dogfish (2′–5′)
Mustelus canis

Spiny Dogfish (to 4′)
Squalus acanthias

The bull shark, *Carcharhinus leucas,* is probably the one large shark that most often frequents the Bay. It has been captured by commercial fishermen well up the Bay into Maryland waters, near Annapolis and the mouth of the Chester River. Bull sharks grow to 12 feet, but those in the Bay are usually less than half that size. Sharks are a primitive fish, differing from the higher fishes in many ways. Major differences are in the structure of the skeleton, which is made of cartilage rather than bone, and the presence of four or five pairs of gill openings rather than a single pair, as in bony fishes. The bull shark is typically a heavy bodied, torpedo-shaped gray shark. Most sharks look much like one another and are difficult to identify. Oftentimes the differences are quite subtle, such as the relative placement of the fins or the length of the nose. Characteristics that distinguish bull sharks from other sharks are the absence of a ridge along the mid-back between the two dorsal fins and a broadly rounded snout. Small bull sharks resemble the sandbar shark, so common over seagrass meadows, but the latter has a distinctive ridge between the two dorsal fins.

Two of the commoner smaller sharks in the Chesapeake Bay are the smooth dogfish, *Mustelus canis,* and the spiny dogfish, *Squalus acanthias.* Both are usually only two to three feet long and travel in schools in deeper waters. Smooth dogfish are gray to brownish with dorsal fins of approximately the same size (rather than a larger first dorsal fin, as in bull sharks and sandbar sharks). Its most distinctive feature, however, is its flat, blunt teeth, which are unlike the sharp, pointed teeth of other sharks. Spiny dogfish are the easiest to recognize, as they have a stout spine in front of each dorsal fin and only one ventral fin (the others have two). Spiny dogfish are slate-colored; younger spiny dogfish of 14 inches or less have white spots scattered over their body.

Skates and rays are essentially wide and flattened sharks whose mouths and gill slits are on their underside. Skates have thickened tails and lay hard, purse-shaped egg capsules (the mermaid purses of the sand beach); most rays have whiplike tails armed with poisonous spines and they bear their young live. In addition to the bluntnose stingray and the cownose ray, two other species are fairly common in waters near the mouth of the Bay: the clearnose skate, *Raja eglanteria,* and the southern stingray, *Dasyatis americana.* Clearnose skates are quickly identified by two transparent patches on each side of a pointed nose. They have the typical heavy skate tail. The back is covered with prickles, and a line of sharp, short spines extends down the middle of the back and tail, which also has a row of spines on each side. Two fins are located at the end of the tail. The

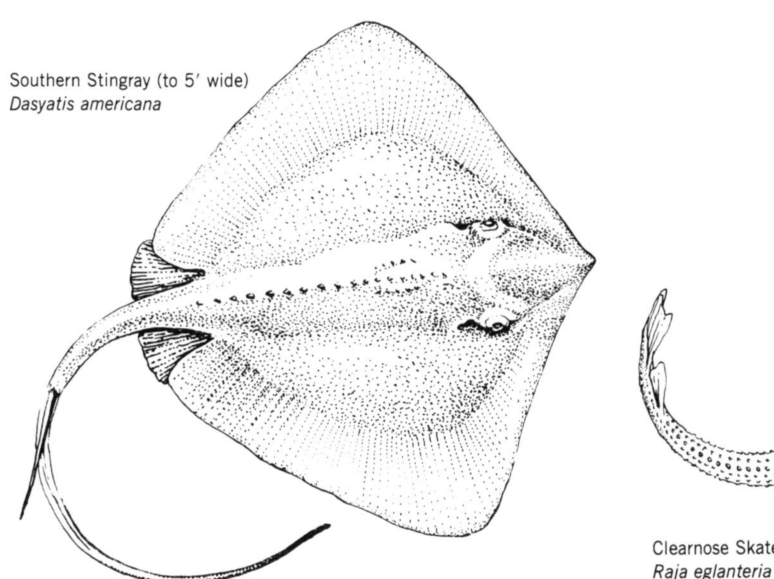

Southern Stingray (to 5′ wide)
Dasyatis americana

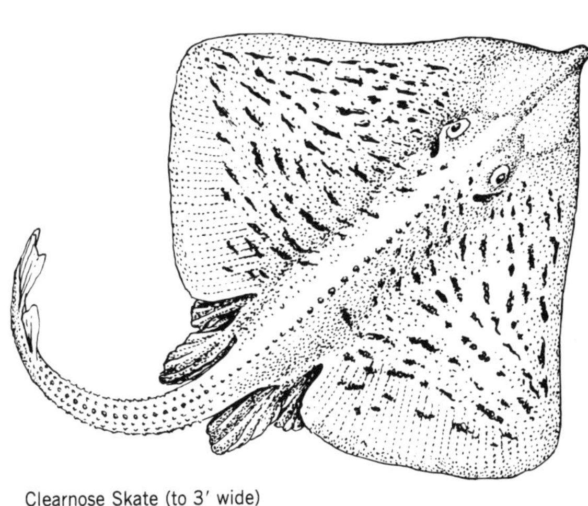

Clearnose Skate (to 3′ wide)
Raja eglanteria

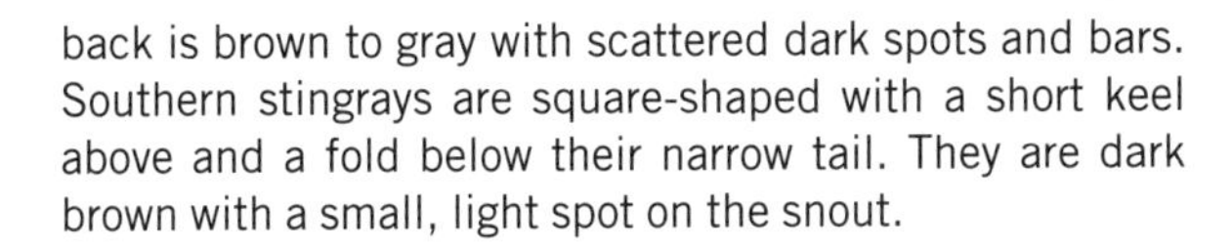

back is brown to gray with scattered dark spots and bars. Southern stingrays are square-shaped with a short keel above and a fold below their narrow tail. They are dark brown with a small, light spot on the snout.

MARINE MAMMALS AND SEA TURTLES OF THE OPEN WATERS

There are many other large and unusual creatures in the Chesapeake Bay and nearby coastal waters that are often unseen or unexpected. The great whales, such as humpbacks, sei, and minke whales, migrate along the coast in a northerly direction in the spring and south toward the West Indies in the fall. Occasionally one of these behemoths will stray into the shallow waters behind the barrier islands and become stranded on the sandy bottom, and almost every year in the Delaware and Chesapeake bays one whale, or perhaps several, will wander into the estuary feeding on fish and other organisms. In fact, humpback whales have been seen above the Chesapeake Bay Bridge near the mouth of the Chester River in recent years. The Atlantic bottlenosed dolphin, *Tursiops truncatus,* is a small toothed whale. Some call these small whales dolphins; others insist on calling them porpoises. There appears to be no clear-cut preference or definition among scientists who study these wonderful marine mammals, although some make a distinction based on tooth shape, considering those with conical or peglike teeth dolphins and those with spade-shaped teeth as porpoises. However, preference as to dolphin or porpoise seems to lie with common usage, which varies from area to area.

The bottlenosed dolphin is a summer inhabitant, particularly in the saltier waters of the lower Bay. They are often seen in small schools feeding in the swift currents near the Elizabeth and James rivers. But they are not confined to the lower Bay and are actually found throughout the Chesapeake—perhaps less predictably than near the mouth of the Bay, but here and there they suddenly appear, often in schools of twenty or more. Every summer bottlenosed dolphins are seen in the upper Chesapeake and tributaries such as the Miles River off St. Michaels, the Chester and Choptank rivers, and along the western shore near Annapolis and Baltimore. They have even been sighted far up the Potomac River near Washington, D.C.

Bottlenosed dolphins are the familiar dolphins seen in marine parks and aquariums. They are the great leapers that seem to have a perpetual smile that everyone loves. These very social animals are found all along the Atlantic Coast, sometimes in pods of a hundred or more but usually in smaller numbers, especially in near coastal waters and in the Bay. They can reach a length of about 12 feet and weigh up to 1,400 pounds, although they usually weigh between 300 and 400 pounds. They have beaklike snouts

and slender, streamlined bodies in contrast to what some refer to as a porpoise, which has a blunter snout and a relatively more stocky body form. They are usually dark or slate gray on the back, lighter on the sides, and somewhat lighter on the belly. Their teeth are used for grasping rather than shearing or biting. Since dolphins are mammals, they are air breathers and give birth to young which are nursed by the mother. The gestation period is 12 months.

Dolphins make puffing and hissing sounds as they exhale moisture-laden air through the single blowhole located on the top of the head. They have been observed sleeping in calm waters just below the surface; with regular, almost imperceptible movements of their powerful tails, they raise their heads above the water's surface and breathe. Bottlenosed dolphins feed on many species of fish and shrimp; in the Chesapeake Bay they prey on menhaden, catfish, eels, mullet, squid, shrimp, and crabs. They are known to eat between 12 and 15 pounds of fish a day in captivity.

One of the least known and unexpected animals found in the Chesapeake is the loggerhead turtle, *Caretta caretta.* Occasionally other sea turtles appear in the Bay as well; the huge leatherback, the world's largest sea turtle, which can weigh more than 1,400 pounds, has been seen north of the Chesapeake Bay Bridge off Kent Island as well as in other areas of the lower Bay. The much smaller Atlantic Ridley sea turtle also wanders in from the ocean to feed on fish, crabs, and mollusks. But it is the loggerhead turtle that regularly moves into the Bay to feed during the summer. Scientists estimate that several thousand loggerheads visit the Bay every year. Loggerheads concentrate in the lower Bay and its tributaries, but they are often seen by boaters in Maryland waters near Kent Island, in the Choptank River, and off Cove Point on the western shore. On languid summer days when the water is "cam as a dish" (local jargon for flat, calm waters), their large round heads are easy to spot as these massive turtles bask on the surface. They have reddish brown, heart-shaped shells, or carapaces, and paddle-like legs marked with yellow; the plastron, or lower shell, is creamy yellow. Loggerheads generally weigh 200 to 500 pounds or more; most of the turtles that visit the Bay weigh in excess of 100 pounds. They are omnivorous, feeding on everything from sponges and jellyfishes to mollusks, barnacles, crabs, and eelgrass.

Loggerheads are great wanderers and roam the Atlantic from Newfoundland to Argentina and even into the Mediterranean Sea. They are the most common sea turtle in the mid-Atlantic and Chesapeake Bay. They nest from late spring through summer on beaches in the Caribbean and from Florida to Virginia. Adults mate in shallow waters off nesting beaches; sometime after, the females lumber ashore, where they dig nest cavities with their hind legs and deposit approximately a hundred leathery-shelled eggs. They fill and cover the nests and return to the sea, often returning at a later time to dig a new nest and deposit more eggs. After about two months the eggs hatch. The young turtles instinctively head down the beach slope, attracted by the reflected moonlight or the rising sun, and enter the surf to swim out to sea. Sometimes they are confused by backlighting from cars and buildings along beach roads; hatchlings may lose their way as they move toward the artificial lights and become vulnerable to waiting predators.

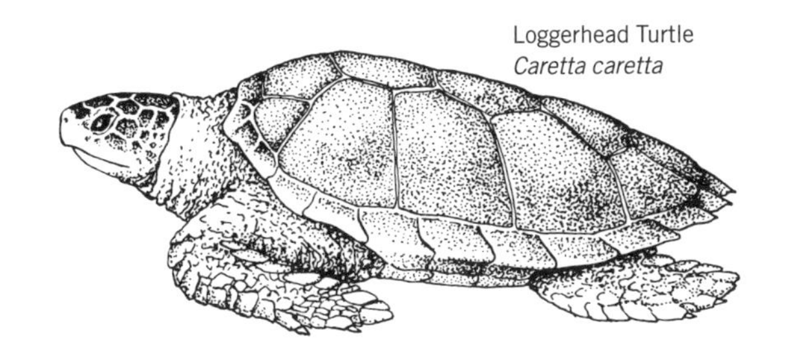

Loggerhead Turtle
Caretta caretta

Those that survive the hazards of the beach and sea, and relatively few do, may live 50 years or more.

BIRDS OF THE OPEN WATERS

The deeper, open waters of the Bay have no obvious physical indicators or barriers to movements by fish, crabs, marine mammals, sea turtles, or birds. Consequently many marine animals move through the open waters, often using this area as a corridor to move freely from one place to another. Sometimes they tarry and feed; yet, some of these animals are consistently found in open waters more often than in other parts of the Bay. Similarly, there are a group of related and unrelated birds, collectively called sea ducks, that typically are found in and over the open waters more often than in other habitats. There are also several species of seabirds: terns, the Bonaparte gull, and gannets, which are also part of this very visible open water assemblage.

The bufflehead, *Bucephala albeola,* the most effervescent and energetic of all the sea ducks, is also the smallest one. These small, chunky butterballs are constantly moving—never motionless, they scurry across the water's surface or bob underneath. The male shows mostly white below; he has a black back and a glossy, greenish black head strikingly marked with a white patch that roughly approximates the space from 12 to 3 on a clock's face. The patch, however, can change size as the crest is spread. The female is dark with lighter undersides and a white cheek patch below and behind the eye.

Most other diving ducks patter across the water and scramble for a takeoff. Not so the bufflehead—it flies straight up from the water like a puddle duck. They are abundant throughout the Bay in various salinities and into fresh water. They do not form huge flocks, but they do raft together in sizable numbers. Usually a sentry remains on alert as the others dive in search of food. Buffleheads prefer water depths of from 4 to 15 feet, where they feed on aquatic vegetation, seeds, mollusks, crustaceans, and fish. Their numbers build in October and peak in November as they arrive from their breeding grounds in Canada, Alaska, and the northern tier of the United States.

The common goldeneye, *Bucephala clangula,* is a close relative of the smaller bufflehead—in fact, both species nest in tree cavities hollowed out by flickers or other woodpeckers. From a distance, the male appears to be

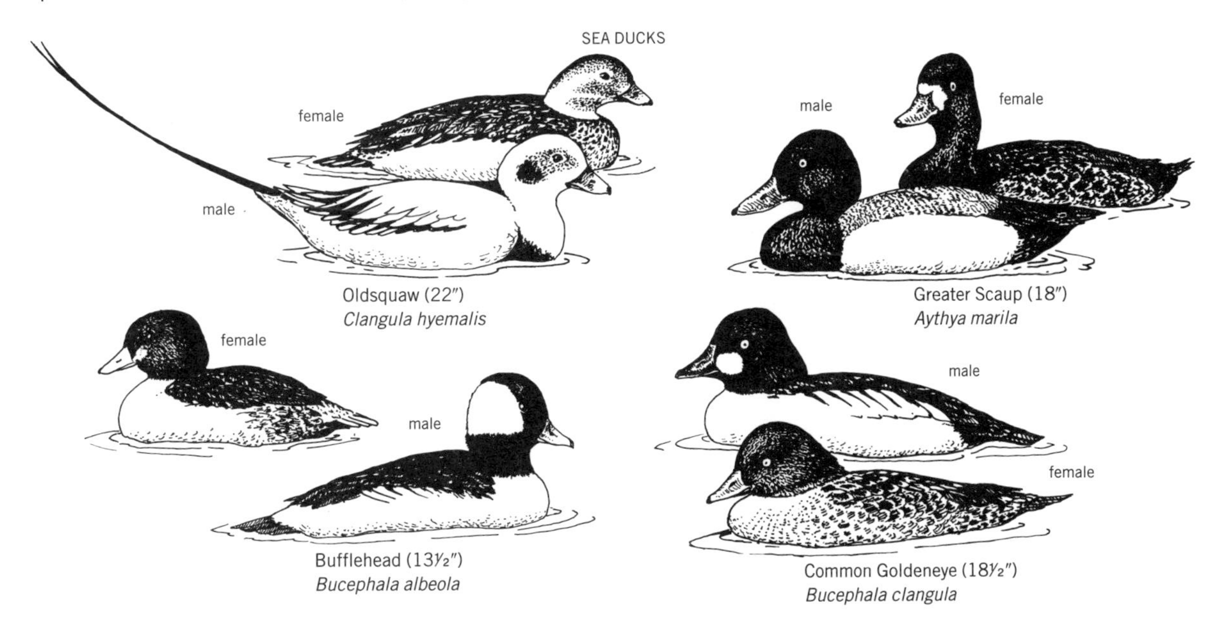

SEA DUCKS

Oldsquaw (22″)
Clangula hyemalis

Greater Scaup (18″)
Aythya marila

Bufflehead (13½″)
Bucephala albeola

Common Goldeneye (18½″)
Bucephala clangula

mostly white bodied with a black back and head. Close observation shows that its rounded head is a dark, glossy green with a characteristic round white patch between the base of the bill and the eye. Typically, the female has a more subdued coloration, usually chestnut brown, with a gray back and a white neck collar. In some locales goldeneyes are called whistlers for the piercing, singing sound their flight feathers make as they pass overhead. They are active and strong fliers and usually move singly or in small flocks.

The peak arrival of goldeneyes in the Bay area is between mid-November and mid-December. They are widely distributed throughout the Bay; they prefer brackish or salt water, although they also feed in fresh water. They seem to concentrate over water depths from 5 to 20 feet, where they feed on crabs, crayfish, water boatmen, the naiads of dragon- and damselflies, snails, clams, and vegetation. They are daytime feeders and raft up at night.

Oldsquaws, *Clangula hyemalis,* have the most seasonably variable plumage of the sea ducks. In fact, unlike most ducks, they do not always have a distinct pattern. Their coloration is a mixture of white, black, brown, and gray. Usually the sides are white or gray, with some white present on the head and neck. A good field mark for the male is its black breast and long, drawn-out tail feathers. The female has dark wings folded over the back, white sides, and a white neck; she lacks the long central tail feathers of the male. Oldsquaws are deep divers and have been reported to forage for food at depths of 50 to 100 feet and more; however, in the Chesapeake they normally dive to about 25 feet in order to reach food. They use their wings, rather than their feet, when swimming underwater and feed primarily on amphipods and other aquatic invertebrates, fish, and some vegetation.

Oldsquaws generally do not associate with other bird species; they may form compact flocks at times, but quite often they swim in small aggregations within a large, loose, almost undefined gathering of several hundred individuals. They are noisy, garrulous birds, constantly uttering various clucking and yodeling calls. Their scientific name means "noisy" and "wintery": they are certainly talkative, and they seem to be at home in the most exposed conditions when the temperatures are bitterly cold and the seas are at their roughest. They are constantly active—oldsquaws will quickly dive, bob to the surface some distance away, and then take off on a careering flight and land on the water close by. Oldsquaws arrive in the Bay in November and are abundant and widely distributed throughout the Chesapeake.

There is a group of primarily dark sea ducks that appear in the Bay every fall and winter in large numbers, sometimes forming flocks of a thousand or more. They are rapid fliers and, like most diving ducks, must patter across the water before becoming airborne. Once aloft, these rather large ducks fly in pulsating, threadlike patterns over the open waters of the Chesapeake Bay. In the summer, these ducks, called scoters and known to some in New England as coots, can be seen far off the ocean beaches, flying swiftly and low over the sea surface in loose, ever changing, stringy formations. They suddenly appear on the horizon and, wraithlike, disappear just as quickly. Except for hunters, who shoot them for sport, but not usually for the table, most do not know these abundant birds. These stocky ducks just do not have the personality and engaging behavioral traits of buffleheads or oldsquaws. Their coloration, for the most part, is somber and unpatterned, particularly when seen from a distance, which is usually the way one sees scoters.

There are three known species of scoters in North America; all three of these species migrate to the Bay and mid-Atlantic waters in the fall. They share some characteristics: all are dark, short-necked, and chunky. They also share the same scientific name, *Melanitta,* which refers to their black coloration. Similar to those of other diving ducks, their legs are set back close to the tail. They are capable of diving to depths often exceeding 40 feet, but in the Bay they usually feed in 25 feet of water or less. Their diets consist mostly of invertebrates, particularly razor clams, mussels, barnacles, crabs, mud snails, and tubeworms, as well as a small amount of vegetation.

The white-winged scoter, *Melanitta fusca,* is the most common scoter in the Bay and is also the largest, often weighing three pounds or more. The adult male is black or very dark brown with a commalike white mark below and behind the eye. They have white wing patches seen more often in flight than when the bird is on the water. The males have a yellow-orange bill with a knobby protuberance at its base. Female white-winged scoters are dark brown with a dark bill and white markings on the head and wings similar to those of the male.

The surf scoter, *Melanitta perspicillata,* also known as the skunk-head coot, is most often seen along the ocean coast feeding in the rough surf, although it is certainly seen in the Chesapeake, often in large numbers. The species name, *perspicillata,* comes from the Latin, meaning "spectacular" or "conspicuous." No doubt, the name was chosen because of the black and white markings on the male's

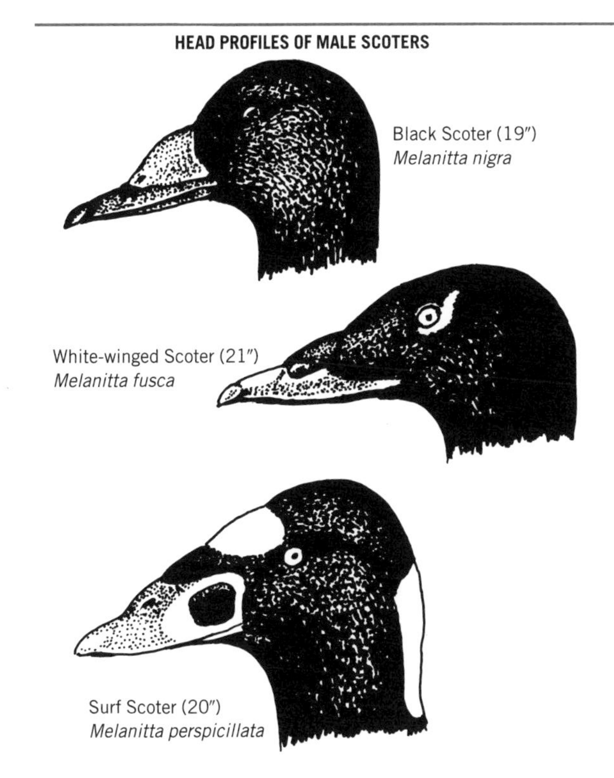

head ("skunk-head"), along with its massive orange, yellow, white, and black-spotted bill. The female is a drab brown with a darker crown and lighter face patches. The female's dark, unmarked bill is similar in size to that of the male.

The black scoter, *Melanitta nigra,* is the smallest of the scoters and is also the blackest of all American ducks. The male has a short black bill with a bright orange-yellow knob at the base, the only touch of color save for the subtle touch of silver on the undersides of its wings. The female is characteristically dark brown with a two-toned head and a dark bill. All three species of scoters arrive in the Bay area in peak numbers around late October to early November. They are creatures of the more open waters of the Bay and rivers, and are widely distributed throughout the Chesapeake.

The greater scaup, *Aythya marila,* and the lesser scaup, *Aythya affinis,* are also diving ducks and are often considered sea ducks. Known by local hunters as bluebills, scaup are typical divers and raft in great numbers in the open waters of the Bay. Lesser scaup are much more common in the Bay than greater scaup, which gather in large flocks in the ocean far from the shore. Adult male scaup, greater and lesser, appear essentially alike. They both have blue bills and a blackish head (the lesser's head is glossed with purple, the greater's with green); their breasts and rumps are black; and they have gray backs with light gray or white flanks. The females of both species are dark brown; their flanks are slightly lighter; and they have a white patch at the base of their blue bills.

Greater and lesser scaup are very difficult to separate from one another, particularly when they are seen at a distance. The greater scaup has a larger bill with a wider black tip than the lesser scaup; but when seen at a distance, which is the usual case, it is almost impossible to make this distinction. The head of the greater scaup is more rounded than the slightly peaked crown of the lesser scaup. Perhaps the light wing stripe on the trailing edge of the wing is the best field mark for separating the species. The light or white wing band extends almost to the tip of the wing on the greater scaup and about halfway on the lesser scaup; these stripes or bands are visible when the birds are in flight. Scaup raft up, sometimes by the thousands, throughout the more open waters of the Bay. In the lower, saltier parts of the Bay, they tend to feed more on mollusks, insects, and crustaceans. In the less saline parts of the Bay system, both species will feed more on vegetation such as seeds, pondweeds, wild celery, and widgeon grass. They seem to linger in the Bay and are among the last ducks to migrate north in the spring.

The common loon, *Gavia immer,* begins arriving in the Bay in late September, and its numbers build in October to, perhaps, 10,000 or more. The males and females arrive in their winter plumage—blackish above, paler below, and with pale gray bills. Immature common loons are similar to adult birds in their winter plumage. As the fall deepens into late October and November, common loons form large, dispersed flocks. They are constantly "peering," which is a term used when loons swim with their heads partially submerged in the water, actively hunting for fish. Loons are powerful swimmers and divers; when they sight their prey they dive beneath the surface and, with the aid of their large, webbed feet, chase down their quarry. They constantly feed during the day. Frequently several loons will cooperate in herding fish, often menhaden, into the shallows, where they glut themselves on the oily fish.

The common loon is a ducklike bird that resembles the double-crested cormorant. Its legs are set well back toward the rump, as in diving ducks, and yet it has a sharp pointed bill that at a first glance looks like that of a cormorant. Like cormorants, loons ride low in the water, but

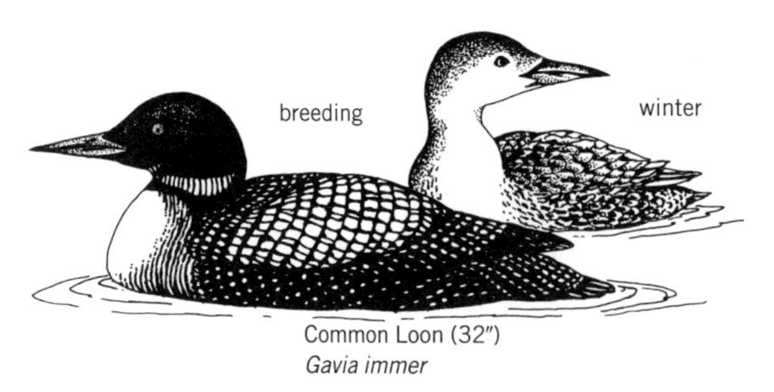

Common Loon (32″)
Gavia immer

they usually hold their straight bills horizontally, in contrast to cormorants, which hold their long, hooked bills pointed upward at an angle.

As winter approaches, the loons begin heading further south. They patter across the water's surface and take off. They gain altitude, and with their large feet extending beyond their tails, their long necks slightly bowed downward, and relatively slow wingbeats, they head south, with most of them overwintering along the Outer Banks of North Carolina. It is in the sounds and open ocean off Cape Hatteras that the common loon begins its winter molt, which renders it flightless for a time. Their insatiable feeding on oily menhaden and other fish from the Chesapeake tides them over until they are able to fly and feed once again. In late winter, February and early March, the loons head north and once again return to the Chesapeake Bay. Now, however, most of the adults are in their handsome breeding plumage, a black bill, a dark black head and neck marked with a black and white necklace, and the unmistakable checkered black and white back. Again the loons begin their heavy feeding on fish as they prepare for their flight to the north and their summer breeding grounds. Generally loons are silent on the wintering grounds, but occasionally on a quiet winter night one will hear their primeval, tremulous yodel.

There are other birds that range over the deeper, open waters in search of food. They spend most of the time flying over the water: terns will often set down on the water to loaf and bob on the waves, but others such as the northern gannet, *Sula bassanus,* hardly ever roost on Bay waters. The northern gannet is a more temperate member of the tropical booby family; its range is from Newfoundland to the Gulf of Mexico.

Gannets are very large seabirds, about the size of a goose, which often make forays from the ocean into the Bay in search of surface schooling fish, particularly menhaden. They frequently make spectacular vertical dives from heights that can be as much as an impressive 90 feet. The impact from such a dive could injure most birds, but northern gannets are protected by a heavy, reinforced skull and air sacs under the skin of the breast, both of which cushion the shock of striking the water from such heights. Adult northern gannets have long, narrow, black-tipped wings; juveniles are speckled with gray and varying shades of white and slate gray until they reach maturity in their third year. They are very sleek birds with a large conical bill that smoothly joins the head. At the other end, its tail is tapered to a point, giving the gannet a streamlined appearance. Northern gannets are never abundant in the Chesapeake, but they do frequent the mid- and lower Bay, particularly in the fall, winter, and early spring.

Two terns are commonly seen in the deeper, open waters of the Bay, the Forster's tern, *Sterna forsteri,* and the common tern, *Sterna hirundo.* When in breeding plumage, both species are black-capped and orange-billed; they are similarly colored and are approximately the same size. The differences are subtle but not impossible to discern. The difficulty, however, in identifying most terns is that they are often seen at a distance; they are usually quite active, flitting here and there; and many of them are similar in appearance and size. To compound the difficulty in identification, terns, like gulls, have different plumage colors and patterns depending on the stage of maturity and the season of the year. In general, an adult Forster's tern in breeding plumage is snow white above and pale gray below; it has a black cap and nape and a black-tipped orange bill. The primaries, the major flight feathers attached to the rear of the wing from the bend of the wing to the tip, are frosty or pearl gray. The common tern also has a black cap and nape; its back is slightly darker than that of the Forster's tern; and its bill is orange or orange-red with a black tip. The common tern has darker primaries, dusky to slate gray, than the Forster's. The color of the primaries is often used as an important field mark to separate the two species. In the winter, it becomes somewhat easier to distinguish the two species both as adults and as immature birds: the common tern has a black stripe beginning at the eye and expanding to a patch around the back of its head; the Forster's tern has a sinuous, broad black stripe that begins in front of the eye and ends at the ear. Both species are graceful fliers, often hawking for insects on the wing or plunge diving for fish and invertebrates.

Both terns breed along the Atlantic coast and in the

A northern gannet, *Sula bassanus* (37 inches), flies behind a Forster's tern, *Sterna forsteri* (14½ inches), with pale upperwings and a common tern, *Sterna hirundo* (14½ inches), with dark wedges on upperwings. A skein of scoters fly low over the water in the distance.

Chesapeake Bay, but the nesting habitats of the two species differ considerably. Forster's terns generally nest with their own kind in salt marshes. Salt marshes are usually near highly productive nursery areas for fish, which provide abundant prey of the right size for the terns. Their nests are usually located in marshy areas vegetated with saltmarsh cordgrass and saltmeadow hay. Apparently Forster's terns prefer to build their shallow nests on deposits of beach wrack, the wave-accumulated deposits of vegetation and other debris. The wrack is usually stranded above the mean high-tide line in long, narrow lines along the shore. The wrack covers the marsh vegetation and provides an elevated ridge upon which the terns build their nests. Females lay an average of three buff and brown eggs, which hatch in about two weeks.

Common terns, on the other hand, nest in colonies on sand beaches of islands, sand bars, and stony or gravelly areas. They are extremely aggressive on the nesting grounds and will readily attack human interlopers by diving at them and inflicting wounds with their bills or feet—or, most memorable of all, diving and, at the low point, defecating on the intruder. Common terns scrape out a shallow depression in the sand and line the nest with grass; the female lays two to three buff and olive-brown eggs in the nest, which hatch in approximately two weeks. There are several nesting sites throughout the Chesapeake Bay from Fishermans Island near the mouth of the Bay in Virginia to Eastern Bay near Kent Island in Maryland.

● The habitats discussed in this book are not as ecologically discrete as the collector might first suppose. It is true that the shifting sands of the beach habitat present an undeniable difference when compared to the pier piling habitat. However, the physical dynamics that influence the beach contours, sort out gravel and sand particles, and transport nutrients to the organisms that live in and on the beach are part of the same physical system that influences all habitats within the Bay.

The various plant and animal communities that occupy the several habitats are inextricably connected one to another. The marsh—the shoreward extension of the estuary—contributes its products to the open waters of the Bay just as surely as the tides transport larval fish from the shallows, across the flooded mud flat, and into the sinuous creek that meanders through the marsh. We encourage you to explore the length and breadth of this great body of water called the Chesapeake—marvel at the beauty of the flowerlike sea anemone and the unexplained mystery of the returning shad; catch a glimpse of a massive sea turtle, a cavorting dolphin, or a heron stalking the shallows.

Glossary, Species List, References, and Index

Alternate Leaves branching singly at different levels on a plant stem.

Amphipod A small crustacean belonging to the phylum Arthropoda, usually with a laterally compressed body.

Anadromous Pertaining to fishes, such as herrings, that ascend from their primary habitats in the ocean to fresh water to spawn.

Aperture Opening into the body whorl in the shell of snails, whelks, and gastropods out of which the head and foot of the animal protrudes.

Apex The first-formed, narrow end of a snail's shell; contains the smallest and oldest whorls.

Axial ribs Ribs or raised ridges parallel to the lengthwise axis of the shell.

Axillaries A group of feathers attached where the wing joins the body, the armpit.

Barbels Fleshy, elongated, tactile projections found under the lower jaw, below the snout, or around the mouth in some fishes.

Basal Leaves arising from the roots.

Basal disk The bottom end of sea anemones and certain jellyfish polyps used for attachment to a substrate.

Beak (umbo) Rounded swellings near the hinge in clams, mussels, and other bivalve shells; the first-formed part of the shell or valve.

Benthos (benthic) Collectively, all animals and plants living in or on bottom substrates in aquatic habitats; the term is often used with reference to animals.

Body whorl Largest and last coil of a snail shell.

Bryozoans Small sedentary colonial animals belonging to the phylum Bryozoa, commonly referred to as moss animals.

Byssus A bundle of fibers released from the foot of mussels and certain other bivalves which attach them to a substrate.

Carapace The hard exoskeleton covering a crustacean's head and thorax. The upper shell of a turtle.

Catadromous Pertaining to fishes, such as the American eel, that descend from their primary habitat in freshwater to the ocean to spawn.

Cardinal teeth Ridges, protuberances, or grooves present on the inner shell surface of clams and other bivalves which hold the shells in alignment.

Catkin A scaly spike of small, usually inconspicuous flowers.

Chondrophore A spoon-shaped indentation in the shell hinge of bivalve mollusks.

Colonial Pertaining to a group of individuals of the same species aggregated together and often structurally interconnected.

Compound leaf A leaf that is composed of two or more leaflets.

Cypris The second-stage, free-swimming larva of a barnacle; metamorphoses into the adult.

Deciduous Shedding, not persistent.

Dextral In snails, having the aperture to the right side of the shell axis when the shell is held with the apex pointing upward and the aperture facing the viewer.

Elytra Thickened, horny forewings of beetles.

Epifauna Animals that inhabit or move over the substrate in aquatic habitats.

Euryhaline Pertaining to organisms that are physiologically adapted for survival in aquatic environments over a wide range of salinities.

Herbaceous Fleshy or nonwoody plant.

Heteronereis The highly modified breeding form of certain species of polychaete worms; adapted for swimming, they swarm in large numbers in the water column from spring through summer.

Hinge The margin of clams and other bivalves where the shells are attached.

Glossary

Hinge teeth Toothlike projections on the inner shell surface of clams and other bivalves which prevent lateral slipping of valves.

Hydroid The polyp form of hydrozoans (phylum Cnidaria); a complicated colonial form often composed of highly specialized feeding and reproductive individuals.

Hydromedusa The sexually reproductive, free-living jellyfish stage of hydrozoans.

Infauna Animals that live within or burrow through bottom sediments of aquatic habitats.

Inner lip The wall of the body whorl, or largest whorl, opposite the outer lip of a snail; also known as the parietal wall.

Isopod A small crustacean belonging to the phylum Arthropoda, usually dorsally and ventrally flattened.

Larva The immature form of animals, usually invertebrates and fishes.

Ligament The strong dark band that attaches the two shells of clams and other bivalves.

Mantle The back of a bird and the upper surfaces of its wings; a fleshy tissue surrounding a mollusk's body that secretes the shell.

Marsupium A specialized pouch in various species used for brooding eggs and young.

Medusa The sexually reproductive stage of certain cnidarians; the jellyfish stage, usually characterized by a gelatinous saucer-shaped bell and trailing tentacles.

Megalopa The second-stage larval form of crabs; metamorphoses into the crab form.

Mesoglea The jellylike layer between the external and internal tissues of a cnidarian.

Naiad The aquatic larva of certain insects such as dragonflies.

Nape The back of the neck.

Nauplii Free-swimming, earliest larval stage of zooplanktonic groups such as copepods and barnacles.

Nekton Pelagic animals capable of swimming with a directed velocity.

Nematocysts Stinging cells of jellyfish and other cnidarians; for injecting poison or capturing prey.

Nymph Larva of terrestrial insects which undergo simple metamorphosis.

Operculum Leatherlike or horny plate attached to the foot in some snails; seals off aperture when animal withdraws into its shell.

Opposite Leaves arranged oppositely on a stem; a paired arrangement.

Outer lip The edge or external margin of the body whorl, or largest whorl, of a snail.

Pallial line A scar line on the interior of a valve or shell indicating the site of mantle muscle attachment in bivalves.

Pallial sinus A shallow, widened depression of the pallial line indicating the siphon-retracting muscles in bivalves.

Parapodia Appendages, frequently paddlelike, found on each body segment of polychaete worms; used in locomotion and respiration.

Parietal wall The wall or region opposite the outer lip; the inner lip.

Pelagic Pertaining to the open waters of the seas or lakes or to the organisms that inhabit these waters.

Perennial A plant that lives more than two years.

Periderm The thin exoskeleton of certain hydroids.

Phytoplankton Free-floating or weakly motile groups of unicellular aquatic plants.

Plankton The community of floating or weak-swimming organisms easily transported by currents and tides.

Plastron The bottom shell of a turtle.

Primaries The outermost and longest flight feathers on a bird's wing.

Pupa The inactive form of an insect between the larval stage and the adult.

Rhizome An underground horizontal or ascending stem.

Salinity The combined weight of certain salts dissolved in 1 kilogram of sea water, usually expressed as parts of salt per thousand parts of water (ppt).

Scute A large scale or plate covering a turtle's shell.

Sepal An individual segment, usually green, surrounding a flower.

Sessile Organisms permanently attached to a substrate; not motile.

Shrub A woody bush, usually 20 feet or less and often with multiple stems.

Sinistral In snails, having the aperture to the left side of the shell axis when the shell is held with apex pointing upward and the aperture facing the viewer.

Spat Juvenile newly attached oysters.

Speculum Iridescent or brightly colored feathers on the trailing edge of the wing.

Spiral coils The turns or whorls of a snail shell.

Stamen The male reproductive organ of a flower.

Suture The juncture between individual whorls of a snail shell.

Swamp A wetland containing water-loving trees and shrubs.

Thorax The intermediate section of an insect's body bearing the wings and legs.

Tuber A fleshy, enlarged portion of an underground stem.

Whorls The turns or spiral coils of a snail shell; the body whorl is the final and largest and includes the aperture from which the head and foot of the animal protrudes.

Zoea The tiny, planktonic, larval stage of crabs; metamorphoses into the megalopa.

Zooecium In bryozoans, the individual covering or chamber within which each animal (zooid) dwells.

Zooid One of the individual animals of hydroid or bryozoan colonies.

Zooplankton Group of floating or weak-swimming animals easily transported by currents and tides.

Species List and Distributions by Salinity Zones

SPECIES	ZONE 1	ZONE 2		ZONE 3
		upper	*lower*	
PLANTS				
Green Seaweeds — Phylum Chlorophyta				
Green-tufted seaweeds — *Cladophora* spp.		▮▮▮	▬▬	▬▬
Hollow-tubed seaweeds — *Enteromorpha* spp.		▮▮▮	▬▬	▬▬
Sea lettuce — *Ulva lactuca*		▮▮▮	▬▬	▬▬
Brown Seaweeds — Phylum Phaeophyta				
Brown fuzz seaweeds — *Ectocarpus* spp.			▬▬	▬▬
Red Seaweeds — Phylum Rhodophyta				
Tapered red weed — *Agardhiella tenera*				▬▬
Banded seaweeds — *Ceramium* spp.			▬▬	▬▬
Graceful red weed — *Gracilaria foliifera*				▬▬
Coarse red weed — *Gracilaria verrucosa*				▬▬
Laver — *Porphyra sp.*				▬▬
Aquatic Weeds and Wetland Plants — Phylum Spermatophyta				
*Red maple — *Acer rubrum*	▬▬			
*Smooth alder — *Alnus serrulata*	▬▬			
Water hemp — *Amaranthus cannabinus*	▬▬	▬▬	▬▬	
*Shadbush — *Amelanchier canadensis*	▬▬			
Saltmarsh aster — *Aster tenuifolius*	▬▬	▬▬	▬▬	▬▬

Distribution of species in the Chesapeake Bay and its tributaries is shown on the table according to the salinity zones drawn on the regional map in Chapter 1 (p. 6). A solid bar (▬▬) indicates general distribution; a broken bar (▮▮▮▮) indicates marginal distribution. Species are similarly distributed by salinity in other mid-Atlantic bays and estuaries. Trees and shrubs marked with an asterisk (*) tolerate only freshwater; however, they are distributed throughout the Chesapeake Bay uplands.

Zone 1 Tidal freshwaters
Zone 2 (upper) Brackish waters of 1–10 ppt salinity
Zone 2 (lower) Moderately salty waters of 11–18 ppt salinity
Zone 3 Salty Bay waters of 18–30 ppt salinity

SPECIES	ZONE 1	ZONE 2		ZONE 3
		upper	*lower*	
Groundsel tree — *Baccharis halimifolia*				
*River birch — *Betula nigra*				
Sea oxeye — *Borrichia frutescens*				
*Buttonbush — *Cepthalanthus occidentalis*				
Coontail — *Ceratophyllum demersum*				
*Silky dogwood — *Cornus amomum*				
Umbrella sedge — *Cyperus strigosus*				
Salt grass — *Distichlis spicata*				
Common waterweed — *Elodea canadensis*				
*Black ash — *Fraxinus nigra*				
Marsh hibiscus — *Hibiscus moscheutos*				
Hydrilla — *Hydrilla verticillata*				
*Winterberry — *Ilex verticillata*				
Blue flag — *Iris versicolor*				
Marsh elder — *Iva frutescens*				
Black needlerush — *Juncus roemerianus*				
Seashore mallow — *Kosteletzkya virginica*				
Sea lavenders — *Limonium* spp.				
Purple loosestrife — *Lythrum salicaria*				
Wax myrtle — *Myrica cerifera*				
Bayberry — *Myrica pensylvanica*				
Eurasian water milfoil — *Myriophyllum spicatum*				
Bushy pondweed — *Najas quadalupensis*				
Yellow pond lily — *Nuphar lutea*				
*Swamp tupelo — *Nyssa sylvatica*				
Switchgrass — *Panicum virgatum*				
Arrow arum — *Peltandra virginica*				
Reed grass — *Phragmites australis*				
*Loblolly pine — *Pinus taeda*				
*Sycamore — *Platanus occidentalis*				
Saltmarsh fleabane — *Pluchea purpurascens*				
Halberd-leaved tearthumb — *Polygonum arifolium*				
Water smartweed — *Polygonum punctatum*				
Pickerelweed — *Pontederia cordata*				
Sago pondweed — *Potamogeton pectinatus*				
Redhead grass — *Potamogeton perfoliatus*				
*Swamp white oak — *Quercus bicolor*				
*Cherrybark oak — *Quercus falcata*				
*Swamp rose — *Rosa palustris*				
Widgeon grass — *Ruppia maritima*				
Big-leaved arrowhead — *Sagittaria latifolia*				
Glassworts — *Salicornia* spp.				
American threesquare — *Scirpus americanus*				
River bulrush — *Scirpus fluviatilis*				
Olney threesquare — *Scirpus olneyi*				
Seaside goldenrod — *Solidago sempervirens*				
Saltmarsh cordgrass — *Spartina alterniflora*				

SPECIES	ZONE 1	ZONE 2		ZONE 3
		upper	*lower*	

Big cordgrass — *Spartina cynosuroides*
Saltmeadow hay — *Spartina patens*
Bald cypress — *Taxodium distichum*
Narrow-leaved cattail — *Typha angustifolia*
Common cattail — *Typha latifolia*
Wild celery — *Vallisneria americana*
Horned pondweed — *Zannichellia palustris*
Wild rice — *Zizania aquatica*
Eelgrass — *Zostera marina*

INVERTEBRATE ANIMALS

Sponges — Phylum Porifera

Boring sponges — *Cliona* spp.
Potato sponges — *Craniella* spp.
Sun sponge — *Halichondria bowerbanki*
Volcano sponges — *Haliclona* spp.
Stinking sponge — *Lissodendoryx carolinensis*
Redbeard sponge — *Microciona prolifera*

Sea Anemones, Hydroids, Jellyfish, and Corals — Phylum Cnidaria

SEA ANEMONES AND CORALS — CLASS ANTHOZOA

Star coral — *Astrangia astreiformis*
Sloppy gut anemone — *Ceriantheopsis americanus*
White anemone — *Diadumene leucolena*
Burrowing anemone — *Edwardsia elegans*
Green-striped anemone — *Haliplanella luciae*
Whip coral — *Leptogorgia virgulata*
Sea onion — *Paranthus rapiformis*

HYDROIDS — CLASS HYDROZOA

Freshwater hydroid — *Cordylophora caspia*
Horn garland hydroid — *Dynamena disticha*
Pink-hearted hydroid — *Ectopleura crocea*
Tube hydroid — *Ectopleura dumortieri*
Rope grass — *Garveia franciscana*
Snail fur — *Hydractinia echinata*
Graceful feather hydroid — *Pennaria disticha*
White hair — *Sertularia cupressina*
Soft snail fur — *Stylactaria arge*

JELLYFISHES — CLASS SCYPHOZOA

Moon jellyfish — *Aurelia aurita*
Sea nettle — *Chrysaora quinquecirrha*
Lion's mane jellyfish — *Cyanea capillata*
Mushroom-cap jellyfish — *Rhopilema verrilli*

SPECIES	ZONE 1	ZONE 2		ZONE 3
		upper	*lower*	
Comb Jellies — Phylum Ctenophora				
Pink comb jelly — *Beroe ovata*				
Sea walnut — *Mnemiopsis leidyi*				
Flatworms — Phylum Platyhelminthes				
Slender flatworm — *Euplana gracilis*				
Oyster flatworm — *Stylochus ellipticus*				
Ribbon Worms — Phylum Rhynchocoela				
Milky ribbon worm — *Cerebratulus lacteus*				
Leech ribbon worm — *Malacobdella grossa*				
Red ribbon worm — *Micrura leidyi*				
Four-eyed ribbon worm — *Tetrastemma elegans*				
Sharp-headed ribbon worm — *Zygeupolia rubens*				
Green ribbon worm — *Zygonemertes virescens*				
Bryozoans — Phylum Bryozoa				
Dead man's fingers — *Alcyonidium verrilli*				
Spiral bryozoan — *Amathia vidovici*				
Hair — *Anguinella palmata*				
Creeping bryozoan — *Bowerbankia gracilis*				
Lacy crust bryozoan — *Conopeum tenuissimum*				
Jointed-tube bryozoan — *Crisia eburnea*				
Coffin box bryozoan — *Membranipora tenuis*				
Freshwater bryozoan — *Pectinatella* sp.				
Cushion moss bryozoan — *Victorella pavida*				
Phoronid Worms — Phylum Phoronida				
Phoronid worms — *Phoronis* spp.				
Segmented Worms — Phylum Annelida				
BRISTLE WORMS — CLASS POLYCHAETA				
Ornate worm — *Amphitrite ornata*				
Opal worm — *Arabella iricolor*				
Lugworm — *Arenicola cristata*				
Elongated bamboo worm — *Asychis elongata*				
Parchment worm — *Chaetopterus variopedatus*				
Fringed worm — *Cirratulus cirriformia*				
Common bamboo worm — *Clymenella torquata*				
Plumed worm — *Diopatra cuprea*				
Freckled paddle worm — *Eteone heteropoda*				
Bloodworms — *Glycera* spp.				
Chevron worm — *Glycinde solitaria*				
Fifteen-scaled worm — *Harmothoe imbricata*				
Capitellid thread worm — *Heteromastus filiformis*				
Limy tube worm — *Hydroides dianthus*				

SPECIES	ZONE 1	ZONE 2		ZONE 3
		upper	lower	
Twelve-scaled worm — *Lepidonotus* sp.				
Red-spotted worm — *Loimia medusa*				
Red-gilled mud worm — *Marenzelleria viridis*				
Common clamworm — *Neanthis succinea*				
Red-lined worms — *Nephtys* spp.				
Fringe-gilled mud worm — *Paraprionospio pinnata*				
Trumpet worm — *Pectinaria gouldii*				
Whip mud worm — *Polydora cornuta*				
Oyster mud worm — *Polydora websteri*				
Fan worm — *Sabella microphthalma*				
Sandbuilder worm — *Sabellaria vulgaris*				
Glassy tube worm — *Spiochaetopterus oculatus*				
Barred-gilled mud worm — *Streblospio benedicti*				
Mollusks — Phylum Mollusca				
BIVALVES — CLASS PELYCYPODA				
Blood ark — *Anadara ovalis*				
Fossil ark — *Anadara staminea*				
Transverse ark — *Anadara transversa*				
Freshwater mussels — *Anodonta* spp.				
Jingle shell — *Anomia simplex*				
Fossil astartes — *Astarte* spp.				
Gould's shipworm — *Bankia gouldi*				
Atlantic mud-piddock — *Barnea truncata*				
Asian clam — *Corbicula fluminea*				
American oyster — *Crassostrea virginica*				
Angel wing — *Cyrtopleura costata*				
Common jackknife clam — *Ensis directus*				
Atlantic ribbed mussel — *Geukensia demissa*				
Gem clam — *Gemma gemma*				
Hooked mussel — *Ischadium recurvum*				
Fossil clam — *Isocardia* sp.				
Freshwater mussels — *Lampsilis* spp.				
Fossil ribbed scallop — *Lyropecten madisonius*				
Fossil broad-ribbed scallop — *Lyropecten santamaria*				
Baltic macoma clam — *Macoma balthica*				
Narrowed macoma clam — *Macoma tenta*				
Hard clam — *Mercenaria mercenaria*				
Little surf clam — *Mulinia lateralis*				
Long-siphoned fingernail clams — *Musculium* spp.				
Soft-shelled clam — *Mya arenaria*				
Dark falsemussel — *Mytilopsis leucophaeata*				
Blue mussel — *Mytilus edulis*				
Ponderous ark — *Noetia ponderosa*				
Fossil oyster — *Ostrea virginica*				
Fossil pecten — *Pecten quinquecostatus*				

SPECIES	ZONE 1	ZONE 2		ZONE 3
		upper	*lower*	
False angel wing — *Petricola pholadiformis*				
Pill clams — *Pisidium* spp.				
Brackish water clam — *Rangia cuneata*				
Little green jackknife clam — *Solen viridis*				
Short-siphoned fingernail clams — *Sphaerium* spp.				
Surf clam — *Spisula solidissima (shells only)*				
Purplish razor clam — *Tagelus divisus*				
Stout razor clam — *Tagelus plebeius*				
Northern dwarf tellin — *Tellina agilis*				
File yoldia — *Yoldia limatula*				
SNAILS — CLASS GASTROPODA				
Barrel bubble snail — *Acteocina canaliculata*				
Grass cerith — *Bittium varium*				
Two-sutured odostome — *Boonea bisuturalis*				
Impressed odostome — *Boonea impressa*				
Knobbed whelk — *Busycon carica*				
Fossil whelk — *Busycon* sp.				
Channeled whelk — *Busycotypus canaliculatus*				
Convex slipper shell — *Crepidula convexa*				
Common Atlantic slipper shell — *Crepidula fornicata*				
Flat slipper shell — *Crepidula plana*				
Four-lined fossil snail — *Ecphora quadricostata*				
Thick-lipped oyster drill — *Eupleura caudata*				
Coolie hat snails — *Ferrissia* spp.				
Hornshell snail — *Goniobasis virginica*				
Solitary bubble snail — *Hamionea solitaria*				
Seaweed snails — *Hydrobia* spp.				
Eastern mudsnail — *Ilyanassa obsoleta*				
Threeline mudsnail — *Ilyanassa trivittata*				
Marsh periwinkle — *Littorina irrorata*				
Spindle-shaped turret snail — *Mangelia plicosa*				
Saltmarsh snail — *Melampus bidentatus*				
Lunar dove shell — *Mitrella lunata*				
Nassa mudsnail — *Nassarius vibex*				
Shark eye — *Neverita duplicata*				
Pouch snail — *Physa gyrina*				
Fossil moon snail — *Polinices* sp.				
Pitted baby-bubble snail — *Rictaxis punctostriatus*				
Black-lined triphora — *Triphora nigrocincta*				
Interrupted turbonille — *Turbonilla interrupta*				
Fossil turret snail — *Turritella plebia*				
Atlantic oyster drill — *Urosalpinx cinerea*				
Sea Slugs — Orders Sacoglossa and Nudibranchia				
Striped nudibranch — *Cratena pilata*				
Limpet nudibranch — *Doridella obscura*				
Rough-back nudibranch — *Doris verrucosa*				

SPECIES	ZONE 1	ZONE 2		ZONE 3
		upper	*lower*	
Kitty-cat sea slug — *Elysia catula*				
Emerald sea slug — *Elysia chlorotica*				
Cross-bearer sea slug — *Hermaea cruciata*				
Ridged-head nudibranch — *Polycerella conyma*				
Dusky sea slug — *Stiliger fuscatus*				
CHITONS — CLASS POLYPLACOPHORA				
Common eastern chiton — *Chaetopleura apiculata*				
SQUIDS, OCTOPUS, AND CUTTLEFISH — CLASS CEPHALOPODA				
Brief squid — *Lolliguncula brevis*				
Jointed-legged Animals — Phylum Arthropoda				
HORSESHOE CRABS — CLASS MEROSTOMATA				
Atlantic horseshoe crab — *Limulus polyphemus*				
SEA SPIDERS — CLASS PYCNOGONIDA				
Long-necked sea spider — *Callipallene brevirostris*				
CRUSTACEANS — CLASS CRUSTACEA				
Water Fleas — Order Cladocera				
Giant water flea — *Leptodora kindtii*				
Barnacles — Order Thoracica				
Fossil Chesapeake barnacle — *Balanus concavus chesapeakensis*				
Ivory barnacle — *Balanus eburneus*				
Bay barnacle — *Balanus improvisus*				
White Barnacle — *Balanus subalbidus*				
Little gray barnacle — *Chthamalus fragilis*				
Opossum shrimps — Order Mysidacea				
Bay opossum shrimp — *Neomysis americana*				
Mantis Shrimps — Order Stomatopoda				
Mantis shrimp — *Squilla empusa*				
Isopods — Order Isopoda				
Slender isopod — *Cyathura polita*				
Mounded-back isopod — *Edotea triloba*				
Elongated eelgrass isopod — *Erichsonella attenuata*				
Baltic isopod — *Idotea baltica*				
Fish-gill isopod — *Lironeca ovalis*				
Sea roach — *Ligia exotica*				
Fish-mouth isopod — *Olencira praegustator*				
Eelgrass pill bug — *Paracerceis caudata*				
Shrimp parasite — *Probopyrus pandalicola*				
Sea pill bug — *Sphaeroma quadridentatum*				
Amphipods — Order Amphipoda				
Small four-eyed amphipod — *Ampelisca abdita*				
Long-antennaed four-eyed amphipod — *Ampelisca vadorum*				

SPECIES	ZONE 1	ZONE 2		ZONE 3
		upper	*lower*	
Narrow-headed four-eyed amphipod — *Ampelisca verrilli*				
Long-antennaed tube-builder amphipod — *Ampithoe longimana*				
Purple-eyed amphipod — *Batea catharinensis*				
Skeleton shrimps — *Caprella* spp.				
House-carrier amphipod — *Cerapus tubularis*				
Slender tube-builder amphipod — *Corophium lacustre*				
Wave-diver tube-builder amphipod — *Cymadusa compta*				
Scuds — *Gammarus* spp.				
Banded freshwater scud — *Gammarus fasciatus*				
Spine-backed scud — *Gammarus mucronatus*				
Mottled tube-builder amphipod — *Jassa falcata*				
Common burrower amphipod — *Leptocheirus plumulosus*				
Bamboo worm amphipod — *Listriella clymenellae*				
Red-eyed amphipod — *Monoculodes edwardsi*				
Sand digger amphipod — *Neohaustorius schmitzi*				
Saltmarsh flea — *Orchestia grillus*				
Beach flea — *Orchestia platensis*				
Beach hopper — *Talorchestia longicornis*				
Shrimps, Crabs, and Crayfishes — Order Decapoda				
Big-clawed snapping shrimp — *Alpheus heterochaelis*				
Blue crab — *Callinectes sapidus*				
Burrowing crayfish — *Cambarus diogenes*				
Jonah crab — *Cancer borealis*				
Rock crab — *Cancer irroratus*				
Sand shrimp — *Crangon septemspinosa*				
Mole crab — *Emerita talpoida*				
Flat mud crab — *Eurypanopeus depressus*				
Short-browed mud shrimp — *Gilvossius setimanus*				
Grooved-wristed mud crab — *Hexapanopeus angustifrons*				
Six-spined spider crab — *Libinia dubia*				
Nine-spined spider crab — *Libinia emarginata*				
Equal-clawed mud crab — *Dyspanopeus sayi*				
Ghost crab — *Ocypode quadrata*				
Coastal plains river crayfish — *Orconectes limosus*				
Lady crab — *Ovalipes ocellatus*				
Banded hermit crab — *Pagurus annulipes*				
Long-clawed hermit crab — *Pagurus longicarpus*				
Broad-clawed hermit crab — *Pagurus pollicaris*				
Grass shrimps — *Palaemonetes* spp.				
Common grass shrimp — *Palaemonetes pugio*				
Common black-fingered mud crab — *Panopeus herbstii*				
Brown shrimp — *Penaeus aztecus*				
Pink shrimp — *Penaeus duorarum*				

SPECIES	ZONE 1	ZONE 2		ZONE 3
		upper	*lower*	
White shrimp — *Penaeus setiferus*		╍╍	╍━	━━
Parchment worm crab — *Pinnixa chaetopterana*			╍╍	━━
Oyster crab — *Pinnotheres ostreum*			╍━	━━
White-fingered mud crab — *Rhithropanopeus harrisii*		╍╍	━━	━━
Wharf crab — *Sesarma cinereum*		━━	━━	━━
Marsh crab — *Sesarma reticulatum*		━━	━━	━━
Red-jointed fiddler crab — *Uca minax*	╍╍	━━	━━	╍╍
Sand fiddler crab — *Uca pugilator*		╍╍	╍━	━━
Marsh fiddler crab — *Uca pugnax*		━━	━━	━━
Flat-browed mud shrimp — *Upogebia affinis*				━━
INSECTS — CLASS INSECTA				
Beetles — Order Coleoptera				
Northeastern beach tiger beetle — *Cicindela dorsalis*			━━	━━
Predacious diving beetles — *Dytiscus* spp.	━━	━━		
Whirligig beetles — *Gyrinus* spp.	━━	╍╍		
Flies — Order Diptera				
Saltmarsh mosquito — *Aedes solicitans*	━━	━━	━━	━━
Deer flies — *Chrysops* spp.	━━	━━	━━	━━
American horse fly — *Tabanus americanus*	━━	━━	━━	━━
Dobsonfly — Order Megaloptera				
Dobsonfly — *Corydalus cornutus*	━━	╍╍		
Dragonflies and Damselflies — Order Odonata				
Green darner — *Anax junius*	━━	╍╍		
Doubleday's bluet — *Enallagma doubledayii*	━━	╍╍		
Twelve-spot skimmer — *Libellula puchella*	━━	╍╍		
True Bugs — Order Hemiptera				
Water boatmen — *Corixa* spp.	━━	━━		
Water striders — *Gerris* spp.	━━			
Sea Stars, Sea Cucumbers, and Brittle Stars — Phylum Echinodermata				
Common sea star — *Asterias forbesi*				━━
Pale sea cucumber — *Cucumaria pulcherrima*				━━
White synapta — *Leptosynapta tenuis*			━━	━━
Burrowing brittle star — *Micropholis atra*				━━
Common sea cucumber — *Thyone briareus*				━━
Fossil sand dollar — *Scutella aberti*			━━	━━
Arrow Worms — Phylum Chaetognatha				
Arrow worms — *Sagitta* spp.			━━	━━
Acorn Worms — Phylum Hemichordata				
Acorn Worm — *Saccoglossus kowalewskii*				━━

SPECIES	ZONE 1	ZONE 2		ZONE 3
		upper	lower	
Chordates — Phylum Chordata				
TUNICATES — CLASS ASCIDIACEA				
Golden star tunicate — *Botryllus schlosseri*				
Sea squirt — *Molgula manhattensis*				
Green beads tunicate — *Perophora viridis*				
VERTEBRATE ANIMALS				
CARTILAGINOUS FISHES — CLASS CHONDRICHYTHYES				
Order Squaliformes				
Requiem Sharks — Family Carcharhinidae				
Bull shark — *Carcharhinus leucas*				
Sandbar shark — *Carcharhinus plumbeus*				
Smooth dogfish — *Mustelus canis*				
Dogfish Sharks — Family Squalidae				
Spiny dogfish — *Squalus acanthias*				
Order Rajiformes				
Skates — Family Rajidae				
Clearnose skate — *Raja eglanteria*				
Stingrays — Family Dasyatidae				
Southern stingray — *Dasyatis americana*				
Bluntnose stingray — *Dasyatis say*				
Eagle Rays — Family Myliobatidae				
Cownose ray — *Rhinoptera bonasus*				
BONY FISHES — CLASS OSTEICHTHYES				
Order Acipenseriformes				
Sturgeons — Family Acipenseridae				
Shortnose sturgeon — *Acipenser brevirostrum*				
Atlantic sturgeon — *Acipenser oxyrhynchus*				
Order Lepisosteiformes				
Gars — Family Lepisosteidae				
Longnose gar — *Lepisosteus osseus*				
Order Anguilliformes				
Freshwater Eels — Family Anguillidae				
American eel — *Anguilla rostrata*				
Order Clupeiformes				
Herrings — Family Clupeidae				
Blueback herring — *Alosa aestivalis*				
Hickory shad — *Alosa mediocris*				
Alewife — *Alosa pseudoharengus*				
American shad — *Alosa sapidissima*				
Atlantic menhaden — *Brevoortia tyrannus*				
Gizzard shad — *Dorosoma cepedianum*				
Threadfin shad — *Dorosoma petenense*				

SPECIES	ZONE 1	ZONE 2		ZONE 3
		upper	lower	
Anchovies — Family Engraulidae				
Bay anchovy — *Anchoa mitchilli*				
Order Salmoniformes				
Mudminnows — Family Umbridae				
Eastern mudminnow — *Umbra pygmaea*				
Pikes — Family Esocidae				
Redfin pickerel — *Esox americanus*				
Chain pickerel — *Esox niger*				
Order Aulopiformes				
Lizardfishes — Family Synodontidae				
Inshore lizardfish — *Synodus foetens*				
Order Cypriniformes				
Minnows and Carps — Family Cyprinidae				
Goldfish — *Carassius auratus*				
Satinfin shiner — *Cyprinella analostana*				
Carp — *Cyprinus carpio*				
Silvery minnow — *Hybognathus regius*				
Golden shiner — *Notemigonus crysoleucas*				
Spottail shiner — *Notropis hudsonius*				
Suckers — Family Catostomidae				
White sucker — *Catostomus commersoni*				
Creek chubsucker — *Erimyzon oblongus*				
Order Siluriformes				
Bullhead Catfishes — Family Ictaluridae				
White catfish — *Ameiurus catus*				
Brown bullhead — *Ameiurus nebulosus*				
Channel catfish — *Ictalurus punctatus*				
Order Batrachoidiformes				
Toadfishes — Family Batrachoididae				
Oyster toadfish — *Opsanus tau*				
Order Gobiesociformes				
Clingfishes — Family Gobiesocidae				
Skilletfish — *Gobiesox strumosus*				
Order Gadiformes				
Codfishes — Family Gadidae				
Red hake — *Urophycis chuss*				
Spotted hake — *Urophycis regia*				
Order Atheriniformes				
Flyingfishes — Family Exocoetidae				
Halfbeak — *Hyporhamphus unifasciatus*				
Needlefishes — Family Belonidae				
Atlantic needlefish — *Strongylura marina*				
Killifishes — Family Cyprinodontidae				
Sheepshead minnow — *Cyprinodon variegatus*				
Banded killifish — *Fundulus diaphanus*				
Mummichog — *Fundulus heteroclitus*				

SPECIES	ZONE 1	ZONE 2		ZONE 3
		upper	*lower*	
Striped killifish — *Fundulus majalis*				
Rainwater killifish — *Lucania parva*				
Livebearers — Family Poeciliidae				
Mosquitofish — *Gambusia holbrooki*				
Silversides — Family Atherinidae				
Rough silverside — *Membras martinica*				
Inland silverside — *Menidia beryllina*				
Atlantic silverside — *Menidia menidia*				
Order Gasterosteiformes				
Sticklebacks — Family Gasterosteidae				
Fourspine stickleback — *Apeltes quadracus*				
Threespine stickleback — *Gasterosteus aculeatus*				
Pipefishes — Family Syngnathidae				
Lined seahorse — *Hippocampus erectus*				
Dusky pipefish — *Syngnathus floridae*				
Northern pipefish — *Syngnathus fuscus*				
Order Perciformes				
Temperate Basses — Family Percichthyidae				
White perch — *Morone americana*				
Striped bass — *Morone saxatilis*				
Sea Basses — Family Serranidae				
Black sea bass — *Centropristis striata*				
Sunfishes — Family Centrarchidae				
Pumpkinseed — *Lepomis gibbosus*				
Bluegill — *Lepomis macrochirus*				
Smallmouth bass — *Micropterus dolomieu*				
Largemouth bass — *Micropterus salmoides*				
White crappie — *Pomoxis annularis*				
Black crappie — *Pomoxis nigromaculatus*				
Perches — Family Percidae				
Tessellated darter — *Etheostoma olmstedi*				
Yellow perch — *Perca flavescens*				
Bluefishes — Family Pomatomidae				
Bluefish — *Pomatomus saltatrix*				
Cobias — Family Rachycentridae				
Cobia — *Rachycentron canadum*				
Jacks — Family Carangidae				
Blue runner — *Caranx crysos*				
Crevalle jack — *Caranx hippos*				
Lookdown — *Selene vomer*				
Florida pompano — *Trachinotus carolinus*				
Grunts — Family Haemulidae				
Pigfish — *Orthopristis chrysoptera*				
Porgies — Family Sparidae				
Sheepshead — *Archosargus probatocephalus*				
Scup — *Stenotomus chrysops*				

SPECIES	ZONE 1	ZONE 2 upper	ZONE 2 lower	ZONE 3
Drums — Family Sciaenidae				
Silver perch — *Bairdiella chrysoura*				
Spotted seatrout — *Cynoscion nebulosus*				
Weakfish — *Cynoscion regalis*				
Spot — *Leiostomus xanthurus*				
Southern kingfish — *Menticirrhus americanus*				
Northern kingfish — *Menticirrhus saxatilis*				
Atlantic croaker — *Micropogonias undulatus*				
Black drum — *Pogonias cromis*				
Red drum — *Sciaenops ocellatus*				
Wrasses — Family Labridae				
Tautog — *Tautoga onitis*				
Mullets — Family Mugilidae				
Striped mullet — *Mugil cephalus*				
White mullet — *Mugil curema*				
Stargazers — Family Uranoscopidae				
Northern stargazer — *Astroscopus guttatus*				
Combtooth Blennies — Family Bleniidae				
Striped blenny — *Chasmodes bosquianus*				
Feather blenny — *Hypsoblennius hentz*				
Gobies — Family Gobiidae				
Naked goby — *Gobiosoma bosc*				
Seaboard goby — *Gobiosoma ginsburgi*				
Green goby — *Microgobius thalassinus*				
Mackerels — Family Scombridae				
Little tunny — *Euthynnus alletteratus*				
Atlantic bonito — *Sarda sarda*				
Butterfishes — Family Stromateidae				
Harvestfish — *Peprilus alepidotus*				
Butterfish — *Peprilus triacanthus*				
Order Scorpaeniformes				
Searobins — Family Triglidae				
Northern searobin — *Prionotus carolinus*				
Order Pleuronectiformes				
Lefteye Flounders — Family Bothidae				
Summer flounder — *Paralichthys dentatus*				
Windowpane — *Scophthalmus aquosus*				
Righteye Flounders — Family Pleuronectidae				
Winter flounder — *Pleuronectes americanus*				
Soles — Family Soleidae				
Hogchoker — *Trinectes maculatus*				
Blackcheek tonguefish — *Symphurus plagiusa*				
Order Tetradontiformes				
Leatherjackets — Family Balistidae				
Orange filefish — *Aluterus schoepfi*				
Planehead filefish — *Monacanthus hispidus*				

SPECIES	ZONE 1	ZONE 2		ZONE 3
		upper	*lower*	
Puffers — Family Tetraodontidae				
Northern puffer — *Sphoeroides maculatus*		▬	▬	▬
Striped burrfish — *Chilomycterus schoepfi*		┅	┅ ▬	▬
REPTILES — CLASS REPTILIA				
Turtles — Order Testudinata				
Loggerhead — *Caretta caretta*		┅	▬	▬
Snapping turtle — *Chelydra serpentina*	▬	▬	┅	
Red-bellied turtle — *Chrysemys rubriventris*	▬	┅		
Diamondback terrapin — *Malaclemys terrapin*		▬	▬	▬
Snakes — Order Squamata				
Northern water snake — *Nerodia sipedon*	▬	▬	▬	
MAMMALS — CLASS MAMMALIA				
Rodents — Order Rodentia				
Beaver — *Castor canadensis*	▬			
Nutria — *Myocastor coypus*	▬	▬	┅	
Muskrat — *Ondatra zibethicus*	▬	▬	┅	
Whales, Dolphins, and Porpoises — Order Cetacea				
Atlantic bottlenosed dolphin — *Tursiops truncatus*		┅	▬	▬
Carnivores — Order Carnivora				
River otter — *Lutra canadensis*	▬	┅		

Birds—Species List and Seasonal Occurrence

		Spring	Summer	Fall	Winter
Common Loon	*Gavia immer*	o	n	c	c
Brown Pelican	*Pelicanus occidentalis*	c	c	c	n
Northern Gannet	*Sula bassanus*	o	o	o	o
Double-crested Cormorant	*Phalacrocorax auritus*	c	c	a	a
Least Bittern	*Ixobrychus exilis*	c	c	u	r
Great Egret	*Casmerodius albus*	c	c	c	r
Snowy Egret	*Egretta thula*	c	c	c	r
Great Blue Heron	*Ardea herodias*	a	a	a	c
Little Blue Heron	*Egretta caerulea*	u	u	u	u
Tricolored Heron	*Egretta tricolor*	u	u	u	u
Cattle Egret	*Bubulcus ibis*	c	a	c	r
Green Heron	*Butorides striatus*	c	c	c	r
Black-crowned Night Heron	*Nycticorax nycticorax*	u	u	u	u
Glossy Ibis	*Plegadis falcinellus*	u	u	u	r
Mute Swan	*Cygnus olor*	c	c	c	c
Tundra Swan	*Cygnus columbianus*	a	n	a	a
Snow Goose	*Chen caerulescens*	a	n	a	a
Canada Goose	*Branta canadensis*	a	c	a	a
Wood Duck	*Aix sponsa*	c	c	c	u
Green-winged Teal	*Anas crecca*	c	r	c	a
American Black Duck	*Anas rubripes*	c	c	c	c
Mallard	*Anas platyrhynchos*	a	c	a	a
Northern Pintail	*Anas acuta*	c	c	c	c
Blue-winged Teal	*Anas discors*	c	c	c	c
Northern Shoveler	*Anas clypeata*	u	r	u	u
American Wigeon	*Anas americana*	c	r	c	c
Redhead	*Aythya americana*	u	u	r	u
Canvasback	*Aythya valisineria*	c	r	c	c
Greater Scaup	*Aythya marila*	a	r	c	a
Lesser Scaup	*Aythya affinis*	a	r	c	a
Oldsquaw	*Clangula hyemalis*	a	r	a	a
White-winged Scoter	*Melanitta fusca*	a	r	a	a
Surf Scoter	*Melanitta perspicillata*	c	r	c	c
Black Scoter	*Melanitta nigra*	a	r	a	a

a = abundant; c = common, likely to be seen or heard; u = uncommon, often present, but not usually seen or heard; o = occasional, seen infrequently during the year; r = rare, not seen every year; n = not present.

		Spring	Summer	Fall	Winter
Common Goldeneye	*Bucephala clangula*	a	r	a	a
Bufflehead	*Bucephala albeola*	a	r	a	a
Red-breasted Merganser	*Mergus serrator*	c	r	c	c
Ruddy Duck	*Oxyura jamaicensis*	r	u	u	u
Osprey	*Pandion haliaetus*	a	a	c	r
Bald Eagle	*Haliaeetus leucocephalus*	c	c	c	c
Clapper Rail	*Rallus longirostris*	u	u	u	u
King Rail	*Rallus elegans*	u	u	u	u
Virginia Rail	*Rallus limicola*	u	u	u	u
American Coot	*Fulica americana*	u	u	u	c
Greater Yellowlegs	*Tringa melanoleuca*	c	c	c	c
Lesser Yellowlegs	*Tringa flavipes*	c	c	c	c
Willet	*Catoptrophorus semipalmatus*	c	c	c	c
Ruddy Turnstone	*Arenaria interpres*	c	u	c	u
Sanderling	*Calidris alba*	a	u	a	u
Dunlin	*Calidris alpina*	a	u	a	c
Short-billed Dowitcher	*Limnodromus griseus*	c	r	c	c
Long-billed Dowitcher	*Limnodromus scolopaceus*	c	r	c	r
American Oystercatcher	*Haematopus palliatus*	c	c	c	r
Black-bellied Plover	*Pluvialis squatrola*	c	r	c	u
Laughing Gull	*Larus atricilla*	a	a	a	c
Ring-billed Gull	*Larus delawarensis*	a	c	a	c
Herring Gull	*Larus argentatus*	c	c	c	a
Great Black-backed Gull	*Larus marinus*	c	c	c	a
Caspian Tern	*Sterna caspia*	c	c	c	r
Royal Tern	*Sterna maxima*	c	c	c	r
Least Tern	*Sterna antillarum*	u	c	c	r
Forster's Tern	*Sterna forsteri*	c	c	c	c
Common Tern	*Sterna hirundo*	c	c	u	r
Belted Kingfisher	*Ceryle alcyon*	c	c	c	c
Barn Swallow	*Hirundo rustica*	a	a	a	n
Red-winged Blackbird	*Agelaius phoeniceus*	a	a	a	n

Selected References

Abbott, R. T. 1968. *A Guide to Field Identification: Seashells of North America.* New York: Golden Press.

Abbott, R. T. 1974. *American Seashells.* 2d ed. New York: Van Nostrand Reinhold Co.

Amos, W. 1966. *The Life of the Seashore.* New York: McGraw-Hill Book Co.

Angier, B. 1978. *Field Guide to Medicinal Wild Plants.* Harrisburg, Pa.: Stackpole Books.

Army Corps of Engineers. 1977. *Wetland Plants of the Eastern United States.* NADP 200-1-1. New York: Army Corps of Engineers.

Arnold, A. F. 1968. *The Sea-Beach at Ebb-Tide: A Guide to the Study of the Seaweeds and the Lower Animal Life Found between Tide-Marks.* New York: Dover Publications.

Attenborough, D. 1979. *Life on Earth.* Boston: Little, Brown & Co.

Behler, J. L., and F. W. King. 1979. *The Audubon Society Field Guide to North American Reptiles and Amphibians.* New York: Alfred A. Knopf.

Borror, D. J., and R. E. White. 1970. *A Field Guide to Insects: America North of Mexico.* Boston: Houghton Mifflin Co.

Boschung, H. T., Jr. 1983. *The Audubon Society Field Guide to North American Fishes, Whales, and Dolphins.* New York: Alfred A. Knopf.

Brown, R. G., and M. L. Brown. 1972. *Woody Plants of Maryland.* College Park, Md.: University of Maryland.

Buchsbaum, R. 1938. *Animals without Backbones.* Chicago: University of Chicago Press.

Buchsbaum, R., and L. J. Milne. 1960. *The Lower Animals: Living Invertebrates of the World.* Garden City, N.Y.: Doubleday & Co.

Bull, J., and J. Farrand, Jr. 1977. *The Audubon Society Field Guide to North American Birds: Eastern Region.* New York: Alfred A. Knopf.

Burk, B. 1976. *Waterfowl Studies.* New York: Winchester Press.

Carreker, R. G. 1985. *Habitat Suitability Index Models: Least Tern.* Biological Report 82 (10.103). Washington, D.C.: U.S. Fish & Wildlife Service.

Carson, R. 1951. *The Sea around Us.* New York: Oxford University Press.

Carson, R. 1955. *The Edge of the Sea.* Boston: Houghton Mifflin Co.

Chapman, B. R., and R. J. Howard. 1984. *Habitat Suitability Index Models: Great Egret.* Biological Report 82 (10.78). Washington, D.C.: U.S. Fish & Wildlife Service.

Daiber, F. C. 1982. *Animals of the Tidal Marsh.* New York: Van Nostrand Reinhold Co.

Dawson, E. Y. 1956. *How to Know the Seaweeds.* Pictured Key Nature Series. Dubuque, Iowa: William C. Brown Co.

Dawson, E. Y. 1966. *Marine Botany: An Introduction.* New York: Holt, Rinehart & Winston.

Duncan, W. H., and M. B. Duncan. 1987. *The Smithsonian Guide to Seaside Plants of the Gulf and Atlantic Coasts, from Louisiana to Massachusetts Exclusive of Lower Peninsular Florida.* Washington, D.C.: Smithsonian Institution Press.

Edmondson, W. T. 1959. *Fresh-Water Biology.* 2d ed. New York: John Wiley & Sons.

Ehrlich, P. R., et al. 1988. *The Birder's Handbook.* New York: Simon & Schuster.

Elser, H. 1950. "The Common Fishes of Maryland: How to Tell Them Apart." Maryland Department of Research and Education Publication (Solomons, Md.) 88:1–45.

Fassett, N. C. 1960. *A Manual of Aquatic Plants.* Madison: University of Wisconsin Press.

Forbush, E. H. 1939. *Natural History of the Birds of Eastern and Central North America.* Boston: Houghton Mifflin Co.

Fotheringham, N., and S. Brunenmeister. 1975. *Common Marine Invertebrates of the Northwestern Gulf Coast.* Houston: Gulf Publishing Co.

Funderburk, S. L., et al., eds. 1991. *Habitat Requirements for Chesapeake Bay Living Resources.* 2d ed. Annapolis: Maryland Department of Natural Resources.

Gosner, K. L. 1979. *A Field Guide to the Atlantic Seashore.* Boston: Houghton Mifflin Co.

Graves, B. M., and S. H. Anderson. 1987. *Habitat Suitability Index Models: Snapping Turtle.* Biological Report 82 (10.141). Washington, D.C.: U.S. Fish & Wildlife Service.

Hancock, J., and J. Kushlan. 1984. *The Herons Handbook.* New York: Harper & Row.

Harrison, P. 1985. *Seabirds: An Identification Guide.* Boston: Houghton Mifflin Co.

Hayman, P., et al. 1986. *Shorebirds: An Identification Guide to the Waders of the World.* Boston: Houghton Mifflin Co.

Hingtgen, T. M., et al. 1985. *Habitat Suitability Index Models: Eastern Brown Pelican.* Biological Report 82 (10.90). Washington, D.C.: U.S. Fish & Wildlife Service.

Hurley, L. M. 1990. *Field Guide to the Submerged Aquatic Vegetation of Chesapeake Bay.* Annapolis, Md.: U.S. Fish & Wildlife Service.

Johnsgard, P. A. 1975. *Waterfowl of North America.* Bloomington: Indiana University Press.

Klingel, G. C. 1951. *The Bay: A Naturalist Discovers a Universe of Life above and below the Chesapeake.* New York: Dodd, Mead & Co.

Kricher, J. C., and G. Morrison. 1988. *A Field Guide to Eastern Forests: North America.* Boston: Houghton Mifflin Co.

LaMonte, F. 1945. *North American Game Fishes.* New York: Doubleday & Co.

Leatherwood, S., et al. 1976. *Whales, Dolphins, and Porpoises of the Western North Atlantic: A Guide to Their Identification.* NOAA Technical Report NMFS CIRC-396. Seattle: U.S. Department of Commerce.

Lippson, A. J., ed. 1973. *The Chesapeake Bay in Maryland: An Atlas of Natural Resources.* Baltimore: Johns Hopkins University Press.

Lippson, A. J., et al., eds. 1981. *Environmental Atlas of the Potomac Estuary.* Baltimore: Johns Hopkins University Press.

Little, E. L. 1980. *National Audubon Society Field Guide to North American Trees: Eastern Region.* New York: Alfred A. Knopf.

McClane, A. J. 1978. *McClane's Field Guide to Saltwater Fishes of North America.* New York: Holt, Rinehart & Winston.

Martin, A. C., et al. 1951. *American Wildlife and Plants: A Guide to Wildlife Food Habits.* New York: Dover Publications.

Martin, R. P., and P. J. Zwank. 1987. *Habitat Suitability Index Models: Forster's Tern (Breeding)—Gulf and Atlantic Coasts.* Biological Report 82 (10.131). Washington, D.C.: U.S. Fish & Wildlife Service.

Martof, B. S., et al. 1980. *Amphibians and Reptiles of the Carolinas and Virginia.* Chapel Hill: University of North Carolina Press.

Meglitsch, P. A. 1967. *Invertebrate Zoology.* London: Oxford University Press.

Meinkoth, N. A. 1981. *The Audubon Society Field Guide to North American Seashore Creatures.* New York: Alfred A. Knopf.

Milne, L., and M. Milne. 1980. *The Audubon Society Field Guide to North American Insects and Spiders.* New York: Alfred A. Knopf.

Miner, R. W. 1950. *Field Book of Seashore Life.* New York: G. P. Putnam's Sons.

Morris, P. A. 1973. *A Field Guide to the Shells of the Atlantic and Gulf Coasts and the West Indies.* 3d ed. Boston: Houghton Mifflin Co.

Musick, J. A. 1979. *The Marine Turtles of Virginia.* Educational Series, no. 24. Gloucester Point, Va.: Virginia Institute of Marine Science.

Niering, W. A., and N. C. Olmstead. 1979. *The Audubon Society Field Guide to North American Wildflowers: Eastern Region.* New York: Alfred A. Knopf.

Nowak, R. M. 1991. *Walker's Mammals of the World.* 5th ed. 2 vols. Baltimore: Johns Hopkins University Press.

Pearson, T. G., et al., eds. 1936. *Birds of America.* Garden City, N.Y.: Garden City Publishing Co.

Pennak, R. W. 1953. *Fresh-Water Invertebrates of the United States.* New York: Ronald Press.

Peterson, A. P. 1986. *Habitat Suitability Index Models: Bald Eagle (Breeding Season).* Biological Report 82 (10.126). Washington, D.C.: U.S. Fish & Wildlife Service.

Peterson, R. T. 1980. *A Field Guide to the Birds: East of the Rockies.* 4th ed. Boston: Houghton Mifflin Co.

Peterson, R. T., and M. McKenny. 1968. *A Field Guide to Wildflowers of Northeastern and Northcentral North America.* Boston: Houghton Mifflin Co.

Peterson, R. T., et al. 1983. *A Field Guide to the Birds of Britain and Europe.* 4th ed. Boston: Houghton Mifflin Co.

Petry, L. C. 1975. *A Beachcomber's Botany.* Chatham, Mass.: Chatham Press.

Potter, E. F., et al. 1980. *Birds of the Carolinas.* Chapel Hill: University of North Carolina Press.

Prose, B. L. 1985. *Habitat Suitability Index Models: Belted Kingfisher.* Biological Report 82 (10.87). Washington, D.C.: U.S. Fish & Wildlife Service.

Robbins, S. F., and C. M. Yentsch. 1973. *The Sea Is All About Us.* Salem, Mass.: Peabody Museum of Salem/Cape Ann Society of Marine Science.

Rue, L. L., III. 1964. *The World of the Beaver.* Philadelphia: J. B. Lippincott Co.

Schultz, G. A. 1970. *How to Know the Marine Isopod Crustaceans.* Dubuque, Iowa: William C. Brown Co.

Schwartz, F. J. 1967. *Maryland Turtles.* Solomons, Md.: University of Maryland, Natural Resources Institute.

Short, H. L. 1985. *Habitat Suitability Index Models: Red-Winged Blackbird.* Biological Report 82 (10.95). Washington, D.C.: U.S. Fish & Wildlife Service.

Short, H. L. 1985. *Habitat Suitability Index Models: Great Blue Heron.* Biological Report 82 (10.99). Washington, D.C.: U.S. Fish & Wildlife Service.

Silberhorn, G. M. 1976. *Tidal Wetland Plants of Virginia.* Educational Series, no. 19. Gloucester Point, Va.: Virginia Institute of Marine Science.

Silberhorn, G. M. 1982. *Common Plants of the Mid-Atlantic Coast: A Field Guide.* Baltimore: Johns Hopkins University Press.

Spendelow, J. A., and S. R. Patton. 1988. *National Atlas of Coastal Waterbird Colonies in the Contiguous United States: 1976–1982.* Biological Report 88 (5). Washington, D.C.: U.S. Fish & Wildlife Service.

Stewart, R. E., and C. S. Robbins. 1958. *Birds of Maryland and the District of Columbia.* North American Fauna, no. 62. Washington, D.C.: U.S. Fish & Wildlife Service.

Taylor, W. R. 1957. *Marine Algae of the Northeastern Coast of North America.* Ann Arbor: University of Michigan Press.

Tinbergen, N. 1960. *The Herring Gull's World.* New York: Basic Books.

Tiner, R. W., Jr. 1987. *Coastal Wetland Plants of the Northeastern United States.* Amherst: University of Massachusetts Press.

Tiner, R. W., Jr. 1988. *Field Guide to Nontidal Wetland Identification.* Annapolis, Md.: Maryland Department of Natural Resources; Newton Corner, Mass.: U.S. Fish & Wildlife Service.

Usinger, R. L., ed. 1956. *Aquatic Insects of California.* Berkeley: University of California Press.

Vana-Miller, S. L. 1987. *Habitat Suitability Index Models: Osprey.* Biological Report 82 (10.154). Washington, D.C.: U.S. Fish & Wildlife Service.

Wass, M. L., et al. 1972. *A Check List of the Biota of Lower Chesapeake Bay.* Special Scientific Report no. 65. Gloucester Point, Va.: Virginia Institute of Marine Science.

Whitaker, J. O., Jr. 1980. *The Audubon Society Field Guide to North American Mammals.* New York: Alfred A. Knopf.

White, C. P. 1989. *Chesapeake Bay, Nature of the Estuary: A Field Guide.* Centreville, Md.: Tidewater Publishers.

Williams, A. B. 1984. *Shrimps, Lobsters, and Crabs of the Atlantic Coast.* Washington, D.C.: Smithsonian Institution Press.

Williams, J. P., Jr. 1993. *Chesapeake Almanac.* Centreville, Md.: Tidewater Publishers.

Zale, A. V., and R. Mulholland. 1985. *Habitat Suitability Index Models: Laughing Gull.* Biological Report 82 (10.94). Washington, D.C.: U.S. Fish & Wildlife Service.

Zim, H. S., and L. Ingle. 1955. *Seashores: A Guide to Animals and Plants along the Beaches.* Golden Nature Guide. New York: Simon & Schuster.

Zim, H. S., and H. H. Shoemaker. 1956. *Fishes: A Guide to Fresh- and Saltwater Species.* Golden Nature Guide. New York: Simon & Schuster.

Index

Boldface numerals designate illustrations.

ALICE JANE LIPPSON is a biological illustrator and marine scientist. Her books include the best-selling *Chesapeake Bay in Maryland: An Atlas of Natural Resources* and the *Environmental Atlas of the Potomac Estuary* (both available from Johns Hopkins).

ROBERT L. LIPPSON is a marine scientist formerly with the National Marine Fisheries Service and was an adjunct professor of invertebrate zoology at Michigan State University.

The Lippsons live and work on the shores of the Chesapeake. Currently they are writing a new book on the natural history of the coastal bays, inlets, and waterways from Norfolk, Virginia, to Key West, Florida.

Library of Congress Cataloging-in-Publication Data

Lippson, Alice Jane.
Life in the Chesapeake Bay / Alice Jane Lippson ; Robert L. Lippson. — 2nd ed.
p. cm.
Includes bibliographical references (p.) and index.
ISBN 0-8018-5476-8 (alk. paper). — ISBN 0-8018-5475-X (pbk. : alk. paper)
1. Marine biology—Chesapeake Bay (Md. and Va.)
I. Lippson, Robert L. II. Title.
QH104.5.C45L56 1997
574.92'147—dc20 96-27103
CIP